Urban Reforms in India
(Fast Track but Slow Motion)

Urban Reforms in India
(Fast Track but Slow Motion)

V. Gnaneshwar and U.B. Singh

KALPAZ

First Published, 2017

Published by

Kalpaz Publications
C-30, Satyawati Nagar,
Delhi – 110052
E-mail: kalpaz@hotmail.com
Ph.: 9212142040

Printed at: G. Print Process, Delhi

Cataloging in Publication Data—DK

Courtesy: D.K. Agencies (P) Ltd. <docinfo@dkagencies.com>

Singh, U. B., 1953- **author.**

Urban reforms in India : fast track but slow motion / Dr U.B. Singh.

pages cm

Includes bibliographical references.

ISBN 9789351282532

1. City planning—India. 2. Urban policy—India. 3. Municipal government—India. I. Title.

HT384.I4S56 2017 DDC 307.760954 23

Contents

List of Tables

List of Figures

List of Charts

Preface

A new urban world is emerging around the world where urban habitats have become not only the dominant spatial forms but also critical centres playing crucial role in entire human development satisfying the human needs and realising the human dreams and aspirations in multiple dimensions. Another important aspect has been the increasing prosperity of the people understood in a broader perspective in multiple dimensional form including economic growth, quality of life, infrastructure, equity and sustainable development. Pursuing prosperity has become main characteristic feature of the cities of today wherein, every city will inevitably find itself on its own specific and unique historic course. The Indian thinking as reflected in the latest twelfth Five Year Plan document also reflects similar ideas on the future of urbanisation and urban development in the country.

On governance front, cities in the globalising world face paradoxical situations and challenges due to involvement of the multiple stakeholders, interdependent resources and actions, shared purposes and blurred boundaries between the public and private, formal and informal and civil society sectors. The success of urban governance today depends on how the city governments deal with tasks of coordinating the complex and fragmented governance landscape; moderating and negotiating cooperation; and integrating and managing diverse networks, among others.

India has been witnessing an unprecedented growth of urban population as also population shift from rural areas to cities and towns. The contribution of urban population which was about 10 per cent a century back increased to over 31 per cent by 2011. The most disturbing aspect is the absolute population numbers which stood at 377 million in 2011 and projected to increase to 473 million by 2021, 600 million by 2031 and 820 million by 2051 adding about 100 million every decade. Further, the Indian urbanisation has a unique characteristic which makes it different from western concept. Urbanisation in the country is resulting into concentration of population in larger urban centres, on one hand, and emergence of census towns in large number, on the other. These census towns contain a good portion of urban content but are beyond the purview of the urban laws.

The institutional arrangement for managing urban affairs in the country is neither enjoy autonomy nor an independent entity in the uni-federal structure of the country, despite the local government units (urban as well as rural) having their genesis from the supreme law of the land. The 'local governments' in the country function on the whims and fancies of the state governments. Many state governments treat the local governments as their field unit. It is against the spirit of the Constitution.

The current urbanisation debate, therefore, is not about rural-urban divide but how to integrate the interests of the urban areas with the rural areas and, core areas with the peri-urban areas in the metropolitan city regions where the focus of growth has been witnessed in recent periods. The urban policies in India should focus on options for accommodating population in municipal and beyond municipal boundaries in terms of housing, jobs, transport and infrastructure, that is, nurturing metropolitan economics. At the other spectrum, environmentalists are cautioning about impending climatic change threats due to metro centered and unregulated urbanisation, and are advocating for sustainable habitat development. The urban poverty is another dimension demanding the attention of policy-makers and administrators. It is in this context, the urban policies and strategies of urban development become crucial to address the emerging problems and developmental goals. India faces critical challenge of putting in place effective governance structures and strong politico-administrative commitment to bring desirable urban transformation in terms of urban form, infrastructure and development in cities and towns in the twenty-first century, the first urban millennium in history.

The policies and strategies to shape urbanisation and solve the emerging problems are being devised in India. The policy resolve and committed action has been missing until late 1990s when the economic liberalisation and reforms occupied central position in the Indian polity. Since then, focused urban development policies and programmes are being launched to meet the urban sector needs in the areas such as housing, poverty alleviation, slum improvement, civic services, urban governance, urban finance, financial management, etc. Formulation of the national level policies like national urban housing and habitat policy, street vendor's policy and urban transport policies are important policy development initiatives of the Government of India to provide proper direction to the urban development endeavours. Urban reforms also received attention of the governments—central and state. They were ad hoc and project specific till 1990s.

An important milestone in the urban local government sector in the Post-Independence era is the Constitution (74th Amendment) Act, 1992. The urban reforms received constitutional backing and support through this Amendment. The Constitution Amendment Act not only provided constitutional status to the Urban Local Government (ULGs) but also provided institutional arrangements for putting the relationships between the states and the ULGs on a firmer footing and provided mandatory provisions like conduct of regular election to ULGs, reservations to SCs, STs and women, State Finance Commissions, District Planning Committees, Metropolitan Planning Committees and also provided a list of municipal functions in the form of twelfth Schedule. Some improvements are visible in the municipal government side after the Constitution 74th Amendment Act. The Government of India has been pressing for implementation of all the mandatory provisions in all the states. The crucial aspects like devolution of functions, powers and finances to ULGs are yet to be implemented in true spirit.

With increased central sector funding, reforming urban governance has become essential to ensure that the funds are properly utilised to achieve the urban development goals. National level urban reforms are being devised and they are being incentivised through central sector funding to motivate the states and municipalities to implement them effectively. The launching of JnNURM in 2005 was the most significant initiative in India to develop critical urban infrastructure and at the same time improve the governing systems through well conceived package of reforms. However, the quality of urban governance remained far from satisfaction. It has been found that in many states the reform agenda was implemented only to be eligible to access the central assistance. The latest intervention i.e. Atal Mission for Rejuvenation and Urban Transformation (AMRUT) has adopted the approach of 'reform', 'perform' and 'transform'. In the last couple of decades, several policy and reform initiatives are being implemented with varying degrees of success. With increasing urbanisation trend in India, urban government reforms are going to play critical role not only in the urban development sector but in the entire national development process as the urban areas are considered generators of economic momentum and growth.

It is in this backdrop, this publication has been taken-up. The central focus of the book is studying urban reforms in Independent India. It covers major urban sector reform areas namely, urban development policies and programmes, urban governance, urban infrastructure development, financial issues of urban development, slum rehabilitation, housing, urban poverty alleviation, inclusive growth, sustainable urban development, urban planning, etc. The publication covers the entire gamut of urban reforms in the above fields. The major focus is on the nature of reforms, implementation dynamics, achievements, failures, dilemmas and future vision.

The field of urban reforms is quite wide encompassing several sub-sectors like urban planning, infrastructure, urban poverty, urban governance, service delivery system management, etc. Sustainable development demands integration of these interdependent reforms in urban areas. It is really a herculean task demanding the cooperation of several organisations belonging to the central and state governments. In view of such limitations, we should be very careful in assessing, commenting and discussing about the implementation of the urban reforms.

The book contains ten chapters of critical significance in the urban development sector in India. Chapter-One highlights the *Urban Policies and Development Strategies*. National urban policies are of recent development in India. Most of the urban development is taking place without any national level directions and guidance. The Five Year Plans provide some policy vision and direction. While drawing from these Plans and other governmental initiatives, the chapter provides a comprehensive vision and direction on the urban policies and urban development strategies and programmes. It critically examines the urban policies since Independence and identifies the emerging trends. Various urban development programmes are analysed from the perspective of urban development needs.

Chapter-Two entitled *Urban Reforms in India--An Overview* critically examines the national urban reform agenda in India. The emergence of the national urban reform agenda and experiences of the implementation are analysed. The focus of reforms, problems and issues and emerging trends are covered comprehensively. Chapter-Three on *Reforming Urban Governance* examines the implementation of the Constitution Amendment and other reforms relating to decentralisation, e-governance, participation, financé and accountability, etc. The urban governance reforms as such received national focus since the 74th Constitution Amendment. This apart, several reforms in the areas of finance, municipal organisations, functioning, e-governance, etc. are under implementation in the municipalities.

Chapter-Four throws light on *Decentralised Participatory Urban Governance*. Decentralised and participatory governance are essential to meet local people needs effectively. The Constitution 74th Amendment Act provides the institution of ward committees to achieve this objective. Unfortunately, this reform has not received necessary attention of the state governments and municipalities. In recent past the Government of India proposed area sabha concept also to involve the local communities in urban governance and decision-making process. This chapter critically analyses the problems and issues in implementing this reform while highlighting the success stories.

Chapter-Five details the *Property Tax Reforms* as Property tax is the main source of income to the municipalities since Independence. Reforms in property tax received attention since 1980s in India. This is the most important area and at the time several reforms are introduced aiming at standardisation and revenue improvement. This chapter traces all the reform initiatives in property tax and critically examines the current reform trends in the light of the reforms agenda envisaged under erstwhile JnNURM and existing mega mission mode action plan based initiative of AMRUT.

Chapter-Six on *Municipal Accounting Reforms* examines the accounting reforms and the status of their implementation. Municipal financial management is the most neglected area in India. But, with the entry of the international agencies like the World Bank, the situation is slowly changing. Further, the increasing investments in the urban sector have given rise to concrete measures in this area. Accounting reforms are most important reforms taken-up to improve the financial systems in the municipalities. Introduction of the accrual based double entry accounting is the major thrust area. Several state and municipal level reforms are underway in this respect. This is one important area where e-governance applications are being used extensively.

Chapter -Seven deals with *Urban Land as Resource.* Innovative resource mobilisation is one crucial area being proposed to mobilise funds from the critical urban infrastructure development in India. The importance of land is being recognised though late in India. Monetisation of the urban land is one great potential area for resource mobilisation. This chapter examines the potential and mechanisms of exploiting land based revenue generation by the municipalities. This is the emerging area and an attempt is made to analyse the basic issues and proposed reforms to utilise urban land as resource for urban infrastructure development and housing. Chapter-Eight entitled as *Reforming Urban Planning* examines the emerging trends in urban planning and the initiatives being take-up at the state and local levels. Urban planning suffers from too much focus on master plan approach in the country since Independence. The developmental focus has been missing. But, with the realisation of the environmental concerns and rising spatial and social inequities, urban planning is being viewed in a broader developmental perspective. Several reforms are being implemented to promote green environment, green buildings, sustainable urban development and the Constitution 74th Amendment provided institutional arrangements in the form of district planning committees to make urban planning people centred.

Chapter-Nine deals with *Housing the Poor—Shelter Reforms.* With increasing urbanisation, the issue of slums received attention of the governments since Independence in India. Several initiatives are being made to provide basic civic and social services and employment opportunities through urban poverty alleviation programmes. But, housing and shelter issues have not received required attention till the National Urban Housing and Habitat Policy has been formulated. The JnNURM and RAY projects gave big push to this area, but nothing significant changed at the ground. The new initiative viz. Housing For All—Prime Minister's Awas Yojana (HFA-PMAY) hence commits to provide housing to all by 2022. New innovative mechanisms in the building technology, financing and land are being devised to address the urban housing problems especially to the poor. This chapter examines the recent policy trends and reforms introduced in this area and identify the emerging trends and opportunities. Chapter-Ten concludes with *Reflections and Way Forward.* This final chapter while reflecting on the reform initiatives in the urban sector in the vital areas, maps out the emerging scenario. The constraining factors and potential opportunities are critically assessed. Finally, it provides directions to achieve the future vision of urban India.

This effort could not have been succeeded in the absence of support and help received from various angles. I wish to convey my gratitude and thanks to all - the bosses and colleagues at the Regional Centre for Urban and Environmental Studies (RCUES), Osmania University, Hyderabad friends and well-wishers.

The long association with the RCUES has undoubtedly provided ample opportunities in gaining required insight into the subject. In fact, this has been the main motivating factor to venture into such a complex area of urban reforms. It is truly the appropriate time and duty to express my gratitude to the successive Directors, beginning from Professor M.A. Muttalib to Professor Bhupatthi Rau; colleagues especially, Professor Malla Reddy and Professor Ashok Kumar and friends at the RCUES. The RCUES has a very good repository of books, journals and other documents pertaining to urban development, always the favourite place for me. The RCUES librarians provided access to various materials most willingly whenever requested for and I thank all of them.

The various officers at the central and state government levels offered opportunities to work in the urban sector and I thank them all for their help and support.

Finally, I express the heartfelt feelings and thanks for the tolerance, affection and support provided by my family members—wife Swayam and daughters—Deepu, Preeti and son Avinash.

Hyderabad **V. Gnaneshwar**

Writing a book, more so critically analysing the governmental interventions, is really an arduous task. But a momentum is gained when one has devoted whole life to the cause of urban issues—as a teacher, a trainer and a researcher. In this venture cooperation, motivation, inspiration, positive criticism, valuable suggestions and assistance were received from various corners and individuals, including colleagues, friends, government officials, people's representatives, academicians, researchers, family members, etc. I am indebted to all. I have consulted a large number of volumes, municipal laws and rules and reports; visited various related websites; and quoted many authors extensively. I express our sincere gratitude to all. Most of the part of this work belongs to Professor V. Gnaneshwar. My contribution was limited to making certain additions and deletions, here and there. I did only editing and updated the manuscript, added a few case studies and thus made a sincere attempt to highlight the issues on national canvas. This little contribution of value additions however does not lessen my responsibility. I am thankful to Professor Gnaneshwar for his generosity and sharing the manuscript for its updating and making need-based modifications.

The book presents a glimpse of issues related to reforming urban governance. The case studies discussed would necessarily help in understanding the implementation status of urban reform agenda in many states and selected progressive municipalities across the country. This will help guide and motivate other states to realise the importance and benefits of reforming grass-roots cutting edge administration. I am sure the volume will be of immense help to all those having interest in providing quality urban governance.

We are thankful to Kalpaz Publications, Delhi for publishing the book in a record time.

Lucknow **U.B. Singh**
Ashwin Navratra, 2016

Abbreviations

ABPAS	:	Automatic Building Plan Approval System
ACA	:	Additional Central Assistance
ADB	:	Asian Development Bank
AMRUT	:	Atal Mission for Rejuvenation and Urban Transformation
APS	:	Automatic Plan Sanction
APUSP	:	Andhra Pradesh Urban Services for the Poor
ARC	:	Administrative Reform Commission
ARV	:	Annual Rental Value
ATR	:	Action Taken Report
BBMP	:	Brihat Bangalore Mahanagar Palike
BPL	:	Below Poverty Line
BREEAM	:	Building Research Establishment's Environmental Assessment Method
BSUP	:	Basic Urban Services for the Poor
74th CAA	:	Constitution (Seventy-fourth) Amendment Act
C&AG	:	Comptroller & Auditor General
CBO	:	Community Based Organisation
CCBP	:	Comprehensive Capacity Building Programme
CPCB	:	Central Pollution Control Board
CSBEE	:	Comprehensive Assessment System for Building Environmental Energy
CSO	:	Civil Society Organisation
CSO	:	Central Statistical Organisation
CV	:	Capital Valuation
DCB	:	Demand, Collection and Balance
DEAS	:	Double Entry Accounting System
DFID	:	Department for International Development

DPC	:	District Planning Committee
DUDA	:	District Urban Development Agency
ECBC	:	Energy Conservation Building Code
EIUS	:	Environmental Improvement of Urban Slums
EMI	:	Equitable Monthly Installment
EWS	:	Economically Weaker Section
FAR	:	Floor Area Ratio
FBAS	:	Fund Based Accounting System
FDI	:	Foreign Direct Investment
FIRE-D	:	Financial Institutional Reforms and Expansion
FSI	:	Floor Space Index
FY	:	Financial Year
GAAP	:	Generally Accepted Accounting Principles
GDP	:	Gross Domestic Product
GEM	:	Generator of Economic Momentum
GHMC	:	Greater Hyderabad Municipal Corporation
GIS	:	Geographical Information System
GoI	:	Government of India
GPS	:	Geographical Placing System
GPRS	:	Geographical Placing Recording System
GRIHA	:	Griha Rating for Integrated Habitat Assessment
GTPUD	:	Gujarat Town Planning and Urban Development Act
HDFC	:	Housing Development Finance Corporation
HFC	:	Housing Finance Companies
HIG	:	Higher Income Group
HK-BEAM	:	Hong Kong Building Environmental Assessment
HMDA	:	Hyderabad Metropolitan Development Authority
HMWS&SB	:	Hyderabad Metropolitan Water Supply and Sewerage Board
HPEC	:	High Power Expert Committee
HRIDAY	:	Heritage City Development and Augmentation Yojana
HUDCO	:	Housing and Urban Development Corporation
ICDS	:	Integrated Child Development Scheme
ICFAI	:	Council of the Institute of Chartered Accountants of India
ICT	:	Information and Communication Technology
IDBI	:	Industrial Development Bank of India
IDFC	:	Industrial Development Finance Corporation
IDSMT	:	Integrated Development of Small and Medium Towns
IHSDP	:	Integrated Housing and Slum Development Programme

IIT	:	Indian Institute of Technology
ILCS	:	Integrated Low Cost Sanitation
IL&FS	:	Infrastructure Lease and Financing Services
IT	:	Information Technology
IUDP	:	Integrated Urban Development Programme
JnNURM	:	Jawaharlal Nehru National Urban Renewal Mission
KMABR	:	Karnataka Municipalities (Accounting & Budgeting) Rules
KMC	:	Kanpur Municipal Corporation
KRSAC	:	Karnataka Remote Sensing and Application Centre
LAA	:	Land Acquisition Act
LCS	:	Low Cost Sanitation
LEED	:	Leadership in Energy and Environmental Design
LICI	:	Life Insurance Corporation of India
LIG	:	Lower Income Group
LPCD	:	litre per capita daily
MEPMA	:	Mission for Elimination of Poverty in Municipal Areas
MIG	:	Middle Income Group
MIS	:	Management Information System
MML	:	Model Municipal Law
MMRDA	:	Mumbai Metropolitan Regional Development Authority
MOA	:	Memorandum of Agreement
MoHUPA	:	Ministry of Housing and Urban Poverty Alleviation
MoUD	:	Ministry of Urban Development
MPC	:	Metropolitan Planning Committee
MPUSP	:	Madhya Pradesh Urban Services Programme
MT	:	Metric Tonne
NAPCC	:	National Action Plan for Climate Change
NBO	:	National Building Organisation
NCU	:	National Commission on Urbanisation
NeGP	:	National e-Governance Plan
NGO	:	Non Government Organisation
NHB	:	National Housing Bank
NITI Ayog	:	National Institutional for Transforming India
NMAM	:	National Municipal Accounts Manual
NMMP	:	National Mission Mode Programme
NMSH	:	National Mission on Sustainable Habitat
NPC	:	National Priority City
NPV	:	Net Present Value

NRW	:	Non Revenue Water
NRY	:	Nehru Rozgar Yojana
NSDP	:	National Slum Development Programme
NSSO	:	National Sample Survey Organisation
NSUP	:	National Strategy for Urban Poor
NULM	:	National Urban Livelihood Mission
O&M	:	Operation and Maintenance
OBC	:	Other Backward Class
OSRTM System	:	Off Site Real Time Monitoring System
PFDF	:	Pooled Finance Development Fund
PMAY	:	Prime Minister's Awas Yojana
PMIUPEP	:	Prime Minister's Integrated Urban Poverty Eradication Programme
PPP	:	Public Private Partnership
PRI	:	Panchayati Raj Institution
PROOF	:	Public Record of Operational Fund
RAY	:	Rajiv Awas Yojana
RCA	:	Rent Control Act
RCC	:	Reinforced Concrete and Cement
RRY	:	Rajiv Rin Yojana
RWA	:	Resident Welfare Association
SAAP	:	State Annual Action Plan
SAS	:	Self Assessment Scheme
SBM	:	Swachha Bharat Mission
SC	:	Scheduled Caste
SEC	:	State Election Commission
SEWA	:	Self Employed Women's Association
SEWS	:	Socially and Economically Weaker Section
SFC	:	State Finance Commission
SJSRY	:	Swarna Jayanti Shahri Rozgar Yojana
SLB	:	Service Level Benchmark
SLIP	:	Service Level Improvement Plan
SPC	:	State Priority City
SPUR	:	Spatial Priority Urbanisation Region
ST	:	Scheduled Tribe
STP	:	Sewage Treatment Plant
SURDPB	:	State Urban and Regional Development Planning Board
SWAS	:	Single Window Approval System

SWM	:	Solid Waste Management
TCPO	:	Town and Country Planning Organisation
TNUDP	:	Tamil Nadu Urban Development Programme
TOD	:	Transit Oriented Development
TQM	:	Total Quality Management
TSC	:	Total Sanitation Campaign
UAV	:	Unit Area Value
UBS	:	Urban Basic Services
UBSP	:	Urban Basic Services for the Poor
UCD	:	Urban Community Development
UDA	:	Urban Development Authority
UIDSSMT	:	Urban Infrastructure Development Scheme for Small and Medium Towns
UIG	:	Urban Infrastructure and Governance
ULCRA	:	Urban Land (Ceiling and Regulation) Act
ULG	:	Urban Local Government
UNDP	:	United Nations Development Programme
URDP	:	Urban and Regional Development Plan
URIF	:	Urban Reforms Incentive Fund
USAID	:	United States' Aid for International Development
VAMBAY	:	Valmiki Ambedkar Awas Yojana
WDP	:	Ward Development Plan

Chapter - 1

Urban Policies and Development Strategies in India

Introduction

India has been witnessing an unprecedented growth of urban population as also population shift from rural areas to cities and towns. The contribution of urban population which was less that 10 per cent a century back increased to over 31 per cent by 2011. This may look better compared to other large developing countries like China (45 per cent), Indonesia (54 per cent), Mexico (78 per cent) and Brazil (87 per cent). But, the most disturbing aspect is the population numbers which stood at 377 million in 2011 and projected to increase to 473 million by 2021, 600 million by 2031, and 820 million by 2051, adding about 100 million every decade.[1] The current urbanisation debate, therefore, is not about rural-urban divide but how to integrate the interests of the urban areas with the rural areas, and, core areas with the peri-urban areas in the metropolitan city regions where the focus of growth has been witnessed in recent periods. The urban policies in India should therefore focus on options for accommodating population in municipal and beyond municipal boundaries in terms of housing, jobs, transport and infrastructure, that is, nurturing metropolitan economies in the words of the World Bank. The unprecedented increase in number of census towns, containing about one-fifth of the urban population and beyond control of any urban law, is a peculiar facet of recent urbanisation trend in the country. At the other spectrum, environmentalists are cautioning about impending climatic change threats due to metro centred and unregulated urbanisation, and are advocating for sustainable habitat development. The rising number of urban poor is another dimension demanding the attention of policy-makers and administrators. It is in this context, the urban policies and strategies of urban development become crucial to address the emerging problems and developmental goals. India faces critical challenge of putting in place effective governance structures and strong politico-administrative commitment to bring desirable urban transformation in terms of urban form, infrastructure and development in cities and towns in the twenty-first century, the first urban millennium in history. This chapter discusses the shifting policy response to critical urban problems in India since Independence and examines the strategies adopted to deal with the emerging problems with the objective of understanding the urban paradoxes and predicaments.

The Metropolitan Challenge

The urbanisation trend in India indicates that the urban population which was 78 million in 1961 increased to 377.1 million by 2011, near about five fold increase in five decades (Table 1.1), but the level of urbanisation increased from 18 per cent to 31 per cent only, less than two times. The differences with respect to definition of urban areas in India compared to other globally comparable measures make the international comparisons misleading. The global criteria like 150 people per sq. km., is far less compared to Indian criteria of 400 people per sq. km. If such criterion is adopted in India, the urbanisation would be 52 per cent or more.[2] The Census 2011 indicates that though 2,774 additional settlements exhibited urban characteristics between 2001 and 2011, only 242 are officially declared urban areas while the others remained as rural areas but called towns for census purposes.[3] This poses difficult problems to the policy-makers and administrators as no proper legal and institutional set-up exists to monitor development in these areas.

The critical factor is net decadal addition of urban population which rose from 30 million in 1971 to 91 million by 2011. The crux of the issue pertains to the shifts in the urban form which is undoubtedly towards larger and metropolitan cities. For instance, the class I cities (cities having a population of one lakh and above) contributed about two-third to total urban population in 2011. The more disturbing aspect is the spectacular growth of million plus and top order cities.

Table 1.1: Urbanisation Trends in India

Census year	*Urban population in millions*	*Net population increase in millions*	*Urban population as per cent of total population*
1961	78.9	-	17.96
1971	109.1	30.2	19.90
1981	159.5	50.4	23.34
1991	217.6	58.1	25.71
2001	286.1	68.5	27.81
2011	377.1	91.0	31.6

Source: JnNURM Directorate, Ministry of Urban Development and NIUA, *India's Urban Demographic Transition, 2011 Census Report.*

Due to feeble policy and administrative response to the emerging imbalance in urban growth in India, the larger and million plus cities continue to grow faster in number as well as in population. The million plus cities which were just five, namely, Delhi, Mumbai, Kolkata, Chennai and Hyderabad, at the time of Independence, grew to 12 by 1981 (Table 1.2). The growth was phenomenal since 1981 as the number almost doubled to 23 in a decade. This trend continued in the following decades too as the number of million plus cities increased to 53 by 2011 accounting for nearly 43 per cent of the total urban population. Today the top 10 million plus cities have a population of 93 million that is, more than 58 per cent of the total population of million plus cities. The metropolises of Delhi, Kolkata and Mumbai contained collectively about 15 per cent of urban population and five per cent of the total population of the country in 2014.[4] The projections indicate that the million plus cities will increase to 87 by 2031 where 255 million people will live, the top 10 million cities will have 140 million people, and the level of urbanisation is expected to rise to 50 per cent by 2039. Another dimension of the urbanisation in India has been faster growth of the newly developing metros against the reduced growth of the primate metropolitan cities like Mumbai and Kolkata as out of ten fastest growing million plus cities, eight are new.[5]

Table 1.2: Growth of Million Plus Cities in India

Census Year	*Number*	*Population (millions)*	*Percentage of total urban population*
1901	1	1.5	5.8
1911	2	2.8	10.6
1921	2	3.1	11.1
1931	2	3.4	10.2
1941	2	5.3	12.2
1951	5	11.7	18.8
1961	7	18.1	22.9
1971	9	27.8	25.5
1981	12	42.1	26.4
1991	23	70.7	32.5
2001	35	107.9	37.6
2011	53	160.7	42.6

Compiled from Census of India, Provisional Population Tables, Rural-Urban Distribution, 1991; JnNURM Directorate, Ministry of UD & NIUA, India's Urban Demographic Transition-2011 Census.

A recent World Bank report observed that the population densities in and around the largest metropolitan areas are extremely high. This trend is challenging as well as a potential opportunity. The average densities are 2,450 persons per square kilometer in the 50 kilometer vicinity of the seven largest metropolitan areas with population above four million in 2001. Further, it is observed that a third of India's new towns came-up in a 50 km neighbourhood of existing cities of more than one million people.[6] The employment angle also demonstrates the metropolitanisation trend in India. The stringent density and land use regulations, rising costs of living and housing prices at the one end, and the liberalised economic policies at the other are pushing poor, middle class and the entrepreneurs into the metropolitan regions beyond 10 km. radius of core cities. The growth of manufacturing employment with 54 per cent during 1998-2005 was fastest in rural areas adjacent to the largest metropolitan areas.[7] As these areas lack basic infrastructure and good connectivity, the World Bank report suggests that the Indian policy-makers and administrators should focus on land reforms, infrastructure development and connectivity provision in these areas in an integrated way.

Urbanisation Policies in India

Urban policies constitute two basic components viz., urbanisation policies and urban development policies. If the urbanisation policy deals with the spatial dimensions of growth of cities and towns of various sizes, the urban development policies primarily deal with the issues and problems arising out of urban growth such as land issues, housing, infrastructure, civic services, job creation, mobility, etc.[8] Urbanisation policies assume importance in the developing countries because the rapid changes in the settlement pattern impacts the economic efficiency and politico-administrative environment of a country. For instance, concentrated urban growth leads to migration of people from rural areas to urban areas and these shifts often become unmanageable for the administrators.

Bertrand Renaud (1981) summarises well the objectives of urban policies as the following: rapid changes in the settlement patterns create primarily four kinds of imbalances, viz., between the rural and the urban sectors, between the level of development of different regions, between cities of different sizes and, between social groups within cities. Correspondingly, national urbanisation policy has four major objectives: the full development of the national resources of the country, the maintenance of national cohesion among various regions, the prevention or correction of excessive concentration of economic activities within the capital region, and the more efficient and more equitable management of growth within the cities.[9]

The Town and Country Planning Organisation, Ministry of Urban Development, Government of India, for the first time prepared a policy paper on National Urbanisation Policy Resolution in 1975. The policy paper identified following as objectives of national urbanisation policy: [10]

- Evolving a spatial pattern of economic development and location of a hierarchy of human settlements consistent with the exploitation of natural and human resources of the region, and ensuring functional linkages inter se;

- Securing the optimum distribution of population between rural and urban settlements within which each region and also among the towns of various sizes;
- Securing the distribution of economic activities in small and medium-size towns and in new growth centres in order to achieve maximum economic growth for the future;
- Controlling and, where necessary, arresting further growth of metropolitan cities by dispersal of economic activities, legislative measures and establishment of new counter-magnets in the region;
- Providing a minimum level of services for improving the quality of life in rural and urban areas and reducing gradually the differences between rural and urban living; and
- Regulation of rapid growth and decongestion of central areas in metropolitan cities.

The concern for fast emerging spatial imbalances was voiced as back as 1970 in India. John P. Lewis,[11] while discussing the emerging disparities in the regional development, warned that unless corrective measures were taken by the end of 1970, the situation could go out of control. While recognising the gravity of the situation, the Government of India constituted a Task Force in October 1975 to study the problems of small and medium towns in India. The Report of the Task Force (1977) clearly indicated that the problems could be solved only by a multi-dimensional approach involving spatial, social and economic facets of development.[12]

In response to the emerging spatial disparities between the million plus cities and smaller cities and towns, the Government of India launched a scheme called the Integrated Development of Small & Medium Towns (IDSMT) during the Sixth Five Year Plan period (1980-85). This was the first administrative response to the unbalanced urban growth in India. The IDSMT scheme was formulated with the objective of developing small and medium towns having population below one lakh with infrastructural and service facilities as counter magnets to arrest migration of people from rural to larger and metropolitan cities. Though this scheme was continued in the subsequent Plan periods with increased outlay covering cities up to three lakh population till the launch of the Jawaharlal Nehru National Urban Renewal Mission (JnNURM) in 2005, its spatial impact has been insignificant. The faster growth of million plus cities during the period 1981-1991 when the scheme was launched and in the subsequent decades clearly proves this point beyond doubt.

The unprecedented growth of million plus cities in India reflects the economic growth taking place in the country for which cities play very important role. This was first acknowledged by the National Commission on Urbanisation (1988). Setting-up of the National Commission on Urbanisation on 4th October, 1985, is a landmark in the development of urban policy in India. The report of the National Commission on Urbanisation (NCU) analysed the urban problems in quite detail in the light of earlier efforts and emerging trends and presented a comprehensive report on urban policies and strategic interventions required. While reiterating the need for decentralised urban growth, the NCU provided certain specific policy directions on various subjects like environment, planning, local government, poverty, transport, land, housing and infrastructure. Close link between urbanisation and economic development was emphasised by the NCU.[13] It stated that "urbanisation can make a positive contribution to economic growth only when cities generate economic momentum, their economic base expands and productivity increases." The concept of 'backward area' was abandoned because it was felt that instead of inducing investments in areas which are backward and have little infrastructure and in which the concessions are likely to be misused, the identified existing and potential urban centres at intermediate levels could be developed to attract the migrants as they are located in closely interrelated regions. To achieve this, the concept of `Generators of Economic Momentum' (GEMs) was introduced. Focused investments on select towns based upon their potential to generate economic momentum were considered a better strategy. In all, 329 cities were identified as GEMs. These cities were again re-grouped into 77 National Priority Cities and 252 State Priority Cities.[14] The NCU recommended for increased outlay (eight per cent from the existing four per cent) at least to meet the growing urban development needs.[15] On institutional support, urban sector specific financing institutions and the state finance commissions were also put-forth by the NCU.

The central Town and Country Planning Organisation prepared a draft National Urban Policy in 1992 while considering the recommendations made by the NCU. The major objectives of the national urbanisation policy identified in this policy were:[16]

- Evolving a spatial pattern of economic activities and population distribution based on regional development and planning considerations, and location of a hierarchy of human settlements consistent with the exploitation of the natural resources in the region;
- Securing a balanced distribution of population among the urban centres of various sizes, so as to maximise economic gain and minimise social costs of urban development;

- Controlling further growth of metropolitan cities by dispersal of economic activities in the new growth centres to prevent them from becoming national liabilities so as to reduce the sector costs in the growth of large cities;
- Priority in development be given to those urban centres which have been identified as prime economic movers in national economic development, such as the National Priority Cities (NPCs), State Priority Cities (SPCs) and Spatial Priority Urbanisation Regions (SPURs);
- Securing the distribution of economic activities in small and medium towns and in new growth centres in order to induce the desired population distribution;
- Improving the efficiency of the urbanisation process and making urbanisation more productive by removing bottlenecks and breakdowns in the supply of urban services which impose costs on workers, businessmen and consumers; and
- The process of urbanisation can and must be used to improve agricultural performance and create localised employment opportunities.

The Eleventh Five Year Plan observed that the time is ripe to formulate a long-term National Urbanisation Policy indicating the emerging pattern of urbanisation and measures to channelise future urban growth in an equitable and sustainable manner. It should encompass the changing economic scenario, reform agenda of Jawaharlal Nehru National Urban Renewal Mission (JnNURM), and the role of urban local governments (ULGs) in the era of globalisation. The objectives of balanced and sustainable development are to be achieved by reducing spatial disparities.[17]

The economic importance of urban areas continues to be the focus of urban thinkers in the country. The Twelfth Five Year Plan comments that the urban areas are engines of economic growth and this is reinforced by the economic data on the urban share of the Gross Domestic Product (GDP) for the Indian economy. As per CSO data, the share increased from 37.7 per cent in 1970-71 to 52 per cent in 2004-05 and the mid-term appraisal of the Eleventh Plan projected the urban share of GDP at 62-63 per cent in 2009-10[18] and this is expected to increase to 75 per cent by 2021.[19] In this context, the Twelfth Plan observes that urbanisation will be central to India's strategy of achieving faster and more inclusive growth because agglomeration and densification of economic activities (and habitations) in urban conglomerations stimulates economic efficiencies and provides more opportunities for earning livelihoods. Urbanisation thus, increases avenues for entrepreneurship and employment compared to what is possible in dispersed rural areas.

The rural-urban linkages were also highlighted by the High Power Expert Committee (HPEC) under the chairpersonship of Dr. Isher Judge Ahluwalia. The Committee (2011) has observed that the fortunes of the agricultural sector are crucially linked to the manner in which growth in the industry and services sectors unfolds. People living in rural areas typically tap the opportunities that cities provide for employment, entrepreneurial avenues, and education. As urbanisation grows demand for food items other than food-grains that is vegetables, lentils, milk, eggs and so on, also grows. This leads to investments in infrastructure, logistics, processing and packaging in rural and peri-urban areas. Such investments and other economic inter-linkages connect and build synergy between rural and urban centres. The rural sector also benefits from good management of neighbouring urban conglomerations.[20]

Apart from these think tank observations on urbanisation policy, the national level comprehensive urbanisation policy has not emerged in India. The urbanisation was neither properly articulated into a comprehensive policy national frame nor managed effectively. The attention of the government has been only on options to meet the immediate developmental demands of the people. We shall discuss about these options and strategies in what follows.

Urban Development Strategies

Rapid and unbalanced urban growth has led to several development problems in urban India. Haphazard development of cities and towns without proper planning and required civic infrastructure and services has been critical issue in the country. On infrastructure front, huge gap in the availability and requirements raises several administrative, financial and social issues. At the other spectrum, continuous trend of centralisation and bureaucratisation had weakened the urban local governments (ULGs) in terms of resources, powers and competencies. The cumulative effect of all these factors has created complex problems and challenges. Key challenges of urban development in India could be summarised as below:

- Polarisation of economic growth leading to metropolitanisation or urban form,
- Weak public institutions and bureaucracy,
- Increasing urban infrastructure gaps leading to wide inequities,
- Rising urban poverty, informal sector and social and financial exclusion,
- Low quality of services due to inefficient delivery systems,
- Poor revenue base of Urban Local Governments, and
- Threatening environmental concerns.

Spatial and social disparities within cities and between cities and, regions within the same country are growing as some areas benefit more than others from public services, infrastructure and other investments. High levels of urban inequality present a double jeopardy. They have a dampening effect on economic growth and contribute to a less favourable environment for investment.[21] The current urban space is becoming complexly dynamic and fragmented with increasing gap between the smart cities which are adapting well and the lagging cities which are not and, emergence of enclaves of rich and poor communities adding to social tensions.

Globalisation at the other spectrum is interlinking the national economies on the principles of free trade, comparative advantage and global competition. Cities of the developing countries are striving to adapt and prosper in the changed scenario. Economic restructuring of regions and cities is taking place where few advantageous cities in terms of attractive location, economic dynamism, political commitment, telecom connectivity, availability of high quality labour and urban infrastructure, are able to succeed in the global competition by innovative development strategies and financing mechanisms.

Balanced growth and integrated development of cities and towns remained pious policy objectives in India since Independence. As no concrete measures were taken on this front, the cities, especially the metropolitan cities, grew purely on economic factors. The economic significance of cities is well recognised but commensurate response has been lacking to develop the cities through proper planning. Whatever administrative measures were taken, they were only to meet the pressing development problems in particular areas. As a result, sector specific urban policies and strategies are being formulated and implemented through various centrally sponsored schemes and reforms. We shall discuss these initiatives by a review of the Five Year Plans in India.

Urban Development Strategies of the Five Year Plans

The experiences of planning era in the country reflects that massive investments were made in the core sectors like industries, agriculture, irrigation, power, transport, etc., but the dimension of urbanisation which is the logical outcome of these investments was ignored. Comprehensive effort was not made to link economic and social development efforts with the growth of urban areas through a well-formulated policy frame. Nevertheless, the Five-Year Plans formulated certain urban development objectives. It was only in the mid-1970s that the ideas on urban policies got crystallised into certain concrete measures. The liberalisation and privatisation drive injected through the economic reforms in the post-1990 periods had profound impact on the urban development strategies in India. In tune with the emergence of cities as major economic nodes of development, urban infrastructure, slum development, urban poverty alleviation and capacity building of the urban local governments (ULGs) received major thrust in the Five Year Plans at various points of time.

It is interesting that the need for planned development of urban areas was recognised in the very First Plan in 1951. But, the thrust of the First Five Year Plan (1951-1956) was on housing problems. The Second Five Year Plan (1956-1961) emphasised the preparation of master plans for existing and new towns (especially in the project areas). The urban policies began to take shape during the Third and Fourth Five Year Plans. The close linkage between industrial location and urban development received due recognition in the Third Plan (1961-1966). It suggested the strategy of industrial diffusion by setting-up new industries as far away as possible from large cities. The concern for rapid growth of metropolitan cities like Mumbai, Kolkata and Chennai was expressed in the Fourth Plan (1969-1974) and it recommended decongestion and dispersal of urban growth from these large cities. Need for an urban land policy at state level was recognised. This apart, the Plan expressed that the urban land policy should aim at[22]: (a) the optimum use of land; (b) making the land available for weaker sections, (i.e., poor people); (c) reducing or preventing the concentration of land ownership, rising land values, and speculation of land; and (d) allowing land to be used as a resource for financing the implementation of city development plans.[6]

The Fifth Five Year Plan (1974-1979) provided a more detailed statement on urban problems, largely drawing from the ideas expressed in the National Urbanisation Policy Resolution of the Town and Country Planning Organisation. The objectives of the urbanisation policy formulated in the Plan document were:

(i) To augment civic services in cities and towns;

(ii) To tackle the problems of metropolitan cities in a more comprehensive and regional perspective;

(iii) To promote the development of small towns and new urban centres including industrial towns; and

(iv) To assist the metropolitan development projects having national significance.

Development of small and medium sized towns received considerable importance in the Sixth Five Year Plan (1980-1985). The Plan stated that the thrust of the urbanisation policy during the 1980s should be the provision of adequate infrastructure and other facilities in small and medium towns which had been neglected so far.[23] The Integrated Development of Small and Medium Towns (IDSMT) scheme was launched to develop small and medium towns.

The Seventh Five Year Plan (1985-1990), while asserting that planning of urban development should be supportive of the economic development in the country, pleaded for channeling of industrial investment into the small and medium towns so that migration to metropolitan cities could be controlled.[24] This apart, the Plan outlined the following objectives for the urban development sector[25]:

(i) Planned and integrated development of small and medium towns and cities along with slowing down of the growth of big metropolises;

(ii) Revitalisation of civic bodies;

(iii) Reform of municipal administration in general;

(iv) Concentration on the improvement of slums and the provision of basic municipal services; and

(v) Regular devolution of funds from the state governments and establishment of the necessary institutional framework for channeling capital funds for the improvement of urban infrastructure.

The strategy suggested was consolidation and operationalisation of spatial and economic dimensions of planning by a spate of measures: involving an integrated hierarchy of rural and urban settlements based upon primary economic functions, and linking the urban development plans with respective district level planning processes including the programmes of various state-level and central departments like agriculture, rural development, environment, telecommunications, industries and other such organisations. Convergence of all related programmes, i.e., IDSMT, housing and infrastructure development programmes of Housing and Urban Development Corporation (HUDCO), Nehru Rozgar Yojana (NRY) and Urban Basic Services for the Poor (UBSP) to create the desired impact in small and medium towns beyond the threshold level; taking legal, organisational and financial measures for enhanced and equitable supply of urban land and promotion of housing; evolving a decentralised framework of urban government with necessary powers and resources and participation of local communities in the planning and implementation process; promoting private-public sector partnerships in the urban development sector; and developing appropriate specialised institutional support at the central and state levels to deal with financing and development of urban infrastructure.

Apart from the urban poverty related issues, strengthening of urban governance received big push during the Eighth Five Year Plan period (1992-97). The enactment of the Constitution 74th Amendment Act, 1992 was a great achievement in this respect. The Ninth Five Year Plan (1997-2002) stressed the need for evolving a long-term national spatial strategy incorporating environmental and sector specific policies while integrating them at state, district and city levels. The Plan also emphasised the need for democratic decentralisation on the lines of Constitution 74th Amendment Act. Treating land as a resource and urban poverty reduction were other strategies identified.[26]

The Tenth Five Year Plan (2002-07), while recognising the role of cities as important nodal centres for providing services like marketing, health-care, education, etc. emphasised infrastructure development, delivery improvement and resource mobilisation through tax and non-tax sources like land. By recognising the significance of a Draft National Slum Policy drawn in 1999 by the Department of Urban Employment and Poverty Alleviation, Government of India identified certain objectives for implementing the slum policy. They were:[27]

(i) To create awareness of the underlying principles that guide the process of slum development and improvement and the options that are available for bringing about the integration of these settlements and the communities residing there with the urban area as a whole;

(ii) To strengthen the legal policy framework to facilitate the process of slum development and improvement on a sustainable basis;

(iii) To ensure that the slum population is provided civic services, amenities, and economic opportunities to enable them to rise above the degrading conditions in which they live;

(iv) To arrive at a policy of affirming the legal and tenurial rights of the slum dwellers;

(v) To establish a framework for involving all stakeholders in the efficient and smooth implementation of policy objectives; and

(vi) Strengthening urban local governments through capacity building and better financial management.

The Jawaharlal Nehru National Urban Renewal Mission (JnNURM), a mega urban infrastructure and governance reform scheme was launched in December 2005. This scheme was the biggest in terms of financial outlay after Independence in the country. The focus of urban development thus, shifted to urban infrastructure development and governance reforms during the Eleventh Five Year Plan period (2007-12). The Plan commented that the Indian cities face certain structural problems. Prominent of these are: the current urban development model starves cities of financing; weak or outdated urban management practices; inappropriate planning systems; unsustainable service delivery models; and lack of focus on urban poor. All these factors are negating the benefits of agglomerations economies of Indian cities. The strategy suggested by the Plan was to increasing the efficiency and productivity of cities by deregulation and development of land. Reflecting the ongoing urban reform initiatives in the country, it advocated strengthening of urban local governments through capacity building and better financial management, and dismantling of public sector monopoly over urban infrastructure and creating conducive atmosphere for the private sector to invest and establishing autonomous regulatory framework to oversee the functioning of the public and private sector. Reducing incidence of poverty and using technology and innovation in a big way were the other objectives outlined in the Plan.[28]

Inclusive growth and sustainable development are the themes echoed by the Twelfth Five Year Plan (2012-17). This reflects the alarming urban disparities and environmental concerns in urban India. The Plan aims at faster, more inclusive and more sustainable growth and envisions that India's urban growth be aligned with the objectives of inclusion and sustainability. Accordingly, it formulated following specific objectives.[29]

(i) Urbanisation should be guided towards inclusive, equitable and sustainable growth of towns and cities with proper civic amenities;

(ii) Urbanisation should become good by ensuring that towns and cities are free from slums and provide adequate opportunities for productive employment and a decent quality of life to all their inhabitants including the poor;

(iii) The cities must provide world class infrastructure and services at affordable costs to give a competitive edge to the economic activities they host by addressing the needs of all people viz., migrant workers, vulnerable sections of society including women and children, and various vulnerabilities like residential, occupational and social associated with urban poverty;

(iv) As an overriding principle, 'people' should be brought to the heart of the urban agenda: for deciding the vision of their city, for choosing the process of reaching that goal, and for taking responsibility for the cleanliness, safety and hospitality;

(v) Future growth should be consistent with cities' natural endowments and the economic potential of the region in which they are situated to achieve environmental sustainability; and

(vi) All cities should be efficient in using available resources particularly energy, water and land.

The snap-shot of the thrust areas identified for the urban sector and initiatives in different Plan periods in India are briefly outlined in Table-1.3.

Table 1.3: Thrust Areas and Initiatives in Five-Year Plans

Five Year Plan	*Thrust Areas*	*Initiatives/Schemes*
I[st] Plan (1951-56)	• Refugee settlement • Housing in metropolitan cities • Curbing land speculation	• House construction for employees and weaker sections • Institution building - Ministry of Works and Housing, NBO, TCPO, etc. • Slum (clearance and improvement) Act, 1956

(Contd...)

IInd Plan (1956-61)	• Preparation of master plans • Housing	• Preparation of master plan for Delhi • Setting-up of Delhi Development Authority • Slum clearance and sweepers housing
IIIrd Plan (1961-66)	• Master plan preparation • Development of State capitals • Land acquisition and development	• Preparation of master plans for major cities • Building of state capitals - Gandhinagar, Bhubaneswar, etc. • Committee on Urban Land Policy
IVth Plan (1969-74)	• Balanced urban growth • Slum development • Financing of urban development projects	• Creation of HUDCO and Urban Development Authorities in metro towns • Launching of Environmental Improvement of Urban Slums (EIUS) scheme in central sector
Vth Plan (1974-79)	• Metropolitan cities • Arresting of migration to metropolitan cities	• Launching of Integrated Urban Development Programme (IUDP) for large cities • Task Force on Small and Medium Towns • Urban Land (Ceiling and Regulation) Act,1976 • Minimum Needs Programme to ensure basic minimum standard of life for all sections of people • EIUS transferred to state sector in1974 • Launch of Integrated Child Development Services (ICDS) scheme
VIth Plan (1980-85)	• Development of small and medium towns • Integrated provision of urban services	• Launching of Integrated Development of Small and Medium Towns Scheme (IDSMT)
VIIth Plan (1985-90)	• Urban poverty alleviation • Urban basic services • Housing	• Setting-up of National Housing Bank (NHB) and Building Materials Technology Promotion Council (BMTPC) • National Housing Policy (1988) • National Commission on Urbanisation (1985) • Launching of Nehru Rozgar Yojana (NRY) (1989) • Initiation of Low Cost Sanitation (LCS) scheme • Pilot project of urban Basic Services (UBS) sponsored by UNICEF
VIIIth Plan (1992-97)	• Employment generation in urban areas • Basic minimum services • Strengthening of urban local governments • Urban poverty alleviation	• Launching of Urban Basic Services for the Poor (UBSP) scheme • Constitution 74th Amendment Act, 1992 • National Housing Policy (1994) • Accelerated Urban Water Supply Programme (AUWSP) focusing on small towns with less than 20,000 population (1993-94) • Night Shelters • Revision of IDSMT guidelines (1995) • Mega City Scheme for Mumbai, Kolkata, Chennai, Bangalore and Hyderabad (1993-94) • Urban Mapping Scheme for preparing Base Maps for towns

(Contd...)

		• Prime Minister's Integrated Urban Poverty Eradication Programme (PMIUPEP,1995 • Enactment of Manual Scavengers (Liberation and Rehabilitation) Act,1993
IXth Plan (1997-2002)	• Development and up-gradation of urban infrastructure • Urban renewal • Housing to poor • Alleviation of poverty • Strengthening municipal governance • Promoting private sector participation in public infrastructure • Cost recovery and commercial approach to urban infrastructure	• Launching of Swarna Jayanti Shahri Rozgar Yojana (SJSRY) and National Slum Development Programme (NSDP) (1997) • National Housing and Habitat Policy (1998) • Two-Million Housing Scheme (1998-99) • Launch of Valmiki-Ambedkar Awas Yojana (VAMBAY) (2001-02) • Draft National Slum Policy (1999) • Rain water harvesting schemes • Solid Waste Management Rules, 2000 for improving urban solid waste management • Model Municipal Law (2003) • Implementation of 74th CAA
Xth Plan (2002-07)	• Public-Private Partnerships • Strengthening of municipalities • Broad based urban reforms • Urban poverty alleviation • Municipal borrowing and debt financing	• Implementation of SJSRY • Integrated Low Cost Sanitation (ILCS) scheme • City Challenge Fund to promote reforms • Creation of UPA cells in municipalities • Launching of JnNURM and its omnibus schemes UIDSSMT and IHSDP • Pooled Finance Development Fund (2006) • National Policy on Urban Street Vendors (2004) • Launch of Prime Minister's Rozgar Yojana (PMRY)

XIth Plan (2007-2012)	• Strengthening of ULGs through capacity building and better financial management
	• Increasing city productivity by deregulation and land development
	• Promoting public-private partnerships
	• Urban poverty alleviation
	• Governance improvement
	• Making cities slum free
		• Implementation of JnNURM and its sub-missions
		• Affordable Housing in Partnerships
		• National Mission Mode Project for E-Governance in Municipalities
		• National Urban Housing and Habitat Policy (2007)
		• National Policy on Urban Street Vendors (2009)
		• Bill on Street Vendors (Protection of Livelihood and Regulation of Street Vending) (2009)
		• National Urban Sanitation Policy (2008)
		• Launch of Total Sanitation Campaign (TSC)
		• Revamping of SJSRY
		• Interest Subsidy Scheme for Housing the Urban Poor (ISSHUP)
		• Launch of Rajiv Awas Yojana (RAY)
		• Urban Statistics for HR and Assessment (USHA) Scheme
		  • National Habitat Mission, a component of National Action Plan on Climate Change
		  • High Powered Expert Committee (HPEC) on Urban Services (2011)

(Contd...)

XIIth Plan (2012-2017)	• Planned, inclusive and sustainable urban development • Alleviating urban poverty • Improving civic service levels • Empowering urban local government and establishing efficient governance structures • Facilitating participatory governance • Effectively managing land resources • Planned urbanisation by preparing a City Development Plan through a participatory process • Promoting financial sustainability and accountability of ULGs and attracting more private investment • Slum rehabilitation and creation of affordable housing • Planned development of peri-urban areas • Incentivising innovation and rapid learning across urban systems • Urban transformation and rejuvenation • Developing Smart Cities • Housing to All	• Preparation of JnNURM II Scheme as a State sector ACA scheme while covering all the cities and sub-suming earlier sub-schemes viz., Mission Mode Project for E-governance, UIDSSMT, BSUP and USHA (abandoned later on) • National Urban Livelihood Mission (NULM), enlarged in scope and Mission Guidelines revised in 2015 • Comprehensive Capacity Building Programme for urban local government • Capacity Building for Urban Development (CBUD) • Implementation of Service Level Bench marking (SLB) in municipalities • Launch of Swachh Bharat Mission (SBM) in 2014 • Launch of Atal Mission for Rejuvenation and Urban Transformation (AMRUT) in 2015 • Initiation of Smart Cities Mission in 2015 • Prime Minister's Awas Yojana (PMAY) • Enactment of The Street Vendors (Protection of Livelihood and Regulation of Street Vending) Act, 2014 • Enactment of Amended Manual Scavengers (Liberation & Rehabilitation) Act, 2015

Source: Planning Commission, Government of India, compiled from Five Year Plan Documents-I to XII; Guidelines of Schemes; U.B. Singh, N. Ashok Kumar, Awadhesh K. Singh(2014): Challenges in Urban Management in India in New Millennium, GRP, New Delhi.

Sector Specific Policies

The Government of India has been launching several urban development and poverty related schemes through the Five Year Plans. Certain sector specific policies are being formulated in recent periods to provide policy guidance for focused implementation of these schemes and projects by the central, state and local governments. Policies are formulated in areas such as housing and shelter; urban poverty alleviation and slum development; democratic decentralisation and empowerment of the local government units; urban sanitation and solid waste management; and urban transportation. These sector specific policies are discussed hereunder for in-depth understanding of the issues.

Housing and Shelter Policies

Housing in urban India was considered a problem of urban poor including slum dwellers. It was probably perceived that the government as welfare measure is responsible for making provisions of housing for those who are economically weak and are living in slums and shanties. It is therefore found that the plan efforts in the form of policy guidelines focused only to that vulnerable class of the society; least realising that housing has never been their top priority.[32] The First Five Year Plan emphasised on institution building and on construction of houses for government employees and weaker sections. The scope of housing programme for the poor was expanded during the Second Five Year Plan. Programmes like rural housing, slum clearance, and sweepers' housing, and middle income group housing were taken up. The general directions for housing programme in the Third Five Year Plan was coordination of efforts of all agencies and orienting the programme to the needs of the low income group. Special emphasis was made on land acquisition and development as this was basis to the success of all housing programmes. However, it was thought that the individual household would be able to build its own house, or at least manage its construction using small contractors or direct labour in the form of masons and carpenters.

The policy of direct provision of housing in urban areas has become difficult in the face of rapidly increasing demand for housing, spiraling land prices and increasing costs of construction. In the circumstances, government's role has per force to be promotional as suggested in the Seventh Five Year Plan. As major effort is expected to come from the private sector, government's attention has been focused on the priority sector, i.e., urban poor. Further, promoting self-help housing, reduction in cost of construction by adopting low-cost technologies, realistic building bye-laws and land use standards, were other policy directives provided in respect of urban housing.

A national housing policy was first formulated in India in 1988 incorporating these emerging ideas. Main objectives of the National Housing Policy, 1988 as laid on the table of both the Houses of Parliament on 12th May, 1988, were: [33]

- To motivate and help all people and in particular, the houseless and inadequately housed, to secure for themselves affordable shelter through access to land, material, technology and finance;
- To encourage investment in housing in order to achieve a sustained growth of the nation's housing stock, and its proper conservation, renovation and up-gradation;
- To create an enabling environment by eliminating constraints and developing an efficient and accessible system for the delivery of inputs to maximise housing efforts;
- To improve the environment of human settlements with a view to raising the quality of life through the provision of drinking water, sanitation and other basic services; and

- To promote vernacular architecture and preserve nation's rich heritage in the field of human settlements.

The policy document outlining the strategy was to eradicate houselessness, to improve the housing conditions of the inadequately housed, and to provide a minimum level of basic services and amenities to all. It recognised that the magnitude of the housing stock needs involvement of various agencies including government at different levels, the cooperative sector, the community, and the private sector. The policy envisaged a major shift in government's role to act more as a facilitator than as a provider.

According to 1991 census, the total housing stock in the country was pegged at 22.90 million units. More than 90 per cent of this shortage was for the poor and the low income category. It required a huge investment and this could not be met from the public exchequer. It logically followed that no housing policy could make any significant headway without massive participation of the private sector. With this background, the government through the National Agenda declared Housing for All as a priority area and decided to focus on the housing needs of citizens. The National Agenda also emphasised that housing activity would be an engine for substantial generation of employment. Building upon past experience, the government evolved a new National Housing and Habitat Policy in 1998. The new policy aimed at:[34]

(i) Creation of surpluses in housing stock either on rental or ownership basis;
(ii) Providing quality and cost effective housing and shelter options to the citizens, especially the vulnerable group and the poor;
(iii) Ensuring that housing, along with the supporting services, is treated as a priority and at par with the infrastructure sector;
(iv) Guiding urban and rural settlements so that there is planned and balanced growth;
(v) Planning of urban transport is to be an integral part of the urban master Plan;
(vi) Development of cities and villages in a manner which provide for a healthy environment, increased use of renewable energy sources and pollution free atmosphere with a concern for solid waste disposal;
(vii) Using the housing sector to generate more employment and achieve skill up-gradation in housing and building activity, which continue to depend on unskilled and low wage employment to a large extent;
(viii) Ensuring that all dwelling units have easy accessibility to basic sanitation facilities and water;

(ix) Removing legal, financial and administrative barriers for facilitating access to land, finance and technology;

(x) Progressive shift from a subsidy based housing scheme to cost sharing or cost recovery-cum-subsidy schemes for rural housing;

(xi) Progressive shift of rural housing strategies from target orientation to a demand driven approach;

(xii) Empowering the Panchayati Raj Institutions and village cooperatives to mobilise credit for adding to the housing stock and also basic amenities in the rural areas;

(xiii) Using technology for modernising the housing sector to increase efficiency, productivity, energy efficiency and quality. Technology would be particularly harnessed to meet the housing needs of the poor and also specific requirements of rural housing;

(xiv) Forging strong partnerships between private, public and cooperative sectors to enhance the capacity of the construction industry to participate in every sphere of housing and habitat;

(xv) Meeting the special needs of SC/ST/disabled/freed bonded labourers/ slum dwellers;

(xvi) Involving women at all levels of decision-making and enabling them in formulation and implementation of the housing policies and programmes;

(xvii) Addressing the special needs of women headed households/single and working women/women in difficult circumstances. The specific requirements of women in terms of providing necessary facilities in homes to lessen their drudgery would be given sufficient attention;

(xviii) Protecting and promoting our cultural heritage, architecture and traditional skills; and

(xix) Establishing a Management Information System in the housing sector to strengthen monitoring of building activity in the country. This would serve as a decision support system.

The central theme of the Policy was on creating strong public-private partnerships for tackling the housing and habitat issues. The government was to provide fiscal concessions, carry out legal and regulatory reforms and create an enabling environment. The private sector as the other partner was encouraged to take up land assembly, housing construction and to invest in infrastructure services.

In order to generate suitable strategies for housing and sustainable development of human settlements, the Government of India formulated a draft National Urban Housing and Habitat Policy in 2005. The draft policy reaffirmed the importance of small and medium sized urban centres which have vast potential for future urban growth and also promoting a regional balance. It focused on building synergy, convergence and integration of housing and related infrastructure interventions. The draft policy aimed at to act as complementary to poverty alleviation, generation of income and employment to achieve overall objective of shelter for all and sustainable development of human settlements.[35]

The National Urban Housing and Habitat Policy, 2007 further improved the ideas on housing. The focus of this Policy is the provision of "Affordable Housing for All" with special emphasis on vulnerable sections of society such as Scheduled Castes/Scheduled Tribes, Backward Classes, Minorities and the urban poor. The Policy aims to complement poverty alleviation and employment generation programmes for achieving the overall objective of "Affordable Housing for All" with sustainable development. The Policy seeks to enhance the supply of houses especially for the disadvantaged, duly supplemented by basic services. In sum, the National Urban Housing and Habitat Policy, 2007 aims the following: [36]

- *Urban Planning:* Regional planning to promote symbiotic development of rural and urban areas;
- *Affordable Housing:* Promoting development of cost-effective, quality approved building materials and technologies with a view to bringing down the cost of EWS/LIG houses;
- *Increasing flow of Funds:* Promoting larger flow of funds from governmental and private sources; suitable fiscal concessions; access to tenure, land, finance and technology; and shifting to a demand-driven approach and from subsidy based housing schemes to cost recovery-cum-subsidy schemes for housing;
- *Spatial Incentives:* Relaxation of Floor Area Ratio for ensuring 20-25 per cent of the FAR are reserved for EWS/LIG;
- *Increased supply of land:* Facilitating accessibility to serviced land and housing with focus on economically weaker sections and low income group categories;
- Special provision for SC/ST/OBC/Minorities/Disabled;
- *Special Provision for Women:* Catering to women's needs;
- *Employment Generation:* Up-gradation of construction skills and giving an impetus to employment generation;
- *Public-Private Partnerships:* Forging strong partnerships between public, private and cooperative sectors;
- *Management Information System:* Developing information and data-base on housing; and

- *Healthy Environment:* Encouraging use of renewable energy resources and ensuring effective solid waste management and protecting cultural heritage and architecture.

The Policy intends to promote sustainable development of habitat in the country. It seeks to promote various types of public-private partnerships for realising the goal of affordable housing for all. The Policy feels that non-availability of housing by economically weaker sections of society and low income families in urban areas is directly linked with the magnitude of urban poverty.

In addition to these national initiatives, the state governments have been formulating, from time to time, their own state specific housing policies.

Urban Land Policies

A closely related area of housing is land. Land plays a pivotal role in the expansion of urban areas. Not only has that there takes place conversion of agricultural uses of land to the non-agricultural or urban uses, but also competition amongst different uses and different users surfaces. This competition leads to unhealthy trends like speculation, increased land prices, lack of access of land to deprived sections of the society and sometimes, uneconomic utilisation of scarce land. Urban land policies hence, need to address to these issues. A Committee on Urban Land Policy was appointed by the Government of India in 1965 to evolve a comprehensive policy frame on urban lands. The Committee framed the following policy objectives:[37]

(i) To achieve an optimum social use of urban land;

(ii) To make land available in adequate quantity at the right time and for reasonable prices to both public authorities and individuals;

(iii) To encourage cooperative effort and bonafide individual builders in the field of land development, housing and construction; and

(iv) To widen the base of ownership, especially to safeguard the interests of the poor and under-privileged sections of urban society.

The Task Force on Planning of Urban Development later reiterated these objectives in 1983 and added two more objectives for optimum use of scarce land in the face of rapid growth, especially in the large cities. They are: (i) to encourage the socially and economically efficient allocation of urban land such that land development is done in a resource conserving manner and that the magnitude of land used is optimal; and (ii) to promote flexibility in land use in response to changes resulting from a growing city.[38] The policy pertaining to urban land still appears to be basically a welfare approach, i.e., acquiring land at a very cheap rate, developing and distributing on no-loss-no-profit basis with cross subsidy as a balancing factor. The popular policy of allocating house-sites free of cost to the poor had led to gross undervaluation of land. This practice, in fact, worked against the sites and services projects devised with cross-subsidies. Further, the Urban Land (Ceiling and Regulation) Act, 1976 led to legal litigations and blocked precious land for development. As this has not served the purpose intended, the Government of India adopted a policy of abolishing the urban land ceiling Acts by the state governments as a part of the JnNURM reforms. In many, states this has been done away. The shelter and land issues and the reforms are discussed elaborately in the chapters which follow.

Urban Sanitation Policy

Urban sanitation is declining over the years in India. The disturbing aspect here is the lack of awareness among the people about the public health dangers of poor sanitation. The other issues are lack of city wide integrated approaches covering all aspects of managing waste viz., safe confinement, treatment and safe disposal. Limited technological options and resource constraints are some other problems.

Urban sanitation is the other important area which received attention of the central and state governments in recent years. With the rapid decline of urban sanitation and poor solid waste management practices, the Government of India has devised a National Urban Sanitation Policy, 2008. The overall goal of the policy is to transform urban India into community-driven, totally sanitised, healthy and livable cities and towns. The specific goals are:[39]

- ***Awareness Generation and Behaviour Change***: Generating awareness about sanitation and its linkages with public and environmental health amongst communities and institutions, and promoting mechanisms to bring about and sustain behavioural changes aimed at adoption of healthy sanitation practices;
- ***Achieving Open Defecation Free Cities***: All urban dwellers will have access to and use safe and hygienic sanitation facilities and arrangements so that no one defecates in the open;
- ***Re-orienting Institutions and Mainstreaming Sanitation toward city wide approach***: Mainstream thinking, planning and implementing measures related to sanitation in all sectors and departmental domains as a cross-cutting issue, especially in all urban management endeavours; strengthening national, state, city and local institutions (public, private and community) to accord priority to sanitation provision, including planning, implementation and O&M management; and extending access to proper sanitation facilities for poor communities and other un-served settlements; and

- ***Sanitary and Safe Disposal*:** 100 per cent of human excreta and liquid wastes from all sanitation facilities including toilets must be disposed of safely.

The state governments have been developing the state level strategies to implement the national sanitation policy. City sanitation plans are being developed at the local levels to develop sanitation at the community level. But, the progress has been very slow in this area.

Solid Waste Management Policies

Waste is mainly a bi-product of consumer-based lifestyles. The term 'waste' has a different meaning for different people. 'Waste is defined as any material that is not useful and does not represent any economic value to its owner, the owner being the generator of waste. Depending on its physical state, wastes are categorised into municipal wastes, hazardous wastes, medical wastes, liquid wastes and radioactive wastes. Main sources of waste generation in Indian cities are: (a) residential premises, (b) commercial and business establishments, (c) street sweeping, (d) institutional premises, (e) trade wastes, (f) construction and demolition material, and (g) stable waste. Municipal solid waste thus encompasses waste generated from these sources.

Solid waste management is a part of public health and sanitation. According to the Constitution of India, it falls within the purview of the State List (List-II). The responsibility for providing this service lies within the public domain. This activity being of a local nature is entrusted to urban local governments as a mandatory function. The provisions necessary for managing the wastes exist in the municipal laws enacted by the respective state governments. Thus, collection, storage, transportation, processing and disposal of solid wastes are the prime responsibility of municipalities. Managing solid waste generally involves planning, financing, construction and operation of facilities for the collection, segregation, transportation, recycling and final disposal of the waste. Thus, it is one of the most visible of urban services. This labour intensive service consumes a large proportion of operational revenue of a city government. As such, effective and sustainable waste management goes hand-in-hand with good local governance and sound city management. There is a great diversity in the nature and standards of waste management services within the country between different urban areas.[40]

The Government of India, in pursuance of decisions taken at the United Nations Conference on the Human Environment held at Stockholm in June 1972, enacted the Environment (Protection) Act, 1986 to provide for the protection and improvement of environment in the country. The Bureau of Indian Standards formulated Indian Standard Solid Wastes-Hospitals-Guidelines for Management, in 1989, prescribing guidelines for management of solid wastes from hospitals. Subsequently, the Government of India formulated the Bio-Medical Waste (Management and Handling) Rules, 1998 for managing and handling of medical wastes; the Municipal Solid Waste (Management and Handling) Rules, 2000 for managing and handling of municipal solid wastes; the Hazardous Waste (Management, Handling and Trans-boundary Movement) Rules, 2008 for managing and handling of hazardous wastes; and the e-Waste (Management and Handling) Rules, 2011 for managing and handling of radio-active and electronic wastes. These rules have since been revised and new Bio-Medical Waste Management Rules, 2016; E-Waste Management Rules, 2016; Hazardous and other Wastes (Management and Trans-boundary Movement) Rules, 2016; Plastic Waste Management Rules, 2016; and Construction and Demolition Waste Management Rules, 2016 have come into force.

Prior to coming into force the Municipal Solid Waste (Management and Handling) Rules, 2000 municipalities across the country were managing the solid waste unscientifically, depending upon their human and financial resources. The Municipal Solid Waste (Management and Handling) Rules, 2000 framed by the Government of India to regulate the management and handling of municipal solid wastes provided a framework for treatment and disposal of municipal solid waste. These rules came into effect from 25th September 2000 and opened up a channel to tackle the subject more scientifically and in a systematic manner. The Rules explain the responsibilities of municipal authority, State Government, Central Pollution Control Board, State Board or the committees with regard to the management of solid wastes.[41] The Rules were the guiding principles for the municipalities to deal with the solid waste management problems. The implementation of these Rules has been very discouraging in the urban areas. The main problems were lack of commitment at the local level, ignorance about the provisions of these Rules among the ground staff and financial problems of urban local government units. In addition, the Rules mandated that "landfill should always be located away from habitation clusters and other places of social, economic or environmental importance", which implied lands outside the city. Therefore, ever increase in quantum of solid wastes had significant impacts in terms of land required for disposing the waste as it gets more difficult to site landfills.

The latest addition is the promulgation of Solid Waste Management Rules, 2016 which supersedes the Rules framed in 2000. These rules prescribe the functional responsibilities of various authorities knowing very well that coordination among authorities has been a major victim in the country. This is the first time that the provisions of these rules have been extended to census towns as well. The Rules underline that the municipal authority shall encourage the use of municipal solid waste by adopting suitable technology which may include: composting, vermin composting, anaerobic digestion with or without energy recovery, co-incineration, or combination of such technologies as appropriate, to make use of waste so as to minimise burden on landfill. The Rules prescribe the criteria for selecting the landfill sites, development of facilities at the site, specifications for land filling operations and closure on completion of landfill, for pollution prevention, monitoring of water quality, ambient air quality monitoring, plantation at landfill sites, post-care of landfill sites, standards for composting and for treated leachates etc.[42]

It is found that all efforts to manage wastes are mandated under the provisions of the Environment Protection Act, 1986; hence only environmental aspect is taken care of while dealing with this human related problem. As a result of the problems associated with the proper management of wastes, there is a dire need for a multi-stakeholder partnerships in proper and integrated waste management at all stages-collection, segregation, transportation, treatment and disposal, especially at the source where it is generated, and integrates it with policies that encourage reduction, reuse and recycling. While elimination of waste in its entirety may not be feasible, it is possible through the systematic application of waste management hierarchy to reach a point within the next few decades were re-use, recycling, recovery and treatment overtake landfills as preferred options for waste management. The corresponding rules and municipal laws should aim at it.[43]

Policies Relating to Urban Poverty

Poverty is one of the main problems which have attracted attention of sociologists as well as economists. Poverty indicates a condition in which a person fails to maintain a living standard adequate for his physical and mental efficiency. It gives rise to a feeling of a discrepancy between what one has and what one should have. Poverty in India is rapidly acquiring an urban face and becoming a huge issue in the course of country's urbanisation. The urbanisation of poverty has altered India's urban face by imposing two distinct dimensions upon cities. Deprived of a proper urban habitat, the poor have had to find residential footholds in slums; and in their search of employment and enterprise, they have been forced to find survival in the informal sector. Urbanisation of poverty has thereby been accompanied with the informalisation of poverty.[44] Urban poverty is a multidimensional phenomenon resulting from a cumulative deprivation caused by several factors. The prominent features of urban poverty are: unstable income dominated by informal employment; fragile or weak asset base; no shelter/insecure shelter; squalid living conditions with no access to better health-care, education, safe water, toilets, etc.; and lack of social safety nets.

Even after seven decades of independence, the India still has the world's largest number of poor in a country. In urban India, 68 million people as per mixed recall period and 80 million people as per uniform recall period are in the BPL category in 2004-05 as per the Planning Commission report.[45]

The Tendulkar Committee (2009) in the light of new methodology thought proper to re-estimate poverty for previous years. By this exercise, the committee re-estimated poverty at all-India level in 1993-94 was 50.1 per cent in rural areas, and 31.8 per cent in urban areas and 45.3 per cent in the country as a whole, compared to the official figures estimated for the years as 37.2 per cent, 32.6 per cent and 38 per cent respectively.[46] The NSSO (66 th Round) survey on consumption pattern mentions that about 60 per cent of population is living on an average expenditure of Rs 35 in rural areas and Rs 66 in urban areas. The report furnishes that level of all-India per capita monthly expenditure is Rs 1054 in rural areas and Rs. 1984 in urban areas. The survey further observes that 10 per cent of the urban population found at the bottom of pyramid is living on an expenditure of Rs 20 per day per head. Further, 10 per cent of poor are living on monthly income of Rs 599 in urban areas.[47]

Urban local governments are entrusted with developmental functions like urban poverty alleviation, slum improvement, planning for economic development and environmental protection by the 74th Constitution Amendment Act, 1992. However, in few states there is a exclusive government department handling the issue of urban poverty alleviation with the assistance of a specialised agency (State Urban Development Agency), and implementing poverty alleviation and slum programmes independently to urban local governments, a situation contrary to the dictates of the supreme law of the land. The role of decentralised governance in poverty alleviation has been stressed by the central and state governments in the post-74th CAA period.

Although, poor are found living in other parts of the city, slums are recognised as the natural habitat (albeit a major one) of urban poor. This myth has guided the fight against urban poverty, and hence earlier, poverty alleviation schemes focused more on slum development and environmental improvements. Though the social front is essential, the crucial part of poverty alleviation is providing employment to the burgeoning urban poor in the country. Focusing on employment generation to urban poor, the Government of India launched Nehru Rozgar Yojana (NRY), followed by another ambitious scheme Swarna Jayanti Shahri Rozgar Yojana (SJSRY) rightly focusing on this critical issue. The latter effort was designed to provide gainful employment to the urban poor by encouraging setting up of self-employment ventures and provision of wage employment opportunities for families below poverty line in urban areas. The scheme followed community development approach aiming at removing voicelessness among the urban poor, especially women.

The Government of India, in collaboration with UNDP, initiated National Strategy for Urban Poor (NSUP) in November 2003. The move had six broad outcomes as enhanced capacity at national level for the analysis of urban livelihood issues; an all-India network on urban poor livelihoods to serve as a forum for dialogue and support between community and NGO activists and for knowledge/consensus building; technical and financial support for innovative and promising livelihood initiatives broadened and deepened; action based research to inform and improve the policy framework both at the central and state levels with direct bearing on urban poor livelihoods and living conditions; a comprehensive review and capacity analysis of urban poor livelihoods particularly relating to living, working and social security concerns of the poor in urban informal sector; target support to community associations and NGOs to promote urban poor concerns and ground interventions to address multiple vulnerabilities. Further, under mega infrastructure development mission of JnNURM, a seven-point charter was conceived for ensuring tenurial rights to urban poor besides providing social security. Keeping in view the aim of inclusive growth under the Eleventh Five Year Plan, the Government of India formulated a National Skill Development Policy and initiated a National Skill Development Mission there under aiming to create opportunities for all to acquire skills throughout life and especially for youth, women and disadvantaged groups. The Policy and Mission have since been revised and broadened. In pursuance, the Government of India has launched a National Urban Livelihood Mission (NULM). It has multi-pronged strategy: (i) enhancing and expanding existing livelihoods options of the urban poor; (ii) building skills to enable access to growing market-based job opportunities offered by emerging urban economy; (iii) training for and support to the establishment of micro-enterprises by the urban poor -self and group; (iv) ensuring linkages to shelter, basic services and empowerment to promote sustainable approaches to poverty alleviation; and (v) building capacity of the urban poor, their institutions and the machinery involved in the implementation of livelihood development and poverty alleviation programmes. This mission approach focuses on increasing financial inclusion of urban poor. The Government of India has strived hard to eradicate (reduce) poverty through focused but sectoral programmes. The administrative failure and political unwillingness coupled with absence of community (urban poor) involvement have not allowed the benefits of economic growth reach to bottom of the pyramid.[48]

In addition, the state governments sponsor their own programmes under different names. International funding is also available in poverty alleviation programmes. The World Bank, DFID, UNDP, ADB, UNICEF and other multi-national agencies are giving top priority to poverty alleviation programmes and providing funds generously. But these programmes generally are not effectively linked at the local levels for optimum benefits. Efforts must be made to involve the community groups and the NGOs in this effort.

National Slum Policy

The urban poverty alleviation efforts continue to suffer from certain administrative and strategic failures. The Tenth Five Year Plan identified following causes:[49] failure to provide specific provisions in the municipal laws for dealing with issues relating to the urban poor, slums and economic planning for poverty alleviation responsibilities which flow from the 74th CAA; failure to provide a coordinating institutional arrangement across variety of institutions like municipalities, development authorities, housing boards, slum boards, etc.; failure to provide for housing, land and economic needs like production and marketing activities of the urban poor and facilities for street vendors and hawkers in the urban plans; and failure to integrate the slum dwellers into the urban mainstream life.

A National Slum Policy was drafted in 1999 by the Government of India to address the above identified issues. The draft policy provides certain objectives to deal with slum development issues in the country. The Slum Policy while recognising the slums as integral part of the urban areas, suggested that the slum development initiatives should basically focus on: creating awareness amongst the public and government of the underlying principles that guide the process of slum development and improvement and the options that are available for bringing about the integration of these settlements and the communities residing within them into the urban area as a whole; strengthening the legal and policy framework to facilitate the process of slum development and improvement on a sustainable basis; and establishing a framework for involving all stakeholders for the efficient and smooth implementation of Policy objectives. The strategies proposed are: slum up-gradation or improvement rather than clearance; provision of basic civic services irrespective of the legal status of the slum area; the planning goal should be cities without slums; ensuring services and land to poor at affordable costs; greater participation of the stakeholders in planning and development; and registration of all slum dwellers for extending all the social welfare benefits.[50]

The Tenth Plan while drawing from the draft National Slum Policy, 1999 recommended a multi-dimensional strategy focusing on the empowerment of the urban poor. The components of this strategy were: increased attention on the community structures by further strengthening and diversifying them including groups like street vendors; convergence of various basic civic services; creation of slum improvement and urban poverty task forces at the state level and urban poverty cells at the city level; increasing the scope of the Swarna Jayanti Shahri Rozgar Yojna (SJSRY) to include contributory social security assistance through various insurance schemes and capacity building for the personnel engaged in the urban poverty alleviation programmes; an integrated urban poverty alleviation strategy linking the four vital segments viz., shelter, access to civic and social services, urban informal sector and livelihoods; and enabling legal, administrative and planning environment.

While continuing the initiatives of the earlier plans, the Eleventh Five Year Plan identified the following objectives for the urban poor:[51]

(a) To provide affordable shelter and decent living and working conditions to the urban poor;

(b) To make adequate provision of land for the poor in the Master Plan itself;

(c) To help in developing self-employment enterprises and job creation for the wage employment earners; and

(d) To protect the economic interest and safety of women and other vulnerable sections of the society.

On slum development, an integrated strategy was emphasised. The strategy basically involved following measures:[52]

(i) Creating and updating database on slums;

(ii) City-wise perspective and integrated slum development plans;

(iii) Augmenting and facilitating access to services for slum dwellers;

(iv) Granting tenure security to slum dwellers to develop their asset base;

(v) In-situ up-gradation and resettlement options for slum improvement without dislocating the slum dwellers; and

(vi) Using land as a resource for housing and shelter development for slum dwellers.

In view of the not so success of the Swarna Jayanti Shahri Rozgar Yojna (SJSRY) scheme, the Eleventh Plan desired that the focus should be changed to following aspects:[53]

(a) Skill development—PPP mode

(b) Promotion of Self Help Groups

(c) Self-employment skills plus facilitation of credit for setting up self-employment.

Emphasis was also laid on the skill training for employment promotion amongst the poor. It was expected that this would reduce unemployment among the urban poor and strengthen employability, productivity and wages especially among women which is considered a step towards inclusive economic growth. A scheme called Urban Statistics for HR Assessment

(USHA) was launched to monitor the information on skill levels of the urban poor. The Plan also focused attention on promoting healthcare among the poor.

The Twelfth Five Year Plan re-emphasised the in-situ slum development and skill development among the poor. The entire slum rehabilitation exercise revolves around the concept of affordable housing promotion and slum rebuilding through the Rajiv Awas Yojna (RAY) scheme launched during the Eleventh Plan period for making the cities slum free within a stipulated period. Further, creation of rental social housing, adopting transparent policies for slum rehabilitation, building affordable housing stock in the peripheral areas, promoting community participation in slum development for developing customised solutions, etc. are other areas identified. The latest initiatives viz., AMRUT, Smart Cities Mission, PMAY and SBM also lay great emphasis on slum redevelopment and provision of services to un-served areas.

National Policy on Street Vendors

Street vending is a historical fact and an expanding phenomenon today in most of the cities and towns in India. Street vendors mostly constitute illiterate migrants in search of livelihood. But, in the elite and planning circles, vendors are seen as a pestilence – crowding pavements, spilling onto busy streets, creating transport bottlenecks, not just adding to confusion and chaos, but a source of filth, disease and crime. Above all, they are seen as squatters on valuable commercial and residential space, both public and private, violating our sense of order. Little wonder that demands for city planning result in drives against hawkers and vendors.[54]

Working conditions of the street vendors are deplorable living always under multiple threats and constraints. The mafia and even the law enforcing authorities rob their earnings under threat. The existing laws in fact are hostile to the street vendors leading to harassment by police and municipal authorities. All municipal laws prevent street vending and permissions for vending are meagre compared to the demand. In this situation, the civic authorities sometimes take very drastic steps that torment the vendors' lives. For instance, the Calcutta Municipal Corporation Act of 1980 prevents any type of vending on the streets. It was under these provisions that on the night of November 16-17, 1996 that the infamous Operation Sunshine took place. More than a hundred thousand hawkers were forcibly evicted from the streets that night and properties worth several crores of rupees were destroyed or confiscated. This action was, by all accounts, is considered the most brutal action taken against the urban poor in any city in the country.[55] The reality is street vendors provide valuable services to the urban population while trying to earn a livelihood. A number of NGOs like SEWA have been fighting for a proper space in the urban life to the street vendors. If the governments can simplify the laws governing the setting up of petty enterprises and make secure the property rights of the poor, this will go a long way in reducing poverty.

Street vendors are considered as unlawful entities and are subjected to continuous harassment by police and civic authorities. But, the Supreme Court gave a positive direction for the mitigation of the street vendors' problems when it ruled that "*if properly regulated according to the exigency of the circumstances, the small traders on the sidewalks can considerably add to the comfort and convenience of the general public, by making available ordinary articles of everyday use for a comparatively lesser price. An ordinary person, not very affluent, while hurrying towards his home after a day's work can pick up these articles without going out of his way to find a regular market. The right to carry on trade or business mentioned in Article 19(1)(g) of the Constitution, on street pavements, if properly regulated cannot be denied on the ground that the streets are meant exclusively for passing or re-passing and no other use.*"[56] In this backdrop, there exists need for developing strategic inter-connections between urban poverty alleviation and the urban informal sector development in India.

The street vendors form significant part of the urban informal sector in India. It has been estimated that in any city street vendors constitute about two per cent of the population and women constitute large segment of the street vendor population. Street vendors provide valuable services to the people through their personal enterprise, limited resources and labour. They suffer from lack of proper education, resources and skills to improve their conditions. Primarily, they lack proper place to conduct their business. As they conduct their business on the road margins, encroach roads and public places, they are subjected to constant threats, therefore, fear of eviction from the authorities. According to the Constitution of India, the right to livelihood is very well linked with the right to life. The Supreme Court of India (1989) in a case observed that "*Street trading being a fundamental right has to be made available to the citizens subject to Article 19(6) of the Constitution. It is within the domain of the State to make any law imposing reasonable restrictions in the interest of general public......In spite of repeated suggestions by this Court nothing has been done in this respect. Since a citizen has no right to choose a particular place in any street for trading, it is for the State to designate the streets and earmark the places from where street trading can be done. Inaction on the part of the State would result in negating the fundamental right of the citizens.*"[56]

In this backdrop, the Government of India while recognising the contributions of vendors to urban economy and market, and problems being faced by them, formulated the National Policy on Urban Street Vendors in 2004. The overarching objective to be achieved was to provide for and promote a supportive environment for the vast mass of urban street vendors to carry out their vocation while at the same time ensuring that their vending activities do not lead to overcrowding and unsanitary conditions in public spaces and streets. The Policy specifically aimed:[57]

(i) to give vendors legal status by amending, enacting, repealing and implementing appropriate laws and providing legitimate hawking zones in the urban development plans;

(ii) to provide facilities for appropriate use of identified space including the creation of hawking zones;

(iii) to eschew imposing numerical limits on access to public spaces by discretionary licenses and instead moving to nominal fee-based regulation of access where market forces would determine the number of vendors that could be sustained;

(iv) to make street vendors a special component of the urban development plans by treating them as an integral part of urban distribution system;

(v) to promote organisations of street vendors to facilitate their collective empowerment;

(vi) to set up participatory processes that involve all stakeholders;

(vii) to promote norms of civic discipline by institutionalising mechanisms of self-management and self-regulation in matters relating to hygiene, cleanliness etc; and

(viii) to promote access of street vendors to services like credit, skill development, housing, social security and capacity building.

Pursuant to this initiative, many state governments framed the policies to regulate the street vending. However, problem gravitated instead of resolving. Hence, the Government of India framed the National Policy on Urban Street Vendors in 2009 and a model bill entitling Street Vendors (Protection of Livelihood and Regulation of Street Vending) Bill, 2009. The Policy aims to foster a congenial environment for the urban street vendors to carry out their activities. It is conceived as part of the national urban poverty alleviation strategy. The primary objective of this Policy is to provide for and promote a supportive environment for the vast mass of the urban street vendors to carry out their vocation, at the same time, ensuring that their vending activities do not lead too overcrowding and insanitary conditions in public spaces and streets.[59] The basic features of the Policy are listed below: [60]

- According legal status to the street vendors by legislating an appropriate law and providing legitimate vending zones in the city master plans, development plans, zoning and lay-out regulations, and ensuring their enforcement;
- Providing necessary civic facilities in the vending zones;

- Adopting transparent regulation of their space allotments;
- Promoting street vendors cooperatives and associations for collective empowerment;
- Adopting participative methods to facilitate their businesses; and
- Promoting access of street vendors to institutional credit, skill development, housing, social security and capacity building.

The state level legal and institutional support is also emphasised to realise the objectives. Every state is expected to bring-out an appropriate legislation. The city/town level Street Vending Committee is provided to regulate and coordinate all the activities relating to street vendors. A system of registration of street vendors is recommended in this respect. Elaborate guidelines are provided to assist the state governments and the local level authorities for effective implementation of various provisions. A number of state governments enacted their own law based on the model bill; however, the pace was slow. Hence, the Government of India enacted a central law entitled Street Vending (Protection of Livelihoods and Regulation of Street Vending Act, 2014 and advised the state governments to implement the provisions for proper management of urban vending.

Though the policy framework is provided for regulating the street vendors' activities in the cities and towns, the implementation has been slow. Effective implementation of the policy objectives requires enormous capacity building and education programmes to the officials. Necessary legal changes are also required in the town planning and municipal laws to facilitate the process. It is expected that the governments give necessary attention to this aspect and the municipalities develop plans and programmes integrating street vending activities with other urban poverty alleviation projects. The success of the Policy largely depends upon the level of convergence achieved among the organisations and programmes at the local levels.

National Urban Transport Policy

The Government of India has recognised the importance of urban mobility issues and formulated a National Urban Transport Policy in 2006 (since revised in 2014) to develop cities economically vibrant and productive. The mega scheme JnNURM provided fund for building transport infrastructure, parking facilities, purchase of buses, and taking up mass public transport projects. The objectives of the Policy are outlined below: [61]

- Incorporating urban transportation as an important parameter at the urban planning stage rather than being a consequential requirement;
- Encouraging integrated land use and transport planning in all cities so that travel distances are minimised and access to livelihoods,

education, and other social needs, especially for the marginal segments of the urban population is improved;

- Improving access of business to markets and the various factors of production;
- Bringing about a more equitable allocation of road space with people, rather than vehicles, as its main focus;
- Encourage greater use of public transport and non-motorised modes by offering central financial assistance for this purpose;
- Enabling the establishment of quality focused multi-modal public transport systems that are well integrated, providing seamless travel across modes;
- Establishing effective regulatory and enforcement mechanisms that allow a level playing field for all operators of transport services and enhanced safety for the transport system users;
- Establishing institutional mechanisms for enhanced coordination in the planning and management of transport systems;
- Introducing Intelligent Transport Systems for traffic management;
- Addressing concerns of road safety and trauma response;
- Reducing pollution levels through changes in travelling practices, better enforcement, stricter norms, technological improvements, etc.;
- Building capacity (institutional and manpower) to plan for sustainable urban transport and establishing knowledge management system that would service the needs of all urban transport professionals, such as planners, researchers, teachers, students, etc.;
- Promoting the use of cleaner technologies;
- Raising finances, through innovative mechanisms that tap land as a resource, for investments in urban transport infrastructure;
- Associating the private sector in activities where their strengths can be beneficially tapped; and
- Taking up pilot projects that demonstrate the potential of possible best practices in sustainable urban transport.

Implementation strategies identified in this respect are:[62]

(i) Integrating land use and transportation planning;

(ii) Equitable allocation of road space among various categories of vehicles- motorised and non-motorised;

(iii) Priority for the use of public transport by increasing the share of public transport from 22 per cent to 60 per cent;

(iv) Quality and reasonable and fair pricing of public transport;

(v) Integrated public transport systems;

(vi) Setting-up of Special Purpose Vehicles for implementing the projects;

(vii) Priority to non-motorised transport;

(viii) Providing proper parking facilities;

(ix) Use of cleaner technologies;

(x) Innovative financing mechanisms with focus on land as resource;

(xi) Association of the private sector; and

(xii) Public awareness measures to enlist public cooperation.

The large projects like building metro rails require huge investments. Hence, the Twelfth Plan suggests either for implementing either through private sector or with Viability Gap Funding or promoting PPP models.[63] The Twelfth Plan emphasises on providing quality and affordable transport to the urban dwellers. The Plan aims to increase the share of public transport to at least 50 per cent. Other points which were emphasised on the lines of the NUTP, 2006 are to dis-incentivise the use of private vehicles and promoting non-motorised transport are dealt with by new policy.

Urban Policies – Perspectives and Paradoxes

Till recently, most urbanisation policy in India has been considered unconscious, partial and uncoordinated and negative to some extent as commented by Brian J.L. Berry: "It is unconscious in the sense that those who affect it are largely unaware of its proportions and features. It is partial in that few of the points at which governments might act to manage urbanisation and affect its course and direction are in fact, utilised. It is uncoordinated in that national planning tends to be economic and urban planning tends to be physical, and the disjunction often provides competing policies. It is negative in that the ideological perspective of the planners leads them to try to divert, retard or stop urban growth, and in particular to inhibit the expansion of metropolitan and primate cities."[64] But, conscious efforts have been made for the last two decades or so to address the issues and articulate the urban policies in a more focused manner covering a wide spectrum of areas like urban infrastructure, urban sanitation, urban housing and slums, urban employment, efficient and accountable service delivery mechanisms and environment protection for achieving the goal of sustainable habitat development. National level policies are result of such efforts.

Decentralisation, integrated development, inclusive growth and people's participation are the recurring themes echoed in the urban policies in India. Till the launching of the ambitious JnNURM scheme in the country, the programmes taken-up to achieve these objectives were very inadequate and ad hoc. As a result, while urban policies remained idealistic and wishful visions, the market and economic forces shaped the urban development process. The market forces even influenced the governmental decisions to the advantage of a few sections of the population. Consequently, several paradoxical situations emerged on urban scene. If setting-up of large industries in and around large cities went against the policy of decentralised development, inability of the state governments to devolve necessary powers and finances and proliferation of urban development agencies worked against the objective of strengthening of urban local governments. Further, if the policy was to promote people's participation, there has been outsourcing and dominance of bureaucracy in the planning process. At the micro level, haphazard growth of the urban areas across the country reveals the intensity of market forces and real estate lobbies in shaping the urban expansion. The weak politico-administrative system aided these trends.

Since 1990's, the policies of liberalisation and privatisation are being pursued vigorously in India. In view of the mind-boggling costs of urban infrastructure development and decreasing public sector performance, innovative policies, viz., cost recovery, privatisation of services like solid waste disposal, sanitation and public transport are advocated. But, their implementation suffered due to several politico-administrative constraints. Inter-sector consistencies and spatial linkages at national, state and local levels are lacking. The dominance of market forces and economic factors led to selective metropolitanisation without proper planning and governance.

In the absence of a national urbanisation policy, certain sector specific attempts have been made to deal with the issues and problems relating to urban slums, poverty, housing, urban infrastructure and civic services through successive Five-Year Plans. The National Urbanisation Commission provided a broad policy-frame to deal with urban problems. The national urban policy-frame today revolves around five basic themes, viz., sustainable urban development, democratic decentralisation, inclusive growth, participative planning and competent urban governance. Major components of this emerging urban policy-frame are:

(i) A more sustainable urban habitat development,

(ii) Increased spending on urban infrastructure and services,

(iii) Effective linkages between spatial, social and economic planning in a regional context for achieving the objective of inclusive development,

(iv) Continued focus on urban governance reforms,

(v) Developing effective institutional linkages at the local levels for convergence,

(vi) Building public-private-people partnerships, and

(vii) Focus on affordable urban housing and livelihood development though community participation methods.

The launching of the JnNURM scheme with reform orientation and mission approach was a good beginning. The provisions of the Constitution (Seventy-fourth Amendment) Act 1992 and launching of national level urban governance reforms through central schemes like the JnNURM have no doubt yielded results though not to the extent aimed at. There are difficulties in framing national urban policy agenda in federal systems like India. In diversified countries like India, it is still challenging because urban problems are vast and amorphous collection of different physical and social conditions existing in society and translating all of them into a common policy at national level would be a herculean task for the policy makers.[65] The criticism received in respect of the JnNURM by some scholars and the difficulties experienced by the municipalities and the state governments especially, in implementing the nationally framed sector specific policies and governance reforms reflect such a policy dilemma in India. These experiences compelled the Government of India to re-structure the reforms shifting from the one size fits all approach to calibrated approach to suit different types of municipalities and local situations by initiating AMRUT.

To conclude, the Twelfth Five Year Plan identified five enablers for effectively tackling of challenges of urbanisation—governance, planning, financing, capacity building and innovation.[66] The new mission mode approach of AMRUT with increased investments, fast track strategies and reforms is expected to optimise these enables in the coming decades.

References

1. Planning Commission: 12th Five Year Plan (Vol. II), p. 318: HPEC: 2011: National Urban Transport Policy, Ministry of Urban Development; and The World Bank (2013): Urbanisation beyond Municipal Boundaries: Nurturing Metropolitan Economics and Connecting Peri-Urban Areas in India, Washington, p. 77.
2. The World Bank: Urbanisation beyond Municipal Boundaries: Nurturing Metropolitan Economics, op. cit., p. 24.
3. JnNURM Directorate, Ministry of Urban Development and NIUA: India's Urban Demographic Transition— 2011 Census Result, p. 4.
4. UNESCAP and UNHABITAT (2015): The State of Asian and Pacific Cities.
5. India's Urban Demographic Transition- 2011 Census Result, op. cit., pp. 1-8.
6. Urbanisation beyond Municipal boundaries, op.cit., p. 1.

7. *Ibid.*, p. 39.
8. Ramachandran, R. (1989): Urbanisation and Urban Systems in India, OUP, Oxford, p. 323.
9. Renaud, Bertrand (1981): National Urbanisation Policy in Developing Countries, OUP, p. 7.
10. Bhattacharya, Mohit (1975): Urbanisation and Urban Problems in India – Some Policy Issues, Nagarlok, Indian Institute of Public Administration, Vol. VII, No. 4, October-December, pp. 1-14.
11. Lewis, John P.: The Quiet Crisis in India, as quoted in R.K. Wishwakarma and Gangadhar Jha (eds)(1983): IDSMT-Problems and Strategic Policy Issues, Indian Institute of Public Administration, New Delhi, pp. 76-77.
12. Government of India, Ministry of Works and Housing (1977): Report of the Task Force on Planning and Development of Small and Medium Towns and Cities, Vol. I, pp. 85-87.
13. Government of India (1988): Report of the National Commission on Urbanisation, Vol. II, pp. 32-39.
14. *Ibid.*
15. *Ibid.*, 37.
16. Town and Country Planning Organisation, Ministry of Urban Development, Government of India, New Delhi.
17. Planning Commission, Government of India: Eleventh Five Year Plan, Vol. 3, p. 395.
18. Planning Commission, Government of India: Twelfth Five Year Plan, p. 318.
19. Eleventh Five Year Plan, p. 394.
20. As quoted in the Twelfth Plan document, *Ibid.*
21. Tibaijuka, Anna K. (2008/2009): State of the World's Cities, HARMONIOUS CITIES, UN-HABITAT.
22. Ramachandran, R., *op.cit.*, pp. 331-332.
23. Planning Commission, Government of India: Sixth Five-Year Plan, pp. 395-396.
24. Planning Commission, Government of India: Seventh Five-Year Plan, Vol. II, pp. 296-298.
25. *Ibid.*
26. Planning Commission, Government of India: Approach Paper to the Ninth Five Year Plan-1997-2002, pp. 89-90.
27. Planning Commission, Government of India: Tenth Five Year Plan, pp. 611-28.
28. Eleventh Five Year Plan, Vol. 3, Box 11.2, pp. 395-396.
29. Twelfth Five Year Plan, pp. 320.
30. Government of India, Ministry of Urban Development (2015): Mission Statement and Guidelines, AMRUT.
31. Government of India, Ministry of Urban Development (2015): Mission Statement and Guidelines, Smart Cities.

32. Singh, U.B., Ashok Kumar, N. and Singh, Awadhesh, K. (2014): Challenges in Urban Management in India in New Millennium, GRP, New Delhi, p. 95.
33. Government of India (1988): National Housing Policy.
34. Government of India (1998): National Housing and Habitat Policy.
35. Government of India (2005): Draft National Urban Housing and Habitat Policy.
36. Government of India (2007): National Urban Housing and Habitat Policy, 2007.
37. Government of India (1965): Report of the Committee on Urban Land Policy.
38. Government of India (1983): Report of the Task Force on Planning of Urban Development.
39. Ministry of Urban Development, Government of India (2008): National Urban Sanitation Policy, pp. 8-9.
40. Singh, U. B. (2014): Managing Urban Waste in India: In Search of a Separate Act and National Strategy, Nagarlok, XLVI (1), pp. 63-76.
41. Government of India (2000): Municipal Solid Waste (Management and Handling) Rules.
42. Singh, U.B. (2014): op.cit. and Solid Waste Management Rules, 2016; published in Government of India Extra-ordinary Gazette, New Delhi, 8th April, 2016.
43. *Ibid.*
44. *Ibid,* p. 130.
45. Planning Commission, Government of India: Tenth Five Year Plan, p. 405.
46. Government of India (2009): Report of the Expert Group to Review the Methodology for Estimation of Poverty (popularly known as Tendulkar Committee).
47. Government of India (2009/10): NSSO Report.
48. Singh, U.B., *op. cit.*
49. Planning Commission, Government of India: Tenth Five Year Plan.
50. Government of India, Ministry of Urban Affairs and Employment (1999): National Slum Policy (Draft for discussion).
51. Government of India, Planning Commission: Eleventh Five Year Plan, p. 406.
52. *Ibid,* p. 408.
53. *Ibid,* p. 408.
54. Street Vendors, Seminar, July, 2000, p. 12.
55. Bhowmik, Sharit K. : Hawkers and the urban informal sector: a study of street vending in seven cities for National Alliance of Street Vendors of India (NASVI).
56. As quoted in the National Policy on Urban Street Vendors, 2009.
57. Supreme Court of India (1989): Sodan Singh others etc. versus NDMC.
58. Government of India, Ministry of Housing & Urban Poverty Alleviation (2004): National Policy on Urban Street Vendors.
59. Government of India, Ministry of Housing and Urban Poverty Alleviation, National Policy on Urban Street Vendors, 2009, pp. 2-9.
60. *Ibid,* pp. 6-9.
61. Government of India (2006): National Urban Transport Policy, pp. 3-4.

62. *Ibid.*, pp. 5-9.
63. Planning Commission, Government of India: Twelfth Five Year Plan, p. 341.
64. Renaud, Bertrand (1981): National Urbanisation Policy in Developing Countries, OUP, pp. 96-99.
65. Lawrence, Eric, Stoker, Robert and Wolman, Harold (2010): Crafting Urban Policy, Urban Affairs Review, Vol. 45, No. 3, January, p. 414.
66. Government of India: Twelfth Five Year Plan, op. cit., p. 324.

Chapter - 2

Urban Reforms in India - Emerging Scenario

Introduction

Urban reforms in India distinctly differ in the post-economic reform period which began somewhere in the late 1980s. The idealistic welfare policies, top-down approach and the tendency to depend largely on the bureaucratic structures of the public service in the delivery of civic services dominated the pre-economic reform period. As against this, a holistic and comprehensive understanding of the problems, realisation that poverty is a multi-causal and systemic problem, need for going beyond bureaucratic and professional divisions and a more targeted and participative approach have become key features of the present urban reform universe. Allan Cochrane (2007) terms this paradigm shift as managerialism in the evolution of urban policy. He feels that this new thinking focuses on coordination and holistic or joined-up thinking; the rise of partnership working; the making of new urban managers and urban professionals; a concern for evidence-based policy; and mobilisation of new technologies of management.[1]

The urban reform movement in India is encapsulated and discussed in chronological way in this chapter. The objective of this chapter is to trace-out and provide a broad frame of the urban reform movement in India.

Urban Reforms in the Pre-economic Reform Period

Socialistic pattern of development, balanced regional growth and provision of housing to the poor and weaker sections dominated the urban policies in India in the period just after Independence. The strategies of dispersal of industries from the metro cities; building of houses—ownership basis or rental for weaker sections; and control over urban land through legislations like the Land Acquisition Act, 1894, Town and Country Planning Acts, Rent Control Acts and finally Urban Land (Ceiling and Regulation)

Act, 1976 were adopted to deal with the urban problems. Financial institutions like Life Insurance Corporation of India (LICI) and the Housing and Urban Development Corporation (HUDCO) were entrusted with the task of providing funds to various public sector housing schemes. The focus was on subsidised financial assistance for the low-income people and liberal terms for the economically weaker sections of the society. Building of the houses for the increasing urban households was soon found beyond the capacity of the governments—central and state. This was clearly brought-out by the Task Force on Housing and Urban Development, Government of India in 1983. With this realisation and the influence of World Bank's shelter policies, the focus shifted towards slum development projects and environmental improvement of urban slums scheme was launched to provide basic civic services in the slum areas. The urban infrastructure development was totally financed by the budgetary allocations and borrowings from the two public sector institutions viz., LIC and HUDCO. During 1980s, the limitations of the governments—central, state and local to finance the urban development projects were clearly brought-out by several commissions including the NITI Ayog (National Institution for Transforming India) (erstwhile Planning Commission). Emphasis on internal mobilisation of revenues by the urban local governments (ULGs) and the concept of user charges especially for the basic urban services like water supply and sanitation and solid waste management was emerging slowly.

On the urban finance front, the shift was towards more internal mobilisation, more financial devolutions from the state as well as the central governments to the ULGs and increased role for private sector and the civil society organisations. This shift could clearly be seen in the draft National Housing Policy framed in 1987 by the Government of India. The policy suggested that the public sector role in housing should be confined to a facilitator rather than provider and giving increased role to the private sector in building houses. The supply of land and finance, however, continued to be in the public sector domain. Separate budgets at the city level also came to exist in the water supply and sanitation sectors to improve accountability and efficiency and it suited the funding agencies too for advancing loans.

The most experimented reforms on the municipal finance front have been reforms undertaken in the municipal property tax. Standardisation of the property tax assessment received most attention of the reformers. If G.V. Ramakrishna developed capital value based assessment, the others focused on rationalising the existing rental value based system so as to make the system of assessment more objective, rational and scientific. But, the property tax reforms received government acceptability only in the late 1990s.

On the urban government side, certain reforms were introduced in the areas like municipal elections, election method and role of the chairpersons, property tax and community mobilisation. On the municipal election front, reservation system was introduced on the lines of the State Assemblies and the Parliament for various weaker sections of the society like scheduled castes (SCs) and scheduled tribes (STs). Some states also provided reservations to the backward classes (BCs) and women too. The level and basis of reservation, however, differed from state to state. Reforms in the municipal governance were also made in respect of the form of municipal government and election of the chairpersons. A unique system of Mayor-in-Council was introduced in Kolkata Municipal Corporation through the Calcutta Municipal Corporation Act, 1980. The body is a powerful authority consisting of the Mayor, the Deputy Mayor and other councillor as nominated by the Mayor. The municipal laws in West Bengal, Madhya Pradesh, Chhattisgarh and Tripura also provide for the Mayor-in-Council/ President-in-Council System with certain variations compared to the Kolkata model. This system is similar to the cabinet system of governance and it is designed to make the mayor more powerful and a real executive enjoying the executive powers as against the widely prevalent system elsewhere in the country where the mayor is considered a nominal or titular head with ceremonial functions. But, the system did not find favour in the other states. Certain states like Andhra Pradesh, Tamil Nadu and Uttar Pradesh have gone with direct election of the chairperson/mayor as against the indirect election of the municipal chairpersons by the council. The direct election of the chairpersons without much change in the powers of the chairperson appears to be a cosmetic attempt than a real reform. The experience also proves this as there were frequent disruptions in the council meeting where majority of the councilors belonged to the party other than the chairperson. This affected adversely the efficiency in the council performance.

Another reform area which received much significance relates to community mobilisation. With the launch of various slum improvement and urban poverty alleviation programmes during 1980s, involvement of the NGOs and formation of various community-based organisations has become a popular strategy of the governments. The role of these organisations in the provision of the various urban basic services in the slum areas has been found to be of immense help to the municipal authorities. The community involvement initiatives were made on experimental or pilot basis in various cities. International institutions notably, UNICEF and DFID (erstwhile ODA) came forward to offer financial assistance to improve slum areas. These institutions promoted community participation by according top priority to the various community-based organisation in the slum improvement programmes. The success of these initiatives in Hyderabad and other select cities and towns, led to a country-wide moment for involving the local communities in the latter periods.

Urban Reforms in the Economic Reform Period

Historically, cities and towns are intrinsically linked to economic prosperity of the countries. The economic significance of the cities and towns was emphasised by the National Commission on Urbanisation in 1988 itself in India. The economic reports also indicated that urban India contributes about 60 per cent of country's GDP. But the concrete measures to integrate economic growth with the urban development endeavours are yet to be crystallised. The integration has to begin at the planning stage itself where spatial and infrastructure development planning is effectively linked with the economic development planning at different levels—national, state, district and local. Finally, all the planning exercises must converge at the local level leading to comprehensive development.

The current urban reforms in India have a close linkage with the macro economic policy reforms of the Government of India initiated during 1991. The economic reforms in the first phase included the components like opening up of the economy to foreign direct investments (FDI), foreign institutional investors and imports of technology and consumer goods. Simultaneously, the policy of economic liberalisation was initiated while relaxing several administrative controls aimed at encouraging privatisation even in the sectors like infrastructure and basic services, which were earlier in the domain of public sector. Evidence shows that the reform process has been very slow due to governance weaknesses and political constraints. The FDIs, which were expected of, did not come through. Though the foreign investments as a proportion of total capital accounts inflow in India increased from a mere one per cent in 1991-92 to 25 per cent in 2000-01, it was not the foreign investments that increased but it was the loan accounts whose share which was 38 per cent in 1990-91 increased to 45 per cent in 2000-01. At the other end, the share of FDI, which was more robust, and a stable investment remained quite low at five per cent in 2000-01 but the portfolio investment which is very unstable and can fly out of the country even at the hint of a small crisis in the economy was dominant with about 20 per cent in 2000-01. The most visible effects of economic reforms have been the flooding of foreign consumer durables and increased debt burden on the governments.[2] These trends continue even today in India.

Aid-driven Urban Reforms

Though the integration of Indian economy with the global economy remains doubtful, the urban sector reforms are moving more on global thinking. It is widely believed that the Indian policy-makers and bureaucrats have shown exceptional commitment to the reform process under the influences of international institutions like the UNICEF, World Bank, ADB, USAID, UNDP and DFID. The role played by these international financing institutions in the Indian urban reform process, an amazing trend, is discussed in the following.

UNICEF

The UNICEF has been focusing mainly on women and children. It influenced the social sector reforms in India especially, the social mobilisation and community development. In the urban sector, the UNICEF played critical role in urban poverty and slum development, with a special focus on problems and issues of women and children in rapidly declining urban environment. The UNICEF collaborated with the governments, ULGs and the NGOs in the provision of urban basic services in slum areas. In fact, UNICEF assistance prompted the Government of India to launch the Urban Basic Services (UBS) scheme in the towns and cities of India during the late 1980s. The emphasis was on organisation of the local people, involvement of the communities, NGOs and other civil society organisations in the urban development process focusing on urban basic services and women and children. The states like Andhra Pradesh and Kerala played very active role in the reform process. The decentralisation movement of Kerala during late 1990, in fact, was inspired by these initial efforts under the UBS scheme.

World Bank

The influence of international funding agencies on the urban sector reforms goes back to the World Bank campaign of 1970s. The earlier projects of the World Bank focused on shelter, urban housing and infrastructure. The urban reforms as such started with implementing the conditionalities of the lending agencies like devolution of more powers to city governments and improving the financial performance of municipalities. In fact, improving the municipal financial performance was the main condition to the Calcutta Urban Development Project (1973-88) financed by the World Bank. It was followed in the second Calcutta Project also. Similar trends are also visible in the subsequent projects financed by the World Bank, namely, Madras Urban Development Project (1977-86), Tamil Nadu Urban Development Project Phase I, II and III, Bombay Urban Development Project (1982-88), and Uttar Pradesh Urban Development Project.

The World Bank's lending policy influenced the governments in developing countries to institutionalise the privatisation or to use a more respectable and appropriate term, Public Private Partnership (PPP), vigorously in recent years. The classic example in India is the Tamil Nadu Urban Development Fund set-up in 1995. Under this, a mix of public and private sources of finance in the nature of a joint venture including the ICICI and IL&FS was instituted to develop municipal infrastructure in the towns.

The influence of the World Bank is also seen in respect of the accounting reforms in city governments. The accounting reforms were initiated during 1981-91 in Mumbai and Chennai. In these mega cities, the double entry accrual accounting system was introduced at the instance of the World Bank. In Mumbai, the reform was confined to water supply and sewerage sector while in Chennai, it was made applicable to all major functional areas. In Gujarat Urban Development Project, the World Bank insisted that accrual based accounting must be introduced in all the six corporations and Anand Municipality which received the financial assistance. The World Bank even finalised the consultants to take up the job during 1990-95. In recent years, the Tamil Nadu has also joined this reform movement with its TNUDP wherein the accrual accounting system was introduced in all the municipalities. However, some of the corporations have initiated accounting reforms even without any international assistance, namely, Bangalore Municipal Corporation under Bangalore Agenda Task Force, Hyderabad, Indore, Ludhiana, etc. Finally, the accounting reforms, which started under the influence of international financial institutions, evolved into a national policy eventually. In 2000, the Government of India took accounting reforms on a national scale and the Institute of Chartered Accountants of India prepared a technical guide, which ultimately resulted in the preparation of a National Accounts Manual. Implementation of this Manual has become major responsibility of national, state and local governments under various urban development programmes and schemes.

The World Bank focused its financial assistance on water supply and sanitation while allocating about 60 of its urban investment funds. Its strategy has been involving the private sector in the urban infrastructure and delivery including aspects like institutional development, project development and regulatory facilities. The reform focus of the World Bank has been to provide autonomy to ULGs so that they can operate in a commercial and market environment. The key reforms promoted by the Bank are: [3]

(i) Municipal finance reform and capacity building,

(ii) Lines of credit for multi-sector municipal investment,

(iii) Establishing new independent regulatory entities,

(iv) Developing institutional framework to promote the creation of commercial water utilities,

(v) Reducing price distortions and increasing tariffs,

(vi) Implementing new strategies to sanitation,

(vii) Vigorous private sector participation in the urban water supply and sanitation sectors.

The World Bank has been focusing on policy reform and institutional change in recent years in India penetrating deeper into areas of regulation, pricing and incentive system covering the entire range of governance issues viz., local government autonomy; resource mobilisation; and patterns of relationships among local government, the private sector, informal organisations and households. The new strategy of the Bank for the twenty-first century calls for a holistic approach to urban development with more democratisation, decentralisation and public private partnership and collaboration among stakeholders.[4] The ultimate aim of the World Bank is stated to promote sustainable cities and towns that fulfill the promise of development for their inhabitants with a view to build national perspective in urban development.[5] The sustainability implies that the cities are livable ensuring decent quality of life and equal opportunities for all residents including the poorest. The cities must be competitive, well governed and efficiently managed and financially sustainable, in other words bankable to achieve this goal. Competitiveness means the cities require buoyancy with growing employment, incomes and investment coupled with appropriate urban equity and social safety nets. The bank is supposed to act as a facilitator for city wide economic analysis and strategy development through a participative process involving a wide range of stakeholders. Good governance implies inclusion and fair representation to all groups in the urban society, accountability, integrity and transparency and management capacities at the municipal level. Bankability reflects the financial soundness of ULGs that decides credit worthiness to borrow and enter into the capital market. This requires specific and internally consistent local revenues and expenditures, transparent and creditable inter-governmental transfers, prudent conditions for municipal borrowing and generally accepted financial accounting, asset management and procurement practices and more importantly adopting a commercial approach to the service and administrative functions keeping in view the social concerns. The World Bank's strategy currently aims to achieve greater impact mainly on:[6]

Replicability - promoting lasting improvements in raising the living standards of the urban poor on a scale commensurate with the demands;

Relevance - mobilising skills and knowledge across a wide range of urban issues and offering assistance with flexible designs, realistic timeframes and appropriate forms of financing; and

Reach - ensuring more local governments can be reached directly or indirectly.

Asian Development Bank (ADB)

The Asian Development Bank (ADB)[7] provides support to urban infrastructure, transportation, housing, basic services for informal sector and environment management in India. ADB projects are underway in Karnataka,

Madhya Pradesh, Rajasthan and Gujarat involving a huge investment. The Bank specifically focused on mobilising private capital to finance urban infrastructure and services. The ADB also emphasised institutional strengthening and financial reforms to make the ULGs financially sound to raise loans and repay them. Recently, the Asian Development Bank came forward to finance urban projects in Hubli-Dharward, Belgaum, Bellary, Bijapur, Shimoga, Gulbarga, Davengere, Raichur, Gadag-Betgeri and Bidar in the backward regions of Karnataka. The ADB is supporting urban financial reforms under the Nirmala Nagar Project. The focus of the ADB appears to be on e-Governance related issues. Main problems of the ADB funding are observed to be no concession for poverty alleviation components. Further, ADB support is not forthcoming to preparatory works in the initial periods of a project.

United States Aid for International Development (USAID)

The USAID has been another international agency, which supports the urban infrastructure development in India. Its focus has been mainly on urban sector reforms primarily through the debt market, a component of Financial Institutions Reform and Expansion (FIRE-D) Project. The project has been supporting municipal, state and central governments in the areas of water, sewerage and solid waste and also on increasing the access of services to the poor. The ULGs were assisted in the tasks like project development and financing, resource mobilisation, decentralisation, capacity building and training. In short, it promoted the concept of commercial viability in the delivery of urban infrastructure services. In the process, it helped the development of urban infrastructure through accessing domestic capital market and reforming financial management and accounts of ULGs. Discussing about the future perspective, the assessment and analysis report of the USAID states that reforms must focus on providing managers autonomy to operate in a commercial environment and identified key reform measures, such as:[8]

(i) Establishing new independent regulatory entities;
(ii) Implementing new approaches towards sustainability;
(iii) Reducing price distortions and increasing tariffs;
(iv) Reforming institutional framework to promote creation of commercial entities; and
(v) Increasing private sector participation in the provision, delivery, operation and maintenance of services.

Department for International Development (DFID)

The UK's financial assistance under DFID started on a pilot basis in Andhra Pradesh, West Bengal, Orissa and Madhya Pradesh focusing on poverty alleviation and slum improvement sector in select cities. In Andhra

Pradesh, it began with Hyderabad and extended to Visakhapatnam and Vijayawada cities later. The pilot projects though successful in provision of the much-needed basic infrastructure and services in the slum areas, did not have the desired impact on the urban poverty, therefore, found to be inadequate. In recent years, its strategy has changed from pilot based assistance to state level assistance and the scope of assistance has also been widened to cover the whole gamut of urban management and governance reform. Of late, the DFID has been providing technical assistance to urban reforms at the national and state levels in India. This changed approach was visible in the Andhra Pradesh Urban Services for the Poor (APUSP), Madhya Pradesh Urban Services for the Poor (MPUSP), and Basic Urban Services for the Poor (BUSP) projects. Apart from its primary focus on slums and poverty reduction, these projects have supported various governance reforms at the state and city levels in the areas of community participation, accounting, internal revenue mobilisation, capacity building, etc.

It is clear from the above narration that international financial agencies have been playing major part in the urban sector reforms in India. The reforms advocated and promoted by these international agencies in turn, influenced the reform agendas of the central and state governments either directly or indirectly.

Emergence of the National Urban Reforms Agenda

Till 1990s, the urban policies in India focused on master plans, integrated development and environmental improvement of slums and provision of basic services. The National Commission on Urbanisation (1988) was the first major policy drive in India. The Commission provided a documentary base and identified the priority sectors of development. The economic role of the cities and towns was first recognised and reforms were initiated over a range of urban governance and development issues.

The Rakesh Mohan Committee Report (1996) suggested privatisation and commercialisation of all infrastructure development including urban. In fact, the Committee had a profound impact on the urban infrastructure development strategy. This is echoed in the Ninth Five Year Plan, which strongly recommended adoption of a commercial and market approach to infrastructure planning, development and management. This approach was incorporated in the ongoing schemes of IDSMT by revising it during 1995. The Mega City Scheme introduced in 1993-94 in the five mega cities, namely, Mumbai, Kolkata, Chennai, Bengaluru and Hyderabad incorporates the provision for the institutional financing including the private sector involvement. The commercial nature of the scheme is visible in the ratio of financing. The three main categories of projects, namely, remunerative, user charge based and non-remunerative or service oriented were to be in the ratio of 40:40:30 to make the projects financially viable. The Ninth Plan suggested for private sector involvement in the projects like city beautification, solid waste management including commercialisation of recycling of solid waste, housing, land development and commercial complexes. Cost recovery was emphasised in the development of basic urban infrastructure.

74th CAA-The Landmark in Urban Reforms

The urban local governments (ULGs) in India have been suffering erosion in their powers, resources and capacities over the years in the absence of a constitutional protection. The much needed constitutional base finally came in the form of the Constitution (Seventy-fourth Amendment) Act, 1992 (74th CAA). The 74th CAA provided a strong democratic base for ULGs and an institutional backing to the urban reforms in India. If the Twelfth Schedule provides a functional framework for the ULGs, the State Election Commission aims to ensure regular municipal elections every five years. The reservations to SCs, STs and women provide a constitutional support to empower the weaker sections of the society. The District Planning and Metropolitan Planning Committees are devised to democratise and integrate local level planning. The State Finance Commission set-up in states recommends several innovative strategies and methods to strengthen the financial status of the ailing ULGs. The 74th CAA envisions an empowered governing system for city governments in terms of resources, powers and expertise. The financial management reforms initiated in several ULGs could be considered a good beginning in this direction. The Ministry of Urban Development formulated a Model Municipal Law (MML), a legal framework, to implement the provisions of the 74th CAA in true spirit. The MML aims to simplify municipal byelaws, bring uniformity in city administration, promote municipal borrowing and facilitate the entry of private sector in the urban infrastructure and municipal services. It is expected to guide urban local bodies in the areas of empowered local governance, taxation, accounting, pricing urban services, resource mobilisation, accountability, public-private partnerships, etc. The Nagar Raj Bill aims to bring the civic administration nearer to people besides increasing involvement of people in local governance.

Central Guidance for Urban Reforms

In 1996, Urban Development Plans Formulation and Implementation Guidelines were formulated. The guidelines provided a broad vision of urban reform strategies. These guidelines have since been revised in 2015. The guidelines suggest that the traditional system of funding based on plan and budgetary allocations be reduced and ultimately withdrawn due to fiscal deficit. Subsidies need to be rationalised and urban development plans and projects need to be placed on a commercial format by designing commercially viable urban infrastructure services and area development projects. Other innovative resource mobilisation measures suggested include using land as a resource, increase in the non-property taxes and using Public-Private Partnership in service delivery.

Accessing the Capital Market

The urban local governments (ULGs) in India are also entering into the capital market in line with the global trend and the Government of India's policy. Several municipal corporations, viz., Bangalore, Ahmedabad, Ludhiana, Nagpur, Nasik, and Madurai took lead and successfully issued municipal bonds for raising resources for urban infrastructure. In order to attract more private sector investment into the urban infrastructure sector, the Government of India had announced tax exemption in case of bonds issued by municipal / local governments. Guidelines were issued by the Ministry of Urban Development in February, 2001 for regulating issue of tax-free municipal bonds. Under the guidelines, such bonds would be issued for raising resources for capital investment in creation of new infrastructure as well as augmentation of existing systems. To supplement these initiatives, the Ministry of Finance had also extended certain fiscal incentives in 2001-02 namely, five year tax holiday for all the income for investments on development and O&M of water supply, sanitation and sewerage projects. Tax free municipal bonds were issued by Ahmedabad, Hyderabad and Visakhapatnam municipal corporations taking advantage of these provisions. The municipal bonds are riddled with several problems, viz., lack of project development expertise at the city level, poor credit rating culture of the ULGs, prudential norms of the public sector banks limit exposure to municipal bonds, and capital markets also suffer from low liquidity, not many buyers for municipal bonds as the city governments are considered highly risky by the capital market, political risks and future risks.

Pooled Financing for Municipal Infrastructure

Traditionally, urban local governments have relied on subsidised funds for providing urban services, which restrict scope for introduction of the user charges and efficient project operation and maintenance. In view of the huge resource gap, accessing the capital market is now an accepted alternate option. But, access to capital market requires effective financial management and enhanced credit rating. Evidence shows that only a few bigger municipal corporations are in a position to take advantage of the resources available in the capital market. Medium and smaller municipalities are unable to do so due to weak financial position and lack of capacity to prepare viable project proposals.

A state level pooled financing mechanism is considered way out to overcome the above hurdles, especially, problems faced by the small and medium municipalities. The objective of a state level pooled finance mechanism is to provide a cost effective and efficient approach for smaller and medium sized municipalities to access the domestic capital markets for urban infrastructure and to introduce new institutional arrangements for mobilising urban infrastructure finance. The success of the Tamil Nadu Urban Development Fund motivated the Government of Karnataka also to move in this direction of pooled urban financing.

The Government of India has also come-up with Pooled Finance Development Fund (PFDF) in 2006 for such pooled financing initiatives at the state level. The aim is to provide credit enhancement to the ULGs to access market borrowings based on their credit worthiness through State-level Pooled Finance mechanism. The broad objectives of PFDF are:[9]

(i) To facilitate implementation of bankable urban infrastructure projects through appropriate capacity building measures and financial structuring of projects;

(ii) To facilitate the ULGs to access capital and financial markets for investment in critical urban infrastructure by providing credit enhancement grants to State-pooled Finance Entities for accessing capital markets through Pooled Financing Bonds on behalf of one or more identified municipalities for investment in the urban infrastructure projects;

(iii) To reduce the cost of borrowing to urban local government units with appropriate credit enhancement measures and through restructuring of existing costly debts; and

(iv) To facilitate development of Municipal Bond Market.

In spite of the initiatives taken-up by some city corporations, few state governments and the Government of India's incentive scheme of pooled financing strategy for raising capital from the market, the municipal bonds concept did not gain acceptability in India. The efforts so far made had very insignificant impact on resource mobilisation by the city governments.

Foreign Direct Investment (FDI)

In May 2001, the Government of India announced the policy of permitting 100 per cent FDI for the development of integrated townships including housing, commercial premises, hotels and resorts. The FDI was also permitted in city level urban infrastructure like roads, bridges, mass rapid transit systems and the manufacture of building materials and development of land. This initiative also has not picked-up in India. FDI in infrastructure investments in India has been very poor so far.

Increasing the Pace of Reforms - Urban Reforms Incentive Fund (URIF)

The Government of India launched a scheme called Urban Reforms Incentive Fund (URIF) to link financial assistance to states as an incentive to accelerate process of urban reforms identified. It was 100 per cent grant assistance and the Ministry of Finance would release funds based on the recommendations of the Ministry of Urban Development. The states were required to enter into a prescribed memorandum of agreement with the central government to avail the assistance under the scheme. It was decided that 50 per cent of the states allocation, which is based on percentage of urban population of the state to the total urban population in the country, released in the year 2003-04 on signing the MOA agreeing for undertaking reforms in the urban sector and the balance would be released in the subsequent years based on the milestones achieved for that year.

The funds could be utilised for housing, urban development and poverty alleviation programmes by the state governments. The scheme was basically a reform-oriented initiative by the Government of India to make the urban local governments financially self-reliant while focusing on core urban areas. The first phase reforms are as identified and the weightages allocated on achievement of these reforms were as below:

(i) Repeal of Urban Land (Ceiling & Regulation) Act – 10 per cent of the state share;

(ii) Rationalisation of stamp duty to bring it down to five per cent by the end of Tenth Plan 20 per cent;

(iii) Reform of Rent Control Laws – 20 per cent;

(iv) Introduction of Computerised Process of Registration - 10 per cent;

(v) Reform of Property Tax like introduction of Unit Based Tax and achieving collection efficiency at least to 85 per cent by Tenth Plan - 10 per cent;

(vi) Levy of reasonable user charges with the objective of full cost recovery on O & M by the end of Tenth Plan – 20 per cent; and

(vii) Introduction of Double Entry System of Accounting-10 per cent.

An Empowered Committee headed by the Secretary, Ministry of Urban Employment and Poverty Alleviation (Ministry of Housing and Urban Poverty Alleviation now), Government of India identified the second/third generation reforms. Some of the reforms already identified were:

(i) Streamlining the approval process for construction of buildings;

(ii) Revision of municipal laws in line with model legislation prepared by the Ministry;

(iii) Initiation of public-private partnership in the provision of civic services;
(iv) Reduction in staff;
(v) Reduction in revenue expenditure;
(vi) Introduction of property title certification;
(vii) Introduction of independent regulators for urban services; and
(viii) Implementation of decentralised measures as envisaged in the 74th CAA.

Guiding the Public Private Partnerships - PPP Guidelines

Radical reform of the urban water supply sector and sanitation is urgently needed to boost economic growth and well-being of the urban population and to create an institutional and policy framework which can attract level of public and private investments needed to fill the rising gap between demand and supply in a sustainable manner. In order to facilitate the reform process and entry of the private sector participation in the water supply and sanitation services, the Ministry of Urban Development has formulated PPP Guidelines. These guidelines are expected to sensitise state governments and ULGs to the policy and procedural issues, which need to be addressed in the PPP environment. In particular, the guidelines seek to:[10]

(i) Embed an evolving role for the private sector into the broader urban sector reform;
(ii) Facilitate a systematic assessment of the issues and options for successful private sector participation;
(iii) Prevent improperly designed and executed PSP transactions.

The guidelines envisage a clearly articulated state sector institutional and policy framework, underpinned by enabling laws. The framework should provide the following: [10]

(a) Defining the path towards greater transparency, accountability and efficiency, by separating service provision from policy-making and regulation initially and consolidating this separation through an optimal mix of public private partnerships in service provision;
(b) Defining a financing strategy for the sector—complementing budget funding with user charges for slowly moving towards cost recovery tariffs and leveraging private investment, through an effective use of robust cash-flow projections and public guarantees;
(c) Introducing specific pro-poor incentives, which ensure low-income consumers also benefit and afford at least a basic level of service;
(d) Mitigating adverse impact of reforms on employees, contractors and stakeholders;

(e) Providing guidance and manage scarce water resources in an equitable manner;

(f) Establishing principles of economic regulation and creating regulatory authority;

(g) The state governments would create the enabling state level legal, regulatory and industrial structures, stipulate regulatory roles and establish a reform facilitation team to guide the reform process;

(h) The ULGs would restructure service providers, assess public supply costs, tariffs and subsidy requirements, decide and execute appropriate form of PSP and manage the PSP contract;

(i) Adequate capacity building at the municipal level to manage the PSP contracts and governs the service provider; and

(j) The PSP transactions must be transparent and designed to maximise competition, using reputable advisors to structure, market and execute the transaction.

Linking Infrastructure Development With Reforms

The Ministry of Urban Development launched the Jawaharlal Nehru National Urban Renewal Mission (JnNURM) in December, 2005 as an ambitious scheme to facilitate focused attention on the urban infrastructure development across the cities and towns in India. The scheme adopted a mission mode and a time bound programme of implementation. The objectives of the scheme were:

(i) Focused attention to integrated development of infrastructure services in the cities and towns;

(ii) Securing effective linkages between asset creation and asset management so that the infrastructure services are maintained efficiently and become self-sustaining over time;

(iii) Pumping adequate investments to bridge the deficiencies in urban infrastructure;

(iv) Scaling up the delivery of civic amenities with an emphasis on universal access to the urban poor;

(v) Taking up urban renewal programmes for inner city development and reducing congestion; and

(vi) Provision of basic services to the urban poor including security of tenure and housing at affordable prices.

Most important aspect of this scheme was the reform thrust in urban development. It was expected that by implementing the reforms mandated, the ULGs would become financially sound and sustainable with enhanced credit rating and ability to access capital market for undertaking new programmes and there would be an improved public-private participation. The state governments and the ULGs were required to agree and implement an agenda of reforms both, mandatory and optional. A Memorandum of Agreement (MoA) between States/Municipalities/Parastatal agencies and the Government of India, a prerequisite for accessing the Central assistance, would spell out specific milestones to be achieved for each item of reform. All agreed mandatory and optional reforms outlined below were to be completed within the seven year mission period.

Mandatory Reforms at the State Level

(i) Implementation of 74th Constitution Amendment Act;
(ii) Repeal of ULCRA;
(iii) Reform of Rent Control Laws;
(iv) Rationalisation of Stamp Duty to bring it down to five per cent;
(v) Enactment of the Public Disclosure Law;
(vi) Enactment of the Community Participation Law;
(vii) Integration of city planning and delivery functions by assigning ULGs with "city planning function" and transferring all civic service functions to ULGs.

Mandatory Reforms at the Municipality/Parastatal Agency Level

(i) Shift to accrual-based double entry accounting;
(ii) e-Governance;
(iii) Reform of property tax for reaching 85 per cent coverage and 90 per cent collection efficiency;
(iv) User charges—100 per cent cost recovery in water supply and solid waste management;
(v) Internal earmarking of funds for services to urban poor;
(vi) Provision of basic services to the urban poor.

Optional Reforms (common to States, Municipalities and Parastatal Agencies)

(i) Introduction of Property Title Certification System in municipalities;
(ii) Revision of building byelaws to streamline the approval process;
(iii) Revision of byelaws to make rain-water harvesting mandatory;
(iv) Earmarking at least 25 per cent developed land in all housing projects for EWS and LIG category;
(v) Simplification of legal and procedural frameworks for conversion of land from agricultural to non-agricultural purposes;
(vi) Introduction of computerised process of registration of land and property;

(vii) Byelaws on reuse of recycled water;

(viii) Administrative reforms;

(ix) Structural reforms;

(x) Encouraging Public Private Partnerships.

The reforms package of the JnNURM in tune with the goals of the Mission primarily focused on the rejuvenation of the urban areas in terms of deeper democratic decentralisation, governance, infrastructure, delivery systems, financial sustainability and inclusive development. However, some of the reforms like reducing the cost on staff and improving the competencies and delivery capabilities appear to be contradictory in nature. Though the PPP approaches are being promoted as an alternative, they are still in the formative stages riddled with several scams and bureaucratic lapses and inefficiencies. The critics of the reforms under JnNURM argue that the Government of India has been attempting to promote the market-based development of urban areas which basically served the interests of elites and private sector. Further, it was put-forth that the implementation of the market oriented reforms like privatisation, ULCRA repeal, property title certification, GIS mapping and full cost recovery, etc., will only further deteriorate the living conditions of the poor and economically weaker localities.[11]

The implementation of the reforms under JnNURM across the municipalities during the seven year period (2005-12) reveals certain important trends. The overall progress appears to be slow. The progress has been, however, better in larger mission cities compared to the small and medium size towns. Further, the performance also varied across the states. Further, it was noticed that the implementation of the state level reforms was better compared to the city level reforms.[12] It has been observed that the reforms like property tax and user charges were found to be difficult due to resistance from the taxpayers and consumers. However, some of the states have done excellent job in implementing the property tax reform. The states like Karnataka, Tamil Nadu and Andhra Pradesh have achieved better success in implementing the accrual-based double entry accounting reform. Some of the states like Andhra Pradesh and Karnataka have pushed through reforms in a more vigorous manner. For instance, the Government of Karnataka has set-up a separate Municipal Reforms Cell to give focused attention on implementation of the urban reforms. The Government of Andhra Pradesh also gave focused attention on reforms through the institution called Andhra Pradesh Urban Services Programme (APUSP) and now through Mission for Elimination of the Poverty in Municipal Areas (MEPMA). Such an institutionalised focus at the state level yielded better results. The broader picture of reforms implementation at the city level reveals that e-Governance, accounting, property tax and to some extent public-private partnership fields received more focus. The implementation process of urban reforms under

JnNURM was slow largely confined to some energetic and well placed metropolitan cities.

The Central Finance Commissions' Approach

In pursuance of the Constitution (Seventy-fourth) Amendment Act, 1992 clause (c) was inserted in Article 280 which makes mandatory for the Central Finance Commission to make recommendations relating to "the measures needed to augment the Consolidated Fund of a State to supplement the resources of the Municipalities in the State on the basis of the recommendations made by the Finance Commission of the State."[13] This clause became effective before the Tenth Finance Commission had finalised its report and thus the Tenth Finance Commission felt obliged to make recommendations in this regard, though no such reference was made to it. The Commission thus recommended a lump sum grant in addition to the normal devolution by the state governments. The Eleventh Finance Commission listed the core civic services which it would support, including primary education, health, drinking water, street lighting and sanitation and indicated that the funds released should be earmarked for operation and maintenance of these services. It suggested improvement in efficiency of collection of property tax as well as assignment of a suitable tax with buoyant revenues in lieu of octroi and levy and periodical revision of user charges. The Commission also recommended review of the accounting heads to ensure clarity; prescription of the format for maintenance of accounts by Comptroller & Auditor General (C&AG); audit of accounts by C&AG, whose report to be placed before a committee of the State Legislature; transfer of functions and schemes; integration of district urban development agency (DUDA) with the ULGs; and revitalisation of district planning committees (DPC).[14] The Twelfth Finance Commission stipulated that municipalities should use the grants to improve service delivery relating to water supply and sanitation. In towns with a population of over 100,000, 50 per cent of the grant was to be earmarked for solid waste management schemes in public-private-partnership (PPP) mode. The urban local governments were also expected to give high priority to expenditure for the creation of database on local body finances and maintenance of accounts through the use of modern technology and management systems. The Thirteenth Finance Commission recommended two types of grant—Basic and Performance. The General Basic Grant would be released in two installments. The incentive framework for General Performance Grant consisted of nine conditions to be complied by the states. These conditions (reforms) were as under:[15]

(i) The state government must put in place a supplement to the budget documents for local bodies, and introduce the double entry accounting system;

(ii) The state government must put in place an audit system for municipalities;

(iii) The state government must put in place a system of independent municipal ombudsman to look into complaints of corruption and maladministration against the functionaries of local bodies;

(iv) The state government must put in place a system to electronically transfer local government grants to the respective local bodies within five days of their receipt;

(v) The state government must prescribe through an Act the qualifications of persons eligible for appointment as members of the State Finance Commission;

(vi) All municipalities should be fully enabled to levy property tax on all types of residential and commercial properties;

(vii) State government must put in place a state level Property Tax Board for assisting all municipalities to put in place an independent and transparent procedure for assessing property tax.

(viii) State government must gradually put in place standards for delivery of all essential services provided by municipalities; and

(ix) All municipalities with a population of more than ten lakhs (2001 census) must put in place a fire hazard response and mitigation plan for their respective jurisdiction.

The Fourteenth Finance Commission has linked performance grants to address the following issues: (i) making available reliable data on municipalities' receipt and expenditure through audited accounts; and (ii) improvement in own revenues. In addition, the municipalities will have to measure and publish service level benchmarks for basic services.[16] Thus, the Central Finance Commissions have linked the devolution of funds with the urban reforms.

Further, one purpose of the Atal Mission for Rejuvenation and Urban Transformation (AMRUT), the latest intervention, is to improve urban governance through a set of reforms, and has delinked the release of funds to completion of reforms and thus has shifted from penalisation to incentivisation. The Mission has purported the package of reforms along with the steps and time line for implementation. The set of reforms as envisaged under the AMRUT is shown in Table 2.1:[17]

The new mission (AMRUT) unlike the earlier mission (JnNURM) has gone to the micro level and has thus made it implementable and measurable.

Table 2.1: Reform Agenda under AMRUT

S.No.	TYPE OF REFORM	STEPS/MILESTONES	TIME-LINE
1.	E-Governance	Digital ULGs:	
		(i) Creation of websites	6 months
		(ii) Publication of e-newsletter Digital India Initiatives	6 months
		(iii) Support Digital India Coverage with E-MAAS	6 months
		• Registration of Birth, Death and Marriage	
		• Water and Sewerage Charges	
		• Grievance Redressal	
		• Property Tax	
		• Advertisement Tax	
		• Issuance of Licenses	24 months
		• Building Permissions	
		• Mutations	
		• Payroll	
		• Pension	
		• E-procurement	36 months
		• Personnel Staff management	
		• Project management	
2.	Constitution and professionalisation of municipal cadre	(i) Establishment of municipal cadre	24 months
		(ii) Cadre linked training	
		(iii) Policy for engagement of interns in ULGs and implementation	12 months
		(iv) Policy for Right-sizing the number of municipal functionaries	36 months
3.	Augmenting double entry accounting	(i) Complete migration to double entry accounting and obtaining audit certificate to the effect w.e.f. 2012-13 onwards	12 months
		(ii) Appointment of internal auditor	24 months
		(iii) Publication of annual financial statement on website	Every year

(Contd...)

4.	Urban Planning and City level Plans	(i) Preparation of Master Plan using GIS	48 months
		(ii) Preparation of Service Level Improvement Plans (SLIP) and State Annual Action Plan (SAAP)	6 months
		(iii) Establish Urban Development Authorities	36 months
		(iv) Action plan to progressively increase Green cover in cities to 15 per cent in 5 years	6 months
		(v) Develop one Children Park	
		(vi) Establish a system for maintaining of parks, playgrounds and recreational areas relying on People Public Private Partnership (PPPP) model	Every year 12 months
		(vii) Policy to implement the parameters given in National Mission for Sustainable Habitat	24 months
5.	Devolution of funds and functions	(i) Ensure transfer of Fourteenth Finance Commission devolution to ULGs	6 months
		(ii) Appointment of State Finance Commission (SFC) and making decisions	12 months
		(iii) Implementation of SFC recommendations within time line	18 months
		(iv) Transfer of all 18 functions to municipalities	12 months
6.	Review of Building Bye-laws	(i) Revision of building bye-laws periodically	12 months
		(ii) Policy and action plan for having a solar roof top in all buildings having an area more than 500 sq.mts. and all public buildings	12-24 months
		(iii) Policy and action plan for having Rain-water Harvesting structures in all commercial, public buildings and new buildings on plot of 300 sq. mts. and above	12-24 months
		(iv) Create single window clearance for all approvals to building permissions	12 months
7.	Set-up financial intermediary at state level	Establish and operationalise financial intermediary—pool finance, access external funds, float municipal bonds	12-18 months

(Contd...)

8.	(a) Municipal Tax and fees improvement	(i) At least 90 per cent coverage (ii) At least 90 per cent collection (iii) Policy to periodically revise property tax, levy charges and other fees (iv) Post Demand Collection Book of tax details on website (v) Achieve full potential of advertisement revenue	12 months
8.	(b) Improvement in levy and collection of user charges	(i) Policy on user charges for individual and institutional assessments (ii) Action plan to reduce water losses to less than 20 per cent and publish on website (iii) Separate accounts for user charges (iv) At least 90 per cent billing (v) At least 90 per cent collection	12 months
9.	Credit Rating	Complete the credit ratings of all municipalities	18 months
10.	Energy and Water audit	(i) Energy (street lights) and Water Audit (including non-revenue water) (ii) Making STPs and WTPs more energy efficient (iii) Optimise energy consumption in street lights (iv) Incentives for green buildings e.g. rebate in property tax or charges on building permission/development charges	12 months 12 months 12 months 24 months
11.	Swachh Bharat Mission	(i) Elimination of open defecation (ii) Waste collection (100 per cent) (iii) Transportation of wastes (100 per cent) (iv) Scientific disposal (100 per cent)	36 months

Source: Mission Statement & Guidelines, AMRUT, Annexure-1, June 2015.

Service Level Benchmarking for Improving Service Delivery

The universe of urban infrastructure and services is turning most challenging to tackle in the background of increasing urban population, haphazard growth and the connected civic service needs. The improvements in productivity and efficiency therefore, are essential to make continued growth possible with growing population, urbanisation and constraining resource base. But, poor levels of service, interruptions, low coverage levels, and other problems can undermine quality of life in city and erode trust in local government. Access and provision of quality services to all the citizens is the critical challenge here.

Various studies and reports in India have brought-out the deplorable status of the basic civic amenities. For instance, as per the 2011 Census only 70.6 per cent of urban population is covered by individual connections compared with 91 per cent in China, 86 per cent in South Africa and 80 per cent in Brazil. Duration of water supply in Indian cities ranges from one hour to six hours compared with 24 hours in Brazil and China and 22 hours in Vietnam. Per capita supply of water in Indian cities ranges from 37 lpcpd to 298 lpcpd for a limited duration. Most Indian cities do not have metering for residential water connections. Seventy per cent of water leakages occur from consumer connections and due to malfunctioning of water meters. Non-Revenue Water (NRW) accounts for 50 per cent of water production compared with five per cent in Singapore.[18]

In respect of sanitation, even partial sewerage network is absent in 4861 cities and towns in India. Almost 50 per cent of households in cities like Bengaluru and Hyderabad do not have sewerage connections. As per 2011 census, about 13 per cent of urban households do not have access to any form of latrine facility and defecate in the open. Census 2011 also revealed that about 37 per cent of urban households are connected with open drainage and another 18 per cent are not connected at all. Less than 20 per cent of the road network is covered by storm water drains. As per the report of the Central Pollution Control Board (CPCB) 2009, only about 20 per cent sewage generated was treated before disposal in Class-I cities and Class-II towns (as per 2001 census). As per CPCB report bought out in 2005, about 1,15,000 Metric Ton (MT) of Municipal Solid Waste is generated daily in the country. However, scientific disposal of the waste generated is almost non-existent.[19]

The JnNURM scheme aimed to give an impetus for economically vibrant, sustainable, and equitable initiatives in the urban infrastructure and services front. This demanded that the service delivery management of the municipalities must undergo radical changes focusing on performance improvement, quality and equitable service. The Ministry of Urban Development, Government of India embarked on reforming the service delivery management through the reform of Service Level Benchmarking (SLB).

The systems for measuring performance and taking further action on them have not been institutionalised in urban agencies in India. Recognising its importance, the Ministry of Urban Development (MoUD), Government of India has launched the Service Level Benchmarking (SLB) initiative covering water supply, sewerage, sanitation, solid waste management and drainage. The Service Level Benchmark is a useful tool in recognising the current status of the city using the standardised set of indicators. Benchmarking is now well recognised as an important mechanism for introducing accountability in service delivery. It involves measuring and monitoring of service provider performance on a systematic and continuous basis. Sustained benchmarking can help utilities to identify performance gaps and introduce improvements through the sharing of information and best practices, ultimately resulting in better services to people.

The framework for implementing Service Level Benchmarking (SLB) has been developed by the Ministry of Urban Development. The framework provides a minimum set of standard performance parameters for the four identified services for adoption uniformly across the municipalities in the country. The monitoring and reporting on these indicators and the method of operationalsing the system were also outlined. The data requirements and the measurements procedure were also provided. The following standard details are prescribed for each of the indicators:[20]

(i) Title, Units and Definition
(ii) Data requirements
(iii) Rationale for the indicator
(iv) Reliability of measurement
(v) Frequency of measurement
(vi) Jurisdiction of measurement

In order to get commitment from the state governments and the ULGs, the implementation of the SLB was made one of the conditions for getting performance based grants under the Thirteenth Finance Commission grant-in aid scheme.

Performance Parameters for Basic Urban Services

Service level performance parameters numbering 28 have been identified for four basic urban services, viz. Water Supply, Sewerage, Solid Waste Management and Drainage.[21]

Water Supply

As water supply is a basic need, emphasis has been laid on performance related to reach and access to quality service; and prevalence and effectiveness of the systems to manage the water supply networks. As financial sustainability is critical for continued effectiveness in service delivery, performance is measured on this aspect too. Key performance indicators and the standards are as below:

(a) Coverage of Water Supply Connections-100 per cent
(b) Per Capita Supply of Water-135 lpcd
(c) Extent of metering of water connections-100 per cent
(d) Extent of Non-revenue water-20 per cent
(e) Continuity of Water Supply-24 hours
(f) Quality of Water Supplied-100 per cent
(g) Efficiency in reddressal of customer complaints-80 per cent
(h) Cost Recovery in Water Supply Services-100 per cent
(i) Efficiency in Collection of Water Supply-related charges-90 per cent

Sewerage and Sanitation

For waste water management, performance related to reach and access of the service, effectiveness of the network and environmental sustainability have been emphasised, apart from financial sustainability of operations. Indicators selected are:

(a) Coverage of toilets-100 per cent
(b) Coverage of sewage network services-100 per cent
(c) Collection efficiency of sewage network-100 per cent
(d) Adequacy of sewage treatment capacity-100 per cent
(e) Quality of sewage treatment-100 per cent
(f) Extent of reuse and recycling of sewage-20 per cent
(g) Efficiency in redressal of customer complaints-80 per cent
(h) Extent of cost recovery in sewage management-100 per cent
(i) Efficiency in collection of sewage charges-90 per cent

Solid Waste Management

Performance related to reach and access, effectiveness of network operations and environmental sustainability has been considered, apart from financial sustainability of operations. Indicators selected are:

(a) Household level coverage of SWM services – 100 per cent
(b) Efficiency of collection of municipal solid waste – 100 per cent
(c) Extent of segregation of municipal solid waste – 100 per cent
(d) Extent of municipal solid waste recovered – 80 per cent
(e) Extent of scientific disposal of municipal solid waste -100 per cent

(f) Extent of cost recovery in SWM services – 100 per cent

(g) Efficiency in redressal of customer complaints – 80 per cent

(h) Efficiency in collection of SWM related user related charges – 90 per cent.

Drainage

Extent and effectiveness of the network are emphasised to assess storm water drainage systems performance. As this service does not yield any direct revenues, financial sustainability is not considered. Indicators selected are:

(a) Coverage of Storm Water Drainage Network-100 per cent

(b) Incidence of water logging/flooding-0

It is evident from the performance indicators selected that the reach, access and network coverage are the important aspects pertaining to all the four services identified. The citizens' grievance redressal and the revenue generation indicators are found to be common in three of the services namely, water supply, sewerage and solid waste management. This shows that the SLB strategy has given utmost importance to financial sustainability of the civic services which was neglected so far in the municipalities.

The Ministry of Urban Development has provided extensive guidelines on the indicators, standards, information requirements, performance improvement plans and implementation monitoring mechanisms. These guidelines, if seriously taken into consideration and implemented by the ULGs, the service levels will improve in the urban areas across the country. Though the standards fixed are a bit higher, the municipalities should strive to achieve them in true sense. In order to commit the state governments, the Government of India has made their notification compulsory under the Thirteenth Finance Commission grant scheme. This yielded results at least in the beginning as the state governments moved fast and notified the SLBs in all the municipalities and other parastatals.

The Ministry of Urban Development (MoUD) has in fact cautioned and expressed concerns to be addressed for effective implementation of the SLBs. They are:(i) systems for capturing key data elements identified for the SLBs are not present in many cases at the field level; (ii) tendency of the officers to measure it through ad hoc systems; (iii) resistance of field staff or other stakeholders to collect and report correct information due to vested interests; and (iv) definition and measurement methodology issues. The Ministry of Urban Development suggested that performance management will be sustainable only if-disclosure, reporting, monitoring and performance management feedback, incentives and disincentives are also brought into the cycle. Else the system of measurement and disclosure of SLBs may not sustain itself.[22] The overall framework must be viewed in a performance improvement model as depicted in Fig. 2.1.

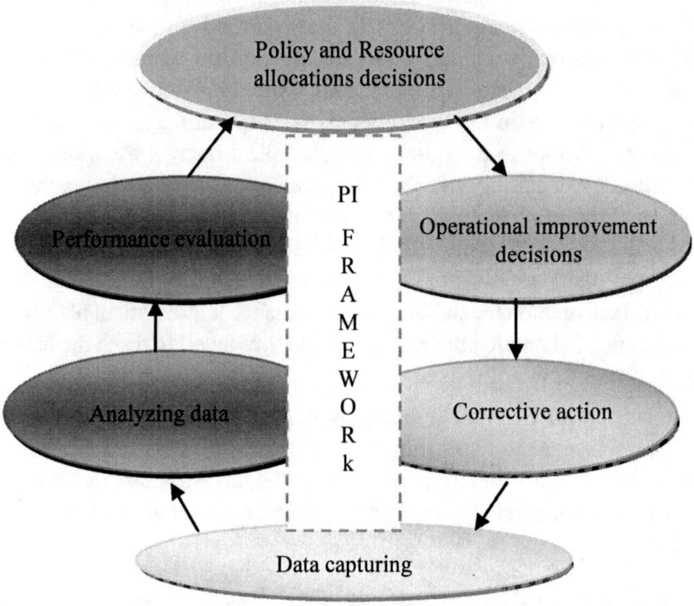

Fig. 2.1: Performance Improvement Framework under SLB Scheme.

The municipalities are facing several teething problems in the implementation of the Service Level Benchmarking (SLB). The bureaucratic lethargy, lack of proper understanding of the purpose and methodology of measurement and, data inadequacies are some constraints impacting implementation of SLBs negatively. A close observation of the implementation process so far reveals that the municipalities are hurriedly providing the information to satisfy the deadlines. The ground level status assessment is not being done properly. The reliability of the data hence, becomes questionable. This will also give wrong signals in rating exercise based on this information. In spite of support provided in the form of training the concerned officials, providing the measuring and reporting formats and making the entire process IT oriented, there appears to be casual approach to the reform. Its significance for service level improvement is yet to be properly understood and carried out efficiently at the municipal level.

Appreciation of the purpose of the SLB by the functionaries concerned is the primary need. The concept of SLB has to be viewed in a broader perspective of analysing the delivery management process of urban services aiming at improving the quality and reach of the services. The officers responsible must understand the parameters, data requirements, measurement and assessment methodology and reporting mechanism to make the implementation purposeful. The data has to be captured at the operational level, the data has to be collated and analysed against the identified indicators and finally, the performance has to be assessed and evaluated. The performance reports of municipalities will support the decisions pertaining to policy decisions, resource allocation, operating decisions aiming at improvement and, again the circle begins with capturing of new data. The entire process works in a cyclic manner as performance assessment is a continuous activity aiming at further improvement. Efforts are being made to devise proper reporting templates covering all the required information while adopting the IT tools. The stakeholder consultation and performance improvement plans are essential in this respect. If stakeholder consultation promotes effective implementation, the performance improvement plans along with the financial plans aid improvements in service levels to reach the national standards on a sustainable basis.

The success of SLB scheme largely depend on the reliability of data, correct measurement to assess the service gaps, effective monitoring and finally, integrating of the SLB data base with the infrastructure development projects so that the levels of services could be improved as per the national standards.

Citizen Charter for Ensuring Quality Urban Services

'Citizen's Charter' initiative is a response to the quest for solving the problems which a citizen encounters, day in and day out, while dealing with the organisations providing public services. The concept was first articulated and implemented in the United Kingdom by the Conservative Government of John Major in 1991 as a National Programme with a simple aim: to continuously improve the quality of public services for the people of the country so that these services respond to the needs and wishes of the users. The First Report on Citizens' Charter that was released by Prime Minister John Major in 1992 stated, "The Citizens' Charter sees public services through the eyes of those who use them. For too long the provider has dominated and now it is the turn of the user... The Citizens' Charter will raise quality, increase choice, secure better value and extend accountability."[23] The programme was re-launched in 1998 by the Labour Government of Tony Blair which rechristened it "Service First".[24] Nine main principles were framed in UK to operationalise the citizen charters. They are: [25] (i) Set standards of service; (ii) Be open and provide full information; (iii) Consult and involve; (iv) Encourage access and promote choice; (v) Treat all fairly; (vi) Put things right when they go wrong; (vii) Use resources effectively; (viii) Innovate and improve; and (ix) Work with other providers.

Later, the concept was implemented in different countries under different nomenclatures and focuses.[26] The Government of Malaysia's Client's Charter emphasised on a written commitment by an agency to deliver outputs or services according to specified standards of quality. The Commonwealth Government of Australia launched its initiative of Service Charter in 1997 as part of its on-going commitment to improve the quality of service provided by agencies to the Australian community to shift focus from bureaucratic approach to customer-focused outcomes. The Total Quality Management (TQM) concept has similar objectives and this has also impacted the quality focus of public service organisations through citizen charters.[27]

The citizen centric administration as a concept has been very old in India. But, its adoption in the modern public administrative systems is of recent phenomenon. The thinking on citizen centric administration is also being generated in India while taking cue from the international practices. In a Conference of Chief Ministers of various States and Union Territories held on 24 May, 1997 in New Delhi, presided over by the Prime Minister of India, an "Action Plan for Effective and Responsive Government" at the Centre and State levels was adopted. One of the major decisions at that Conference was that the Central and State Governments would formulate Citizen's Charters. [28] Later, the Department of Administrative Reforms and Public Grievances in Government of India initiated the task of coordinating, formulating and operationalising Citizen's Charters. The Second Administrative Reforms Commission (ARC) articulated the strategies in its Twelfth Report. It identified several strategies like right information, grievance redressal system, active citizen's participation, etc and one of the strategies identified was citizens' charters.[29]

Citizen Charter is a document presented by a government institution explaining the services and its standards it promises to provide to the citizens. Before preparing the citizen charter, adequate groundwork must be done by any organisation. The Second ARC in India identified some very important steps to make the exercise successful and purposeful. They are:[30] "(i) Define all services which you provide and identify your clients, (ii) Set standards and norms for each service, (iii) Develop capability to meet the set standards, (iv) Perform to achieve the standards, (v) Monitor performance against the set standards, (vi) Evaluate the impact through an independent mechanism, and (vii) Continuous improvement based on monitoring and evaluation results."

As municipalities are democratic bodies and established to provide services to the urban communities, it is their primary responsibility to implement Citizen Charter. Citizen charter basically focuses on service delivery- its promptness, citizen satisfaction and finally achieving the objectives of the ULGs for which they are constituted. Citizen charters become significant from the point of efficiency in the provision of services and also to achieve public accountability which is one of the basic tenets of good governance. Citizen charter is a commitment of the ULG to the citizens for delivering of quality and efficient services in time with helpful attitude and by transparent method. The Charter is a commitment of municipalities to provide all residents of urban areas a range of standard services like adequate and safe drinking water, proper roads, drains, adequate street lighting, prompt registration of births and deaths and, issue of various permissions and licenses.

Often, it is observed that citizen charters are kept in public offices simply to observe the rules and regulations and their implementation suffers from lack of commitment among the officers, inability to provide services in time due to financial and staff problems. Following steps are needed to implement citizen charter effectively:

(i) The local governments need to make public commitments of what services it will provide to the citizen's as per entitlements with specific reference to the informal settlements, with clearly stated information about standards and quality;
(ii) If the services are not provided, there must be a clear cut grievance redressal process;
(iii) Establishing the process of public hearings to identify priorities for taking up projects, their costs and contractual arrangements and periodic reviews of progress; and
(iv) Charter should spell out areas of municipalities working with citizen's groups, the various activities in which it will work in partnership with people.

Timely service and promptness in response are essential to achieve the objectives of citizen charter. The municipalities should put in all efforts to deliver the services as per the citizen charter standards. Otherwise, ULGs have to pay a reasonable compensation on day-basis to the applicants towards loss of their valuable time, if the time schedule is not adhered to in according permissions and issuing certificates. This amount may be collected by the municipality from the person responsible for the delay in rendering the service. This step will definitely go a long way in adhering to the time schedule by the municipalities. In Andhra Pradesh, penalty provisions are introduced for timely implementing the citizen charters. The municipalities in India have been implementing the citizen charters but the effectiveness varies from city to city. If the ULGs perform this task with commitment, the results would be far better. Two closely related reforms in this regard viz., citizen grievance redressal system and the service centres are discussed in the following.

Citizen Grievance Redressal Mechanism

The citizen charter implementation or any other civic service related aspect cannot be implemented successfully unless a grievance redressal mechanism is put in place in the municipalities. A public grievance in simple terms is dissatisfaction expressed by the citizens about the services being rendered by the government agencies. Whether it is imagined, right or wrong, written or oral, it should be redressed. If citizen charter provides, the promises, the grievance redressal system ensures that the citizen charter and other services are provided as per the expectations of the people and as promised by the ULGs. In municipalities, some mechanisms are provided to look into the complaints of the people.

The citizen grievance redressal is increasingly becoming important in the ULGs where the citizens receive basic services necessary for daily living. The JnNURM scheme gave much importance for implementing efficient grievance redressal mechanism in the municipalities and the parastatal organisations like slum clearance boards and water supply and sewerage drainage boards. In Karnataka, for instance, the grievance redressal system has been standardised at the state level for implementation in the municipalities across the state. The e-governance tools are being utilised for effecting operation of the system. Multiple access channels - Internet, Phone, E-mail and Paper Form are being used for registration of complaints. With the introduction of web based systems, the service is made accessible to the citizens 24×7. Some NGOs are also involved in the process. Similarly, the grievance redressal systems are being run on e-Governance mode in most of the cities especially, the Mission cities under JnNURM.

The Hyderabad Metropolitan Water Supply and Sewerage Board has re-modelled the grievance redressal system and put it on on-line. Similar to Karnataka, model, multiple channels are opened to receive and register complaints. The redressal process is being monitored on-line. The benefits achieved by introducing the IT-based system clearly indicate that the citizens are provided with citizen friendly services in a transparent manner.

Service Centre/Facilitation Centre

The foremost requirement to ensure quality service to its citizen is to minimise the interaction of the citizen with officials and employees and at the same time providing services at the accessible locations for the convenience of the people. It can be achieved if, Service Centrrs are opened in the office, where people would present their applications for payment of taxes, permissions, certificates, etc., and collect the required acknowledgements/receipts, permissions or certificates from the same place on a designated day. These centres should be manned by trained, competent and knowledgeable persons. They are expected to guide the citizen in the procedural aspects and payment aspect of fees/charges. Leaflets have to be provided about the procedure to be followed for obtaining various permissions / certificates. A time bound disposal system has to be made.

The effective operational mechanism of the service centres is as described below:

(a) The citizen who wishes to avail the services has to submit an application at the Service Centre duly enclosing required documents in the forenoon hours. (Leaflets explaining the procedure for obtaining permissions, certificates, sanctions and the relevant fee/ charges have to be made available to the applicants at Service Centres free of charge).

(b) The executive manning the Service Centre has to carefully scrutinise the applications received and satisfy himself that all the documents and information required to be attached are submitted by the applicant (The executive must be thoroughly oriented of the procedure and systems in the office).

(c) The executive manning the Service Centre has to give an acknowledgement to the applicant duly indicating the date and time at which he can obtain permission / certificate from the Service Centre as per the time frame indicated above.

(d) Permissions / certificates etc. have to be issued from the Service Centre in the afternoon hours.

(e) Separate registers are to be maintained for various services.

(f) The heads of concerned service departments have to review the register daily and satisfy themselves that the time frame fixed is followed strictly in issue of permissions, certificates and sanctions.

(g) The Commissioner has to review the registers once in every week to ensure effective functioning of Service Centre.

(h) Feedback from the applicants may be gathered after handing over permissions /certificates about the functioning of the Service Centre.

Emergence of Urban Financial Institutions in India

Several financial institutions have come-up on the urban infrastructure front in India. These institutions are not only providing the necessary capital and financial support to ULGs, but also playing key role in the urban reform process. Prominent of these are Infrastructure Development Finance Company Limited (IDFC), ICICI Limited and Infrastructure Leasing and Financial Services Limited (IL & FS).

The Infrastructure Development Finance Company (IDFC) was incorporated in 1997 and the Reserve Bank of India, Industrial Development

Bank of India (IDBI) and Government of India are the shareholders of this Company. Major objectives of this Company are to[31]: lend private capital to commercially viable infrastructure projects in India; provide policy advice to the government to met the above objective; and develop and strengthen the linkages between markets, institutions and infrastructure projects.[32] Its focus areas are transportation, energy, telecommunication, information technology and urban infrastructure. IDFC has collaborative agreement with Asian Development Bank (ADB) to provide line of credit for financing urban infrastructure in India and it is participating in setting-up of Infrastructure Funds like Gujarat Infrastructure Fund. The IDFC promotes private infrastructure financing with credit enhancements such as guarantees, bridging finance, risk participation and refinancing. It lends private capital to viable commercial infrastructure projects while developing linkages between markets and urban institutions. IDFC also plays an important role in institutional capacity building. In short, it is for building a commercial frame of mind in the urban infrastructure development.

The ICICI Ltd.[33] is a commercial financial institution aiming to become the market leader among the private sector financial intermediaries, which finance the urban infrastructure projects in the medium term future. It was set-up in August 1999. Though it started as a project-based lending institution to manufacturing sector, it changed with the times and widened its scope to corporate and infrastructure financing. In the urban sector its activities include: providing direct financing of large projects to the municipalities; participating in setting-up state level intermediaries to finance projects; and facilitating municipalities to access capital market. It has promoted Tamil Nadu Urban Development Fund to finance municipal projects, established a joint venture with Kerala Industrial Infrastructure Development Fund for developing private sector infrastructure projects, and established a joint venture with West Bengal Industrial Corporation for infrastructure development.

The Infrastructure Leasing & Financial Services Ltd. (IL&FS) implements infrastructure projects without government support to budgetary allocations. It works with state governments and ULGs in the commercialisation of urban infrastructure projects. The major areas include water supply, power, telecommunications, transport, etc.

Looking Ahead

It may look paradoxical that if national issues, problems and needs shaped the policy-making and dominated the urban development and governance reforms process in India in the pre-1990 period, it is the global forces and international financial agencies that are playing key role in the current decade of reform movement in India. Privatisation, commercialisation, decentralisation, sustainable and inclusive development and good urban governance are the selling governance concepts today attracting the attention of the national, state and urban local governments. Though the Constitution 74th Amendment Act, 1992 initiated the governance reform in India with a broad objective of democratic decentralisation and empowering ULGs, it is the financial issues like state financial devolutions to municipalities, self-reliance, resource mobilisation, debt financing and financial management, which dominate the reform process today. Introduction of the double entry accounting system; reduction of subsidies and user charges to make urban services self-sufficient; repeal of land ceiling legislation; and dominance of the parastatal institutions on the urban front despite pronounced policy of strengthening of ULGs - all indicate the urban governance paradox in India. The Central Government has been playing very effective role in this process through a number of centrally sponsored schemes. Though these reforms are criticised on the grounds of their thrust on commercialisation and privatisation, a balanced view reveals that certain reforms focusing on decentralisation, resource mobilisation, accountability and efficient performance will certainly facilitate strengthening of the urban governments if pursued with commitment and vigour. It is observed that there is a stiff resistance at the state and local levels for a radical reform agenda being pushed through by the international financing agencies overlooking the prevailing situation and realities. A balanced perspective is therefore, needed to meet the emerging challenges arising out of rapid urban growth, globalisation and increasing urban turmoil. Vigorous educating, resource and institution building efforts are essential at the local level. While the central government should continue to play a catalytic role, the state governments must take the lead role to realise the urban development goals.

The entire urban reform movement in India finally converged on the following core themes:

(i) Local level resource mobilisation through a combination of conventional and innovative strategies;

(ii) Innovative financing schemes including setting up of development funds;

(iii) Accounting and budgetary reforms for improving financial management;

(iv) Privatisation and public private partnership (PPP) in the provision of urban services and infrastructure development;

(v) Good governance focusing on decentralisation, capacity building, peoples' participation, transparency and accountability;

(vi) Improving urban civic service levels and delivery management; and

(vii) Sustainable development while promoting inclusive urban growth.

In respect of the SLB reform for improving the delivery management and performance, the municipalities are notifying service level standards. The major problems were found to be defective data and reliability of the information. Citizen's charters and grievance redressal system are slowly picking-up.

The reforms implementation through the JnNURM Mission reflects the Government of India's commitment for developing the urban areas in terms of the governance and service delivery. The implementation pace was better in some of the municipalities. The experiences indicate that the pace in general was slow ridden with various administrative issues and constraints like inadequate municipal involvement and initiatives, lack of political commitment and peoples' resistance for some of the reforms like user charges. The bigger challenge now is sustaining of the reform momentum.

It is argued that the partial success of reform implementation has prevented the ULGs emerging strong and competent entities in spite of the commendable success achieved by some municipalities. In this situation, the Government of India is contemplating to rationalise the system across the city sizes. The Twelfth Five Year Plan[33] commented that "a significant interstate variation in completion of reform underscores the need for their careful calibration on the basis of city size and capabilities." Accordingly, the Twelfth Plan while emphasising the need for completion of the pending reforms has done some rationalisation exercise and identified reforms for cities having more than five lakh population and reforms for metropolitan cities. As some of the reforms identified were found to be difficult due to local conditions, the mechanism of incentive fund (10 per cent of the overall budget) is provided in the AMRUT to motivate the ULGs to implement these reforms. The reforms identified for the AMRUT are thus, a consolidation of all the earlier reforms being carried-out in urban India.

Looking ahead, the Twelfth Plan envisions faster, inclusive and sustainable Indian cities. Further, the cities will be the locus and engines of economic growth over the coming decades. In order to achieve these goals, the cities must be made more competitive, livable, bankable, inclusive and more sustainable. Truly, this realisation should guide the reform crafting and administrative interventions of the central and state governments.

References

1. Cochrane, Allan (2007): Understanding Urban Policy— a critical approach, Blackwell Publishing, Oxford, p. 32.
2. Mahadevia, Darshini (2003): Globalisation, Urban Reforms and Metropolitan Response, India, Manak Publications Private Limited, New Delhi, pp. 68-73.
3. USAID (2001): Assessment and Analysis of the Status of Urban Service Delivery and Urban Infrastructure Sector Reform in India – Draft Final Report, November 2, New Delhi, India.

4. The World Bank (2000): Cities in Transition-World Bank, Urban and Local Government Strategy, Washington.
5. *Ibid.*
6. The World Bank (2000): Cities in Transition – World Bank, Urban and Local Government Strategy, Washington.
7. USAID (2001): Op.cit.
8. *Ibid.*
9. Planning Commission, Government of India: Eleventh Five Year Plan, Vol. 3, p. 397.
10. Government of India (2004): Guidelines for Sector Reform and Successful Public-Private Partnerships-Foreword and Executive Summary, January.
11. *Ibid.*
12. Chandran, Deepa (2013): Developmental Transformation of Urban India-JNNURM Showcase in the South, Kalpaz Publication, Delhi, p. 128.
13. Constitution of India.
14. Government of India: as quoted by the Thirteenth Finance Commission, pp. 149-152.
15. Government of India: Thirteenth Finance Commission, pp. 178-180.
16. Government of India: Fourteenth Finance Commission, p. 113.
17. Government of India (2015): Mission Statement and Guidelines AMRUT, pp. 29-30.
18. Ministry of Urban Development, Government of India (2012): Compendium of Projects and Reforms under Jawaharlal Nehru National Urban Renewal Mission, Vol. I, July, p. 8.
19. Government of India: Twelfth Five Year Plan, Box No. 18.2, p. 321.
20. *Ibid.*
21. Ministry of Urban Development, Government of India: Handbook on Service Level Benchmarking, p. 10.
22. *Ibid.*
23. *Ibid,* pp. 9-10.
24. Cabinet Office, U.K. (1992): as quoted by the Second Administrative Reforms Commission, Government of India (2009), Twelfth Report, Citizen Centric Administration, February, p. 33.
25. Centre for Good Governance (2008): Citizens Charter Handbook, pp. 1-2.
26. Second ARC Report, op. cit., p. 34.
27. e.g., Australia (Service Charter, 1997), Belgium (Public Service Users' Charter 1992), Canada (Service Standards Initiative, 1995), France (Service Charter, 1992), India (Citizen's Charter, 1997), Jamaica (Citizen's Charter 1994), Malaysia (Client Citizen's Charters - A Handbook Charter, 1993), Portugal (The Quality Charter in Public Services, 1993), and Spain (The Quality Observatory, 1992) (OECD, 1996)
28. Citizen Charter Handbook, op. cit.
29. *Ibid.*

30. Second Administrative Reforms Commission, Government of India (2009): *op.cit.*, p. iii.
31. *Ibid.*, pp. 54-55.
32. USAID, *op. cit.*
33. The ICICI Ltd. was later converted into a private sector bank.

Chapter - 3

Reforming Urban Governance

Introduction

The paradox of the current urban scenario is while the cities are becoming important territorial nodes in the face of globalisation, there is also an increasing trend of fragmentation in terms of their communities, economy, politics and administration. At the other end, city governance, to-day, has become a complex process involving multiple stakeholders, interdependent resources and actions, shared purposes and blurred boundaries between private and public, formal and informal and state and civil society sectors demanding coordination, negotiation and consensus. Urban governance, therefore, must address three key tasks: coordinating a fragmented governance landscape; steering interdependent activities through bargaining, confederating, moderating and negotiating processes; and integrating and managing diverse networks rather than simply focusing on internal affairs.[1] In order to perform these tasks successfully, the urban local governments and the governing processes should undergo a paradigm shift from the traditional 'government-centred' approach to the broad based 'governance-centred' approach. This chapter while tracing the historical roots and causes of decline of urban local governments discusses the current reform initiatives in India aimed at improving the urban governance.

Historical Background

Historically, local government evolved earlier to the state and national governments through small communities. India has a rich local self-government tradition which could be traced as back as to Vedic times. Though local government in different forms existed in India especially, in the villages, the modern form of the local government as it exists to-day was evolved during the British Rule periods. The first local authority was formed in Chennai in 1687 and the system of local government was evolved by various

Acts and Resolutions. The Charter Act of 1793, Act of 1850 and Resolution of 1874 were the early steps for encouraging people to involve in the local affairs. Dominated by the nominated members, several forms of local bodies like Improvement Trusts, Port Trusts, District and Taluka Committees and Local Fund Boards were constituted.

Initially, these institutions were administrative units with a mayor and mostly British born councillors and performed certain judicial powers too along with other local functions like roads, conservancy, lighting, etc. The elected system of local governance as such began with the Lord Mayo's reforms. The system was further developed by Lord Ripon in 1882. The urban local governments were designed during Lord Ripon's period on democratic lines with the predominance of the elected representatives. In fact, the Lord Ripon's Resolution is considered the *Magna Carta* in the evolution of the local self-government in India. The local self-government was transferred to Provincial control. Indirect state control, endowing local bodies with sufficient financial resources commensurate with functions entrusted and local body control over the local government officials are the other features of the Ripon's ideology. Main features of Resolution are:

- local bodies should have mostly elected non-governmental members;
- state control should be indirect than direct;
- local bodies must be endowed with sufficient financial resources to carry-out their functions; and
- local government officials should work under the control of the local bodies.

The Royal Commission on Decentralisation, 1906 also reiterated these views and recommended that the local bodies should have elected members, they elect their own president/chairman, they should have authority to determine taxes and prepare budgets and outside control should be in the nature of advice and suggestions. The Government of India Act 1919 led to further democratisation of local government and passing of local Acts viz., Calcutta Municipal Act, 1923, the U.P. District Boards Act, 1922 and the Bombay District Boards Act, etc. Finally, the Government of India Act, 1935 paved way three fundamental trends in the field of local self-government in India:[2]

- abolition of nomination to the local bodies and introduction of adult suffrage;
- separation of deliberative and executive functions; and
- rule making powers of the state governments for functioning of the local governments.

Types of Municipal Governments in India

After Independence, the rural and urban local governments developed on different lines based on the requirements. Various forms of urban local government were constituted to serve different purposes. Main categories of urban local governments are: municipal councils and municipal corporations, notified area committees, town area committees, and cantonment boards. Except cantonment boards, which were established in capital and main cities with domination of nominated members by the Central Government, all others are under the control of the state governments.

Municipal corporations and municipal councils are truly democratic local self-government formed for the towns and cities. Both are constituted under separate state Acts (in some states like Bihar, Jharkhand, Sikkim, Mizoram, Tripura, etc a single law has been enacted for managing all types of municipalities). The corporations are generally constituted in the major cities or capital cities and municipalities in the other cities and towns. However, both are democratically elected bodies with provision for ex-officio members mainly, Members of Legislative Assembly/Councils and Members of Parliament.

Municipal corporations and municipal councils are set-up under state Acts. Generally, they are constituted based on certain criteria viz., minimum population, density of population, development factors, etc. Corporations are constituted for the large urban areas and they enjoy more powers and autonomy compared to the municipalities. As the area and population are bigger, they are given higher powers and autonomy. Earlier corporations like Bombay (now Mumbai), Calcutta (now Kolkata), Madras (now Chennai), Bangalore (now Bengaluru), Guwahati and Hyderabad are under the direct control of the state secretariat departments. But, recently constituted municipal corporations are placed under the control of the Directors of Municipal Administration along with the municipalities.

Municipal corporations and municipal councils perform same set of functions. There may be small variations depending on the local conditions. By and large they have same functional responsibilities. They perform three types of functions: civic functions, regulatory functions, and social and developmental functions. Civic functions are providing drinking water supply, street lighting, drains, waste disposal and conservancy, roads, public conveniences, public parks and gardens, etc. Regulatory functions include registration of births and deaths, regulation of offensive and dangerous trades, taking measures to control spread of epidemic diseases, controlling building activities, giving licenses for various trades and professions, etc. Social and developmental functions are providing and maintaining hospitals, dispensaries, women and child welfare centres, rest houses, tree plantation, providing housing especially for the poor people, providing market facilities, etc.

Cantonment Boards

Cantonment Boards are unique to India and are centrally administered by the Ministry of Defence. These areas acquired under different treaties from the erstwhile princes and were mainly inhabited by the military personnel. But, slowly, these areas were also populated by the civilians due to requirements. This necessitated municipalisation of the boards through an Act called Cantonments Act 1924 (this Act has been replaced by a new Cantonment Board Act, 2006). Cantonment Boards are constituted under this Act comprising of the Officer-in-Command of the station, a magistrate, health officer, executive engineer, military officers—all nominated and some elected members. The president is the commanding officer-of-the station, some are ex-officio members and the elected members hold office for three years. The vice-president is elected by the elected members.[3]

The functions of the cantonment boards are similar to the municipalities. They have certain obligatory functions and discretionary functions. Obligatory functions include: street lighting, cleansing streets and drains, regulating offensive trades, removing dangerous buildings, maintaining the places for the disposal of the dead, constructing and maintaining streets, drains, markets, slaughter houses, etc., supply of pure drinking water, plantation, registering births and deaths, maintaining public dispensaries and hospitals, protection against fire and establishing schools, etc. The discretionary functions include construction of public works, taking census, making survey, relief measures on spread of epidemics, provision of electricity, provision of public transport, etc.

In order to perform these functions they are provided with powers to raise money through taxes and non-taxes similar to the municipalities in India. In 1953 through an amendment, the powers were enhanced to include building control, issue of trade license, etc. In spite of the changes in recent years, their functioning basically remains similar to that of the municipalities except that they are controlled by the Central Government and official domination with nominated members.

Notified Area Committees

A notified area committee was constituted for a transitional area which did not fulfill all the conditions necessary for formation of a municipality but assumed importance due to its peculiar developments. They were mostly created for growing towns. Generally their formation was notified in the government gazette, hence, the name of notified area committee. These institutions existed in states like Bihar, Gujarat, Haryana, Madhya Pradesh, Karnataka, Punjab, Uttar Pradesh, Himachal Pradesh, Odisha, etc. Notified area committees performed similar functions of the municipalities but which were specifically notified. They were primarily nominated bodies. Its council, members and the chairman were nominated by the state government. These units have since been converted as Nagar/Town Panchayat as a consequence of the Constitution (seventy-fourth) Amendment Act, 1992.

Town Area Committees

Town area committees were constituted for small towns. They were formed under specific state Acts but in Assam they were formed under the Assam Municipal Act, 1956. They were constituted in Assam, Kerala, Madhya Pradesh, Uttar Pradesh, West Bengal and Himachal Pradesh. The town area committees were either total nominated or partly nominated and partly elected. The district collector controlled these local government units. They performed limited civic functions like street lighting, drainage and conservancy. These units have since been rechristened as Nagar/Town Panchayat as a consequence of the Constitution (seventy-fourth) Amendment Act, 1992.

Forms of Municipal Government

The form of modern urban local government as it exists in Independent India is a British legacy. It is not an exaggeration to say that the basic functional and financial structure of the Urban Local Governments (ULGs) inherited from the British period remains almost the same even today. Though there have been certain changes in the composition of the municipal councils like representation to the weaker sections, committee system, personnel policies, involvement of the local communities and financial management, the core functions and the financial resource base remained static. This is the area of foci of the reforms in urban governance in India.

Form of urban local government is a mechanism of sharing the political and administrative powers between the deliberative and executive wings. The form, therefore, does not alter the total functions and powers domain of the urban local governments. Forms of urban local government evolved in different countries based on historical and local politico-administrative conditions prevail in particular country. As a result, each country has its own form of urban local government.

India opted for a federal system of government inspired by the American system but its functional structure remains centralised following the British model. The local governments are put under the State List. Hence, they are totally controlled by the state governments. Their form, functions and finances are derived from the state Acts.

Corporation Form

Corporations are formed in the larger cities and capital cities reflecting their elevated positions in the habitat form. The corporation is body corporate having perpetual succession. They are constituted as per the State Municipal

Corporation Acts. In the corporation form, the entire powers of the urban local body are shared by the corporation and the standing committee and the commissioner appointed by the state government. The corporation comprises of three kinds of members- elected members, ex-officio members and the nominated/co-opted members. Generally, the commissioner in the corporations belongs to I.A.S. cadre, but non-I.A.S. officers are also appointed as commissioner in some corporations. Usually, in smaller corporations non-I.A.S. officers are appointed. In the corporation Acts, corporation, standing committee and the commissioner are the statutory bodies. The Acts specify powers of the corporation, standing committees and the commissioner.

The corporation enjoys all the deliberative powers vested by the Corporation Acts. The political executive in the municipal corporations is called Mayor in India. The institution of mayor remains a ceremonial head of the corporation form of urban government in India. Traditionally, mayor is considered the first citizen of the city/town. There are two methods for election of the mayor. In the states like Tamil Nadu, Uttar Pradesh, Jharkhand, Uttarakhand, Mayor is directly elected by the registered voters of the municipal corporation area. In most of the other states like Andhra Pradesh, Karnataka, Bihar, Odisha, Madhya Pradesh, Chhattisgarh, etc., the mayor is indirectly elected by the elected members of the municipal corporation. The corporation can be dissolved by the state governments for any financial or other factors.

Municipality Form

The municipality form is similar to corporation form. It consists of the council and executive officer (in few states he is known as commissioner). The council comprises of three types of members viz., elected members, nominated member/co-opted members and ex-officio members. The elected members are elected from demarcated electoral wards. The nominated or co-opted members are appointed to include the expertise element also in the deliberations of the council. The MPs and the MLA/MLCs representing municipal areas are made ex-officio members.

The council enjoys all the deliberative powers vested in the municipalities by the Act. However, the government can suspend the resolution of the council for specific reasons. The council can be dissolved by the state government for any financial crisis or if the council finds difficult to function due to internal political problems or due to lack of majority.

The executive officer is head of the executive wing and exercises control over the entire administration of the municipality. He is appointed by the state government and he can be transferred by the state government. The office of the executive officer is not a term post. He executes all the decisions taken by the municipal council. He wields considerable administrative powers for carrying-out the duties allocated. The commissioner performs all administrative functions like managing the municipal funds, supervision and control over all the officers and other employees working, and maintains better relations with the councillors and helps in the smooth functioning of the council meetings. He can also take disciplinary action over the employees.

Cabinet Form

The credit of introducing a totally new model after Independence goes to West Bengal in India. This model is known as Mayor-in-Council system. According to the Calcutta Municipal Act 1980, Mayor-in-Council model was adopted in the Calcutta and Howrah Municipal Corporations. The municipal government was restructured into the mayor, the chairman, the mayor-in-council and the commissioner. The mayor and chairman are elected by the council from amongst its elected members. The chairman presides over the meetings of the council and acts like the speaker. On the other hand, the Mayor-in-Council is similar to the cabinet at the state and central levels. The Mayor-in-Council constitutes the mayor, deputy mayor and about ten members nominated by the mayor. The mayor-in-council is collectively responsible to the council and it is headed by the mayor.

The Mayor-in-Council form, a microcosm of state and central level governments at the municipal level, appears to be more suitable in large cities and structurally it is more balanced compared to the existing corporation or municipality models. The Mayor-in-Council system of government was introduced and certain changes were brought out in tax administration. Under the scheme, the commissioner is no longer a statutory authority and he has to work under the mayor. The mayor is elected by the council and the mayor in turn elects his team from amongst the elected councillors. A separate chairman is also elected by the council to chair its sessions as mayor is not made the chairperson of the council. The Mayor-in-Council is the real executive at municipal level and Borough Committees are provided to aid the civic administration. The Mayor-in-Council gets dissolved once the mayor is removed. The mayor can be removed either by resignation, disqualification as member of the council or by a special resolution supported by the majority of the total number of councillors.

This cabinet system in urban local governments is in existence in Madhya Pradesh, Chhattisgarh, Tripura and whole of West Bengal. In other municipalities it is called President-in-Council.

Committee Form

In many states like Bihar, Jharkhand, Odisha, Sikkim, Mizoram an Empowered/ Standing Committee is constituted under the chairperson. The members, unlike cabinet system, are elected by and from amongst the elected members of the municipality. All the executive powers vest in the committee, which is collective responsible to the municipality. In Uttar Pradesh and Uttarakhand, an Executive Committee, consisting of members elected on proportional representation of political parties present in the municipal corporation, is constituted under the chairmanship of the Mayor. This committee is executive body and collectively responsible to the corporation.

Decline of Municipal Government

Decentralisation has been the proclaimed philosophy behind development process in India since Independence. People-centred development and planning from below are the twin principles advocated to achieve decentralised development. In the governance parlance, decentralisation entails building democratic governing institutional structures with adequate functions, resources and powers at the local and grassroots levels.

The Indian Constitution did not accord an autonomous status to the urban local governments (ULGs) in Indian federation. The subject of local government falls within the jurisdiction of the state governments. That is why they are considered as creatures of the states.[4] Their entire functional and financial powers, therefore, emanate from the statutes of the state governments. The state governments in general preferred to strengthen their own arms rather than promoting ULGs as self-governing institutions. As a result, there has been a continuous trend towards centralisation of powers and unwarranted state government interference in the local affairs and several other problems relating to urbanisation challenge, rising people's expectations and inadequate municipal responses due to stagnant low financial resource base and competency levels among the staff, all adversely impacting the working of the municipalities.

The Central Government, however, initiated some thinking action for bringing-out improvement in the status of municipalities by appointing several commissions and committees during the post-Independence periods. Prominent of these are: Local Finance Enquiry Committee (1951), Taxation Enquiry Committee (1953-54), Rural-Urban Relationship Committee (1963-64) and Committee of Ministers on Augmentation of Financial Resources of Urban Local Bodies (1963). It is interesting to note here that most of these commissions and committees were focusing on financial aspects of the ULGs. This shows the government concern on the declining revenues of the ULGs. In fact, much of the problems arose because of the abolition of the local taxes like motor vehicle tax, octroi and other lucrative taxes which were earlier levied by the ULGs in India.

There has been, however, rethinking and reinforcement of the principle of decentralisation in the governance of urban areas and role of the ULGs in the development of cities and towns recently, especially in the post-economic reform periods in India. The report of the National Commission on

Urbanisation (1988) paved way for such a shift among the think tank in India. The Urbanisation Commission report is a comprehensive document covering a range of areas like strategies for the development of cities and towns, urban poverty, housing, transport, energy, water and sanitation, and financing and governing. The recommendations of the Urbanisation Commission were discussed widely in various forums viz., the Conference of Housing Ministers and Ministers of Urban Development (December,1988), the Meeting of the Central Council of Local Government and Urban Development and the Executive Committee of All India Council of Mayors (1989). These deliberations finally converged on the need for amending the Constitution of India and providing a constitutional status to the ULGs resulting in the Constitution 74th Amendment Act (74th CAA), 1992.

Constitution 74th Amendment Act- The Harbinger of Urban Governance Reforms

The most significant step has been the enactment of the Constitution (seventy-fourth) Amendment Act, 1992 (74th CAA) in the area of decentralisation and strengthening of local governments as self-governing institutions. The 74th CAA, considered as a mile-stone in the evolution of local self-government in India, primarily focuses on democratic and decentralised urban governance and development. Main objectives of the 74th CAA are:

(i) Endowing the municipalities with such powers and authority as may be necessary to enable them to function as institutions of local self-government; and

(ii) Providing for adequate devolution of powers and finances to the urban local governments to perform the functions entrusted including those listed in the Twelfth Schedule.

Major features of the 74th CAA are as follows:

- Providing for three main categories of ULGs: Corporations for larger urban areas; Municipalities for smaller urban areas; and Nagar Panchayats for transitional areas;
- Ensuring local democracy through conduct of regular elections by the statutorily constituted State Election Commissions and making municipal elections mandatory while limiting the State Power to do away with the democratically elected local governments;
- Representation to weaker sections viz., SCs, STs and Women. In case of SCs and STs, the per cent of reservation is in proportion to the respective populations to the total population in the urban area concerned. It is one-third, in respect of women;

- Constitution of ward committees in municipalities with a population of three lakhs and above leaving it optional in other cases to promote popular participation at the grass-roots level;
- Providing a framework for transfer of certain functions to municipalities through the Twelfth Schedule;
- Fiscal empowerment of ULGs through the statutorily constituted State Finance Commission and enlargement of the mandate of the Central Finance Commission under Article 280 to include the aspect of local government finance and the issues related; and
- Providing for the constitution of the District Planning Committees and the Metropolitan Planning Committees to prepare draft development plans at the district and metropolitan levels respectively.

The 74th CAA did not alter the respective financial and functional jurisdiction of the state governments as enshrined in the Constitution. It provides for devolution of functions and finances to ULGs out of the powers and functions vested in the State. Consequently, it required conformity legislation at the state level. Almost all state governments have already enacted the conformity legislation.

Without disturbing the existing federal structure of the state, the Constitution 74th Amendment Act aimed at three-tier administrative structure wherein the central, state and local governments operate with specified functional and financial jurisdiction. The institutional provisions are made for the regular conduct of elections through the State Election Commission, devolution of functions through the Twelfth Schedule and devolution of finances through the State Finance Commissions to the urban local governments.

The 74th CAA is not a panacea for all the ills of the urban local governments. But, it propelled certain positive trends and reform initiatives especially, in the areas like local government autonomy both, financial and functional, democratic local governance and decentralised development. Some of the prominent on-going reform initiatives in the country are:

(i) Regular conduct of local body elections;

(ii) Review of the local government functions in the light of the Twelfth Schedule of the 74th CAA;

(iii) Extensive survey and analysis of municipal revenue and expenditure patterns and the entire gamut of municipal financial administration including the legal provisions;

(iv) Search for alternative strategies and mechanisms to bridge the fiscal gaps at the municipal government level;

(v) A sound and comprehensive statutory base for urban governance;

(vi) Integrated planning and administration of urban development through District/Metropolitan Planning Committees;

(vii) Administrative decentralisation at the local levels through Ward Committees;

(viii) A re-look at the delivery management of the municipal services.

The Central Government has been playing critical role in urban governance reforms by way of generating new thinking on urban governance issues and formulating national reform agenda. The central government funding is also increasing through channels like the Central Finance Commission and the sponsoring of central schemes, the most recent being the Atal Mission for Rejuvenation and Urban Transformation (AMRUT). The crucial element here is that the Central Government is trying to push through the national urban reform agenda through these funding mechanisms inviting the central-state controversies. Many a time, the state governments are resisting this trend asserting to retain total control over the ULGs. Cumulative effect of all these has been undermining the role of the ULGs on development map of the country.

Functional Domain of the Urban Local Governments

The functional domain of the ULGs in India indicates that they have similar roles and responsibilities across the country. The municipal enactments of respective states provide for their functions and powers. The municipal functions traditionally are of two types: obligatory functions and discretionary functions. But, some Municipal Acts do not make such differentiation at all. For instance, the Municipal Acts in Andhra Pradesh, Tamil Nadu, Bihar, Sikkim, Mizoram and Jharkhand do not make such differentiation. In fact, all the municipal governments perform almost same functions though the Acts in some states do not specify clearly. By and large, the ULGs perform the following important functions:

(i) Provision and maintenance of protected water supply;

(ii) Construction and maintenance of the streets;

(iii) Street lighting;

(iv) Scavenging, removal and disposal of solid wastes;

(v) Provisions and maintenance of roads and drains;

(vi) Regulation of offensive or dangerous trades;

(vii) Securing or removal of dangerous buildings;

(viii) Establishment and maintenance of hospitals, dispensaries, and maternity and child welfare centres;

(ix) Measures for prevention of dangerous diseases and spread of epidemics;

(x) Vaccination;

(xi) Registration of births and deaths;

(xii) Naming of streets and numbering houses;

(xiii) Construction and maintenance of public parks, gardens, libraries, museums, rest houses, leper homes, orphanages, rescue homes for women, slaughter houses, etc.;

(xiv) Tree plantation;

(xv) Town planning and building regulation;

(xvi) Organising public receptions, public exhibitions, public entertainment, etc.; and

(xvii) Slum development/improvement.

The municipalities in India are entrusted with several tasks and responsibilities by the respective acts, rules and regulations evolved over time. The emphasis has been, however, on the regulatory and maintenance roles. The developmental role of the ULGs has been less significant. Further, some of the functions of the Pre-Independence era like watering of streets, which have become irrelevant are still being continued, at least, in the statute books. Some of the functions like taxation, health and medical aid, planning, prevention of food adulteration, etc, need cooperation from other state departments. In respect of functions entrusted to the municipalities, there could be a distinction between some core and essential functions, which are generally termed as obligatory functions, and non-core functions termed as discretionary functions. The basic issue relating to the functions is that the ULGs are not in a position to carry out their responsibilities in the face of rapid urban growth and increasing demands. The ULGs needed strengthening efforts by the state governments in terms of finances, powers and expertise to meet the emerging needs. But, the state governments on the other hand, preferred to set-up special purpose nominated bodies to take-up specialised functions that required high level of expertise and competence.

It is true that the state governments in general show reluctance to transfer functional responsibilities to the municipalities unless there are compulsions. Growth of special purpose agencies viz., urban development authorities, slum clearance boards, and water supply and drainage/sewerage boards, state urban development agencies were set-up to carryout local functions in India demonstrates this tendency. It is also argued that when the ULGs are not in a position to carry out the existing functions either due to lack of manpower or financial resources, additional devolution has no meaning unless functional devolution is accompanied by the necessary financial and manpower support.

Strengthening of the ULGs is thus related more to the extent of authority, powers and finances devolved rather than just providing a list of functions in the Municipal Acts. General feeling at the municipal level has been that they lack adequate authority, powers and finances to perform their functions to the fullest possible extent. They have to depend on the State authorities for various sanctions, expertise and finances.

The ULGs are not able to maintain even the existing civic services in the rapidly growing urban areas due to financial crunch. It is not an exaggeration to state that many ULGs are not even able to pay salaries to their employees for months together. The internal revenue generation efforts are far inadequate compared to the increasing financial demands. Further, some of the functions, viz., urban poverty alleviation, planning for socio-economic development and environment protection are very broad-based. They spill over a wider area cutting across the existing municipal administrative jurisdictions. In this respect, the municipalities can play a key role in planning and implementing development programmes only in collaboration with several other departments and agencies. However, coordination has always been a victim in Indian administration of the day.

The ULGs also suffer from obsolescence in terms of managerial procedures and employee skills/expertise. Majority of the employees, especially new entrants, lack basic knowledge about the functional responsibilities of the ULGs, their individual duties, powers and responsibilities due to absence of pre-entry training and very meager post-entry training opportunities. This was one of the main factors, which undermined the importance of the ULGs in India. They should, hence, be reoriented and strengthened through appropriate staff training to take-up the challenges of fast urban growth and manage the consequential demands in a planned manner.

The functions listed in the Twelfth Schedule represent broad areas, most of which are already being performed by the ULGs. Some of the functions like protection of the environment, tanneries, fire services and economic planning are being performed by state level agencies, which need to be devolved to the ULGs. Part of these functions like poverty alleviation, protection of the environment and economic planning are very broad in nature performance of which need support of the state and central governments could be devolved to the ULGs. The question therefore, is not whether the list of functions is sufficient or not, but it is the extent of functional responsibilities devolved on the ULGs. The aspect of devolution of functions and powers to the ULGs is a complex area. The state governments are reluctant to devolve more functions and powers to the ULGs. As the functions enumerated in the Twelfth Schedule are only indicative and not mandatory, the reluctance of the state governments to devolve more functions is, therefore, understandable.

The Twelfth Schedule incorporated in the 74th CAA provides a list of functions for devolution to the ULGs and it is left to the discretion of the state government concerned to decide on the matter. There is no point in prescribing a list of broad general functions and blaming either the local governments for non-performance or the state governments for their reluctance to devolve the functions. In the constraining circumstances mentioned above, the municipal functions should be defined in specific terms rather prescribing generalised functions like town planning, economic planning, poverty alleviation and environmental protection, and it should be made mandatory to devolve such specifically defined functions to the ULGs along with the necessary authority, powers, expertise and finances. This appears to be essential in the context of complex nature of the process of urban development and the need for involving various agencies on partnership basis to achieve the goals of urban development at various levels.

There are, thus, two main issues pertaining to the devolution of the functions listed in the Twelfth Schedule to the ULGs, viz., discretion of the state governments and incapacity of the ULGs in terms of finances, human resources and administrative powers. Interestingly, in the piece of legislation intended to extend the Part IX A of the Constitution to the Scheduled Areas, it is made obligatory to devolve certain functions to the ULGs in the Scheduled Areas. Such an obligatory provision is necessary for devolution of functions listed in the Twelfth Schedule as the states prefer to take a casual approach in this respect while carrying-out the mandatory provisions. A clearly specified 'Local List' (in addition to Union List, State List and Concurrent List) of functions appears to be a necessity at the present juncture, which requires further amendment to the Constitution. Any devolution of functions to the ULGs should always accompany the necessary resources- financial and human, and administrative powers. Unfortunately, such a holistic transfer of functional responsibilities and resources is not picking-up in the states in general. It requires true commitment on the part of the state governments and enormous capacity building efforts at the municipal level. Until then, the demand for devolution of functions and powers may remain a far cry.

Urban Governance

In public domain, governance involves a process that goes much beyond government and its institutions integrating the governing institutions and the governed that is, the people at large. Governance is not government. Governance includes government, the private sector, and civil society. Governance emphasises "process". It recognises that decisions are made based on complex relationships between many actors with different priorities.[5] The United Nations defines governance as 'the exercise of political, economic and administrative authority in the management of a country's affairs at all levels. It comprises the mechanisms, processes and institutions through which citizens and groups articulate their interests, exercise their legal rights, meet their obligations and mediate their differences'. In this perspective, governance is not government and it is a broader concept which recognises that power exists inside as well as outside the formal authority and institutions.[6]

In democracies, governments are formed by the people to realize their common needs. If the governments ignore people's needs, such governments cannot survive and sustain. The governments, therefore, should evolve such processes through which people realise their objectives. Urban governance is not dissimilar and should follow this fundamental principle. Urban governance in simple terms means the process through which the urban local governments strive to meet the common needs of the people living in urban areas. It implies that the policies, programmes and activities of the governments must be people centred meeting their needs. The process of urban governance could be mapped as shown in the Chart 3.1.

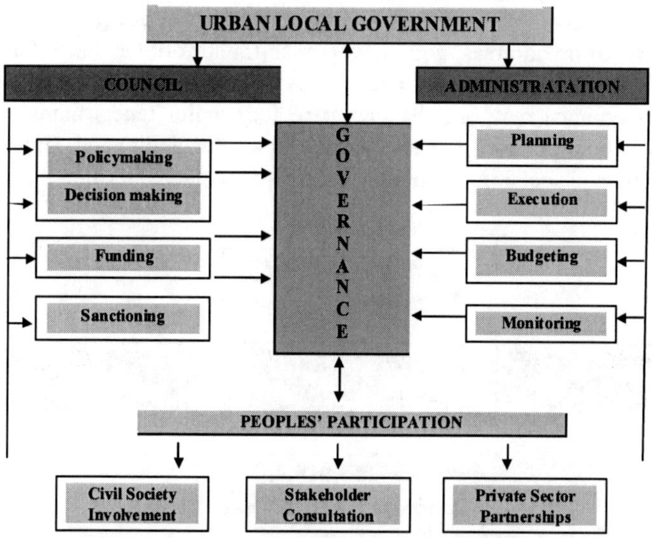

Chart 3.1: Process of Urban Governance.

The growing poverty, inequities and urban exclusion which are threatening the urban areas across the world have given rise to global thinking on the emerging issues and finding solutions to confront the challenges. The inequities are observed to be more prominent in the developing countries like India. As the inequities become dominant in society, it leads to exclusion on the development front.

UN-Habitat launched the Global Campaign on Urban Governance in 1999 to support the implementation of the Habitat Agenda goal of "sustainable human settlements development in an urbanising world." The campaign's goal is to contribute to the eradication of poverty through improved urban governance. There is a growing international consensus that the quality of urban governance is the single most important factor for the eradication of poverty and for prosperous cities.[7]

The campaign aims to increase the capacity of local governments and other stakeholders to practice good urban governance. It focuses attention on the needs of the excluded urban poor. It also promotes the involvement of women in decision-making at all levels, recognising that women are one of the biggest levers for positive change in society.

Good governance in the urban context envisages improving the quality of life in cities through improved local governance by reinventing a city as an inclusive city. Such a city provides space and voice to all its stakeholders through inclusive decision-making, since such decision-making is at the heart of good governance. The inclusive city allows full flow to women's thoughts and initiatives as women are one of the biggest levers of positive change in the society. In urban areas, good governance framework has been defined differently by different lending agencies. All these characteristics of good urban governance may broadly comprise four major tenets namely, (i) decentralisation, (ii) transparency, (iii) social accountability and, (iv) civic engagement and social integration of the poor as depicted in Chart 3.2.

Chart 3.2: Tenets of Good Urban Governance.

These broad areas are further elaborated for clarity. The good governance is the system of governance which promotes: participation of a wider community representing various sections in the community in the governing process; society develops respect for law for common good; public agencies adopt transparent methods of decision-making; public authorities behave in a responsible way in implementing the decisions and meeting the expected needs of the people; issues and problems are resolved in a consensus manner; public policies and decisions are made to achieve equity while including all sections of the people; public actions should focus on effective results; and ultimately public accountability is the prime principle of good governance.[8]

Action Areas

The main action areas identified for good urban governance are well articulated by the UN Habitat. These aspects are further discussed here under.[9]

Decentralisation

Important areas for achieving decentralisation are: devolution of functional and financial powers to urban local governments; transfer of power for elected representative to be coupled with right to recall; and defining State Municipal Relationship. In the urban context, decentralisation connotes the establishment of a local representative government endowed with administrative and financial powers to deliver mandated services to its citizens.

Integration of the Poor and Marginalised

The poor, differently abled, women, children—all disadvantaged citizens should be mainstreamed and integrated. Some important action areas are: shelter for all; promoting better understanding among all stakeholders; city-wide action plan to integrate the poor and marginalised population into the city with special attention to land tenure, basic services, housing and employment needs, including support for informal enterprise of the poor, particularly women; partnership with non-government and community-based organisations; and ensuring access to education, health and other social services.

Transparency and Civic Engagement

Transparency connotes the conduct of public business in a manner that affords stakeholders wide accessibility to the decision-making process and the ability to effectively influence it. Some important action areas are: identifying major bottlenecks through consultative processes; discretionary powers in favour of classes rather than individually; code of conduct for municipal executives and elected representatives; public-private partnership; networking; transparent engagement of private sector; and right to information on all matters pertaining to local government.

Social Accountability

The ULGs should install adequate social accountability measures to ensure that the citizens' problems are solved and needs fulfilled. Some action areas are: implementing citizen charters; providing grievance redressal mechanisms; and introducing measures like social audit.

The Tenth Five Year Plan emphasised on the need for strengthening of the urban governance in India. It observed that "urban governance today is characterised by fragmentation of responsibility, incomplete devolution of functions and funds to the elected bodies and ULGs, unwillingness to progress towards municipal autonomy, adherence to outmoded methods of property tax and reluctance to levy user charges. State governments continue to take decisions on such matters as rates of user charges, property tax, octroi, role of parastatals in water supply and sanitation services, etc., with little reference to the ULGs that are affected by these decisions. Far from strengthening the constitutional role of the elected ULGs such developments only reinforce the perception that ULGs are subordinate entities under the day-to-day control of the state governments, beholden to them not only for the development of the cities but often for their very survival. There appears to be a lack of confidence that many of the ULGs are capable of meeting their obligations as institutions of local self-governance. In the present set up, initiatives for local developmental activities rarely come from the ULGs. Experience shows that functional autonomy becomes a reality only when it is accompanied by financial independence. State governments, therefore, need to strengthen the autonomous functioning of the municipalities through positive measures, and in particular, ensure their financial self-reliance."[10]

The role of the ULGs in the immediate future will be:

- To be responsive and accountable to the community
- To develop cities with standards of service comparable to the best in that particular category.

The Twelfth Five Year Plan also makes inclusive development as the central running theme. Participative planning involvement of the stakeholders, accountability, decentralisation and transparency which are the pillars of good governance campaign at the global level—all find their proper place in the national reform agenda of the Government of India specifically, through AMRUT initiative. These reforms are discussed more elaborately in the Urban Reforms chapter and other chapters. In this chapter, however, e-governance focusing on achieving efficiency, transparency and accountability is discussed.

E-governance Reforms in Municipalities

Information and communications technology are transforming the way of life, but they are most popular in certain highly entrepreneurial technology intensive enterprises. The not-for-profit and public sector organisations are however slow in picking-up. Information and Communication Technology (ICT) is all about connectivity. Connectivity brings proximity which makes the delivery of services offered by governments nearer to people. The ULGs deserve maximum attention as they are at the cutting edge of the administration and hence immensely affect the daily lives of the citizens. The term e-Governance means the use of a range of modern information and communication technologies such as internet, local area networks, mobiles etc. by government to facilitate effective and efficient delivery of services though transparent methods. The strategies of introducing ICT in municipalities may be:

(i) As Information Technology rests on connectivity, a stand-alone computer can offer only minimal utility, and networking therefore would be essential to maximise the gains;

(ii) The larger goal is to maximise citizen satisfaction about the delivery of civic services and the design of the system would therefore keep citizens at the centre-point and every process would be designed keeping them in view;

(iii) The design of the system would be such that it is amenable to web enabling as web offers the easiest method of citizens gaining access to the information they need;

(iv) To the extent possible, the project may be designed on a public-private partnership model that would involve cost and revenue sharing with the private sector; and

(v) The insiders possessing computer skills be encouraged to develop the model for the computerised system.

Mission Mode Project of E-governance in Municipalities

The Government of India has formulated the National e-governance Plan (NeGP), part of which includes a National Mission Mode Programme (NMMP) for E-governance in Municipalities. This NMMP intends to roll out e-governance in municipalities on a nation-wide basis and was as part of the erstwhile mega infrastructure development project JnNURM[11] and now forms an integral part of the much ambitious mission mode scheme AMRUT also.

Objectives of NMMP

(i) Focus on clearly identified list of citizen services that would be covered with clearly laid down service levels and outcomes that would be achieved;

(ii) Improve efficiency and effectiveness in interaction between local-government and its citizens and other stakeholders (i.e. non-governmental organisations (NGOs), community based organisations (CBOs), residents welfare associations (RWAs), private sector, etc);

(iii) Improve quality of internal local-government operations to support and stimulate good governance;

(iv) Bring about transparency and accountability in the governance of urban local bodies;

(v) Enhance interface between urban local bodies and citizens; and

(vi) Help improve delivery of services to citizens.[12]

Services Covered under NMMP[13]

(i) Registration and Issue of Births/ Deaths Certificate;

(ii) Payment of Property Tax, Utilities Bills and Management of Utilities that come under the municipality;

(iii) Grievances and Suggestions;

(iv) Building Approvals;

(v) Procurement and Monitoring of Projects;

(vi) Health Programmes like Licenses and Solid Waste Management;

(vii) Accounting Systems;

(viii) Personnel Information System.

The E-governance Initiatives in Municipalities

Some of the IT tools that are being initiated in municipalities are:[14]

(a) Online payment of municipal taxes, fees, charges, etc;

(b) Online filing and settlement of complaints and grievances;

(c) Online registration of births and deaths;

(d) Online tracking of building plan status;

(e) Infrastructural works monitoring;

(f) Instant issuance of birth and death certificates;

(g) Online tracking of garbage lifting;

(h) e-tendering;

(i) Accounting package including budget control;

(j) Online legal cases monitoring;

(k) Water supply surveillance and monitoring;

(l) Web based building permission processing;

(m) Monitoring and reporting of service level benchmarking implementation.

The launch of Digital India is certain to give a pace to all the initiatives undertaken in this regard. Most of the municipalities are computerising the municipal accounts and records and rendering online services like payment of property tax, downloading applications, applying for certificates such as birth and death certificates. Some cities like Greater Hyderabad Municipal

Corporation (GHMC) introduced advanced financial management system of ERP which is successfully running.

Off Site Real Time Monitoring System in Greater Hyderabad Municipal Corporation (GHMC)

The GHMC introduced Off-Site Real Time Monitoring System (OSRTM system) to track various municipal operations to ensure 360 degree accountability of municipal body in rendering services and facilitate planning and monitoring of various daily operations. Its major features are:

- Innovativeness of the OSRT mechanism is founded on the integration of accountability tools and processes by relying on the eye of the mobile phone.
- Availability of real time reports / data / images in public domain.
- Utilising the easy-to-use mobile phone eye (camera) for civic governance.
- Reviews based on reports generated with two seconds latency.
- Auto sms to citizens upon registration of grievance and after redressal.
- Status verification of grievance through website.
- Making municipal managers more responsive and responsible and making citizen responsible.
- Single point availability of various civic activities and outputs in public domain.
- Usage of GPS / GPRS.
- Satellite captured Latitude / Longitude and date/time stamp on all images.
- Utilising Photo comparison and perimeter binding technologies.
- Transparency in administration.
- OSRT is in public domain and citizens have access to all its features like official of GHMC.

The OSRT system of GHMC has earned national awards under E-Governance Best Practices category by the Ministry of Urban Development, Government of India. Dr. Isher Judge Ahluwalia commented:[15] *"Even as the wrongdoings in the telecom sector generate anger and dismay, Hyderabad's city governance initiative using mobile phones provides a breath of fresh air on what the telecom revolution can do for India. The Off-Site Real Time Monitoring (OSRT) system is a unique but simple mobile-based IT initiative which uses a combination of GPS (Global Positioning System) and GPRS (General Packet Radio Services) technologies through cell phones for managing civic amenities in the city in a transparent and efficient manner."*

Towards Strengthening of Urban Governance

The philosophy of democratic decentralisation has been the spirit behind planned development in India since Independence. In the governance parlance, decentralisation entails building democratic institutional structures with adequate functions, resources and powers at the sub-national and local levels. As against this pronounced policy of decentralisation, there was a continuous trend of centralisation and bureaucratisation that weakened the urban local governments (ULGs) in terms of resources, powers and competencies. There has been steady erosion in the functional and financial powers of the municipalities at a time when the pressures of urbanisation demanded stronger municipalities.

Local governments constitute the most democratic means for achieving the twin objectives of good governance and local development. But, the local governments—rural as well as urban—are unfortunately caught in a web of complex interrelated problems: lack of employment opportunities in the rural areas; consequential migration to urban areas; resultant pressure on available services and amenities in the urban areas; haphazard expansion of the urban areas due to ineffective development control; increasing poverty; and threatening environmental pollution. This apart, globalisation has been adding its own set of pressures, viz., demand for changes in the policies and institutional structures, emerging marketing approach among the governments leading to dominance of the economic interests over social interests in the decision-making, higher investments of scarce resources in the globally attractive mega projects, growing inequalities in services and development of different regions and alienation of disadvantaged and the poor. In recent years, there has been a marked shift in the attitudes of urban governments across the world from a mere management approach to entrepreneurship showcasing the city as a product to be marketed globally. At the other end of the spectrum, the state governments extended sub-optimal support to local governments leading to the neglect of the local democracy, a gradual deterioration in their functions and erosion in their administrative and financial authority. The problem here is not lack of knowledge to address these issues, but inability to build governing structures to address these problems more effectively.[16]

In a federal system of government, what is needed is a right balance between top-level control and local autonomy. Each country has to evolve its own balance depending on the political, social and economic factors prevailing. The development strategy, therefore, must be integration of macro national and sub-national goals with the needs of local communities. The 74th CAA provides the necessary constitutional backing and an institutional framework in this respect. With this added advantage, the institutional capacity building efforts must address the specific problems confronting the ULGs viz., mismatch between functional responsibilities and the financial resources, distancing of the local communities from the ULGs, unwarranted state government interference in the local governance, the crippling official and non-official relationships at the municipal level, increasing bureaucratisation of urban governance and declining managerial competencies and skills among the municipal personnel.

Nearly about two decades long experience of implementation of the 74th CAA across the states shows that the ULGs are better placed now and are provided with a healthy institutional backing consisting of the SFC, SEC, Wards Committees, District Planning Committees and the Metropolitan Planning Committees. On the positive side, there is an increased awareness about the role of the ULGs in the nation building process, improved scope for women's participation in urban governance and local democracy has been restored. As a result, several reforms are underway in the ULGs across the country which is expected to yield tangible results, if continued with some more vigour and pace. Still, it is the ground reality that the governance is the weakest and most crucial link which needs transformation so urgently needed in India.[17]

The 74th CAA is only a beginning, a first step, towards a lengthy journey of refurbished and strengthened urban local bodies. However, there are many hurdles and blockages in between. To mention a few:

(i) There is unwarranted mistrust and suspicion between the local government functionaries and the state government authorities. The foremost need therefore, is to develop a sense of partnership and involvement in the urban development process.

(ii) The relationship between the non-officials i.e., councillors, chairpersons and mayors, and the bureaucratic functionaries has been the most troubled aspect in the municipal governance in India. If the former is immersed in the narrow political interests, the latter is increasingly becoming indifferent and distanced from the local developmental problems and local communities. The crucial issue, therefore, is how to change the perceptions and understanding of the complementary roles of the officials and non-officials to manage the municipal affairs?

(iii) The urban local governments have been witnessing constant erosion in their functional and financial powers in the Post-Independence period in India. The negative trend must be arrested. It requires considerable efforts to strengthen the financial and functional bases of the urban local bodies. The mechanisms and solutions are debatable and solutions may differ from state to state.

Apart from the 74th CAA, certain other reforms are also being implemented to strengthen urban governance. The initiative of e-governance in municipalities, community participation laws, and a host of reforms initiated under the JnNURM and AMRUT intend to move in this direction. In spite of some improvements, though selective, there is much to achieve as the implementation of the governance reforms has been very slow, lacking political and administrative commitment at the state and local levels. The following measures are essential to make the urban governments and the urban governance stronger enough to meet the emerging challenges of urban development successfully:

- **Strengthening of financial base of the ULGs:** Necessary financial avenues-tax and non-tax must be provided to the municipalities in such a way that a balance of functional responsibilities and the finances is maintained. The objective should be to make municipalities financially sustainable.
- **Refurbishing the management:** The delivery management improvement of basic civic services like solid waste management, sanitation, engineering and planning needs special attention here. Necessary reforms should be brought-out in the budget and financial management to make it performance oriented and purposeful. The tax administration is another area, which requires streamlining in most of the ULGs.
- **Focusing on HRD requirements:** The human resource aspect has been a much-neglected aspect in the urban government system in India. The pathetic aspect is that there are no HR policies at the state level. As a result, the personnel in the municipalities are often improperly deployed based on the pressing demands and pressures. The personnel in ULGs must be well equipped and properly trained to meet the emerging challenges in terms of planning, execution, financing, coordinating and involvement of the people. But, in the absence of proper HR policies in place at state and local levels, the entire exercise of capacity building goes in vain.
- **Building information system:** In the context of emerging information technology, the municipalities should be well equipped with ICT tools to meet the needs of the urban communities. Urban data on various dimensions must be documented regularly and database should be computerised for its multiple applications and effective monitoring. Though the governments are collecting different types of data for different purposes, the integrated approach and building reliable data base are essential to make the information system effective.

- **Performance improvement:** Municipalities carry a very bad image among the urban communities due to their inabilities, inefficiency and corruption. Their performance, therefore, should be improved on par with the peoples' expectations. This will also facilitate more devolution of functional responsibilities to the municipalities. The techniques like service level benchmarking which is already introduced at the national level provide good opportunity in this regard to make urban governance competitive and competent.
- **Fostering partnerships and networking:** In the age of globalisation and privatisation, the present catch-words across the globe, ULGs must develop appropriate strategies for working together with private entrepreneurs, non-governmental and voluntary organisations and with other sister public organisations at the one end, and the local communities and self-help groups at the other end. In the resource scarce situation of the developing countries, this has become inevitable and demanding.
- **Coping with natural disasters:** In the context of increasing global warming and environment related problems, urban areas are also subjected to frequent disasters like floods, cyclones and erratic rains like cloud bursting. The ULGs have to play a key role in managing the natural disasters like cyclones, floods and earthquakes. They should be well equipped in terms of resources and techniques for providing temporary shelter, food, health-care, etc., to meet such crisis situations efficiently.

The aggressive public sector reforms, decentralisation, privatisation, and the involvement of the civil society organisations have fragmented the system of governance in recent years in India necessitating a re-knitting of the institutions in a mutually interdependent way to meet the emerging challenges. In essence, it is basically a question of managing the networks. Public management must be improved through innovation rather than by mere imitation of the apparently attractive efficiency models of private sector. Rhodes feels:[18] 'governance refers to self-organising, intergovernmental networks characterised by interdependence, resource-exchange, rules of the game, and significant autonomy from the state.' This approach is truer in respect of urban governance today.

The irony is cities are becoming important territorial nodes in the face of happening globalisation but, there is an increasing trend of fragmentation in terms of their communities, economy, politics and administration. Predominance of the economic and financial factors is the unique about the on-going urban development initiatives. The urban governance universe has become a complex web with multiple stakeholders- international, national and local, and blurred boundaries between private and public, formal and informal, and state and civil society sectors. Urban reformers should understand these trends for evolving suitable strategies.

References

1. UNCHS (2001): Global Report on Human Settlements.
2. Muttalib, M.A. and Khan, Mohd. Akbar Ali (1982): Theory of Local Government, Sterling Publishers Limited, p. 18.
3. Maheshwari, S.R. (1976): Local Government in India, Orient Longman, p. 212.
4. The Constitution of India which came into force on 26th January, 1950, was silent on the creation of municipal governments as against the Article 40 which gave powers to the states for constitution of panchayats. The only reference to municipal governments could be found in the two entries viz., Entry 5 of list II of the Seventh Schedule and Entry 20 of the Concurrent List. If the former talks about constitution of local government (including municipal governments), the latter, indirectly mentions about the functions of local government.
5. What is Good Governance? United Nations Economic and Social Commission for Asia and the Pacific, www.unescap.org/huset/gg/governance. htm
6. UNDP (1997): Governance for Sustainable Human Development, New York, pp. 2-3.
7. UN Habitat (2002): United Nations Human Settlement Programme, Global Campaign on Urban Governance-Concept Paper, Nairobi, March, p. 5.
8. What is Good Governance? United Nations Economic and Social Commission for Asia and the Pacific, www.unescap.org/huset/gg/governance. htm
9. UN Habitat: Action Plan for Good Urban Governance-Recommendations for National Action Plan for Good Governance, www.UNhabitat org/2105_30168_India_Plan_recommendations.doc
10. Planning Commission, Government of India: Tenth Five Year Plan, p. 613.
11. The Ministry of Urban Development, Government of India: National Mission Mode Programme (NMMP) for e-Governance in Municipalities.
12. *Ibid.*
13. *Ibid.*
14. *Ibid.*
15. Ahluwalia, Isher Judge (2011): Report of High Power Committee on Infrastructure and Services.
16. UN Habitat (2003): As quoted by Susan Chaplin E. (2007) in Annapurna Shaw (ed.), Indian Cities in Transition, Orient Longman, Hyderabad, p. 85.
17. HPEC (2011): Report on Indian Infrastructure and Services, Ministry of Urban Development Government of India, p. xxv.
18. Kjaer, Anne Matte (2004): Governance, Polity Press, p. 3.

Chapter - 4

Decentralised Participatory Urban Governance

Introduction

The top-down approach and bureaucratic style of functioning dominate the municipal governance in India in spite of the pronounced policy of democratic decentralisation. The underlying causes are gross neglect of democratic methods by overemphasising bureaucratic role and absence of community interface. Although the municipalities are traditionally considered local self-governing institutions, the political leadership driven by more party interests provided little scope for peoples' participation. The parallel functioning of civil society organisations, at the other spectrum, led to fragmented efforts. With the failure of municipal democratic processes, the gap between expectations of urban communities and performance of these grassroots level government units in respect of decision-making, delivery of services, grievance redressal, planning and implementation of development projects widened leading to crisis in urban governance. The crisis deepened further with unfolding of globalisation and liberalisation effects demanding efficient management, market friendly response and cost effective approaches for sustaining urban services. The urban local governments (ULGs) are faced with the critical challenge of balancing the market demands and the people's needs. Decentralised and participatory municipal organisations are essential to achieve this delicate balance while bridging the gap between local communities and ULGs. The ward committee and area sabha system is devised aimed at institutionalisation of decentralised participatory urban governance in India. This chapter examines the ward committee and area sabha system in the background of the initiatives taken-up after the Constitution 74th Amendment Act (74th CAA), 1992 and the community participation law initiatives.

Institution of the Ward Committees

In Indian cities, especially in larger ones, there exists wide contrast in the level of civic services and infrastructure facilities. There are slums with no proper shelter and basic services, depicting a picture of poverty. There are newly developing areas with grossly inadequate services as against posh and rich enclaves having very good infrastructure facilities comparable with developed countries. The universe of civic services in mega cities depicts a pathetic state of inequity. Though the entire municipal area is divided into circles, zones and wards, they are basically field offices, not democratic institutions. The attempts of community participation in urban areas are mostly confined to slum development. The UN HABITAT identified civic engagement in the local governance as one of the important norms of good governance. If truly decentralised governance is to be established, representative democratic system is not enough, but participatory governance at the grass-root levels is essential to bring proximity between the local communities and the urban local governments.[1]

A determined revival process started in the post-1990 years in urban-India with the passage of 74th CAA in 1992. Decentralisation of powers and functions to ULGs is one of the objectives of the Amendment. As the existing laws in the states did not have provisions for the regular and effective participation of the local people in the municipal governance, there had been a wide gap between expectations of the urban community and performance of the local body. With a view to bridge the gap between local communities and ULGs, and to enable participatory decision-making and governance, the 74th CAA provided for the institution of Wards Committees. The formation of wards committees to enable participatory decision-making and decentralised municipal governance was mandated as below:[2]

(i) constitution of wards committees in cities having a population of three lakhs or more;

(ii) councillor representing a ward in the area of the wards committee should be a member of the wards committee; and

(iii) councillor, if the wards committee consists of one ward, or one of the councillors elected, if the wards committee consists of more than one ward will be its chairperson.

The 74th CAA left it to the discretion of the state legislature to make provisions on the aspects viz., territorial division and number of wards committees to be constituted in a city; composition of a wards committee; manner of filling the seats in the wards committees ; and functions and powers of the wards committee. The Constitution Amendment provided for reservation of seats to women, SCs and STs and, representation to the civic society organisations. The wards committees are expected to address the local needs like provision of civic services, revenue collections, beneficiary selection, etc., by active peoples' participation. The state governments amended the municipal laws to form wards committees. Although constitution of the wards committees is mandatory in the ULGs having a population of three lakh and above as per the provisions of the 74th CAA,[3] in several states, they are constituted but are not made operational. In Kerala, ward committees are formed in all categories of ULGs but in other states they are constituted only in corporations or large cities with above three lakh populations. There has been reluctance on the part of governments to constitute and make the wards committees effectively functional. For example, the wards committees were constituted in Bengaluru, only after the court's intervention in 1999 though elections were held in 1996 to the Bangalore Municipal Corporation. Similarly, though subsequent election was held for the municipal corporation in 2001, wards committees were not constituted until July, 2003. Though wards committees were constituted, every state evolved its own pattern in making them functional. The constitution, composition and powers of the wards committees differ from state to state. If the Kerala, Uttar Pradesh, Jharkhand, Bihar, Madhya Pradesh and Chhattisgarh constituted ward committees in every ward, each wards committee comprised of 10 wards in Andhra Pradesh. It is observed that each wards committee comprise of large populations e.g., eight lakhs in cities like Delhi where they include more than one ward. In Kerala where ward committees are constituted in each ward, the populations are lesser with few thousands.[4] In Maharashtra, the Wards Committees are in reality zonal committees assigned with some specific functions.[5] As wards committees were constituted for large populations and with no civil society representations in some states like Andhra Pradesh, the spirit has been diluted and they failed to serve as effective platform for peoples' participation. Their working suffered due to lack of commitment for decentralised governance at the grass root level. Their functional role was not clearly laid down and the wards committees lacked powers to supervise the municipal works, prepare and propose projects for development of ward concerned. The studies brought-out that the ward committee system has been more effective in two states viz., Kerala and West Bengal and cities like Mumbai.[6] The ineffective wards committees could not achieve the goal of participatory governance at the local level in India.

Revamping the Ward Committee System

The citizen groups voiced against the ineffective functioning of the wards committees in various states. A non-governmental organisation called Janaagraha Centre for Citizenship and Democracy in Bangalore took initiative and prepared a reformed Model which was discussed at various forums. In view of its importance for effective community participation in local governance, the Ministry of Urban Development, Government of India accepted the Model. In order to revamp the ward committee system in the country, the Bill called Model Nagara Raj Bill, 2008 was circulated to the state governments for adoption or enacting such laws with similar features.[7] The Model Nagara Raj Bill virtually suggested for three-tier municipal governance with council, ward committee and area/ward sabha as constituent units. The most important feature of this Model Bill was introduction of a new local government organ called 'area sabha' as the lowest tier of self-government in the municipal areas. This institution is envisioned as the ultimate people's participatory mechanism at the grass-root level. The Model Bill suggested that the area sabha will consist of all the people registered as voters in the area concerned. The area sabha members will elect area representative who chair the meetings and represent the area sabha in the ward committee. It was also suggested that the elections of the area representatives ideally need to be conducted under the supervision of the State Election Commission. But, it was not an obligatory provision. The Model Bill thus provided scope for nomination process for selecting the area representative. The Model Bill also assigns the role, functions, duties and responsibilities to area sabhas viz., identification of local priority needs and preparation of the lists of most eligible persons for welfare schemes, assistance to the public health centres, tax mapping, mobilising voluntary labour, suggesting the location of public amenities within the area and pointing out the problems and deficiencies in the water supply, street lighting, etc., in the area sabha jurisdiction.

Above the area sabha, ward committee was suggested as an intermediary level below the corporation/ council. The ward committee will be chaired by the ward member or elected councillor. The composition of the ward committee is so modelled in the Bill that majority members, at least two-third of the members, should be area representatives from different areas in the ward concerned. Provision was also made for nomination of a maximum of 10 members from the civil society organisations from the ward by the corporation or municipality. The main functions entrusted are similar to the area sabhas but with focus on preparation of ward plans, preparation of the ward budget, encouraging PPP methods in local administration, supervision over local level works, etc. The Bill also provides for the constitution of a ***Ward Finance Committee*** to prepare the annual ward level budget and a report on the financial transactions, and a ***Ward Information and Statistics Committee*** to compile, maintain and update information pertaining to the economic scenario, land use and infrastructural developments and requirements.

In the meantime, the JnNURM was launched by the Government of India, in December 2005, to develop infrastructure facilities and to make urban governance more democratic, vibrant and accountable. Apart from other reforms, effective implementation of the provisions of the 74th CAA and enactment of the community participation law on the suggested lines of the Model Nagara Raj Bill, 2008 to involve local communities in the decision making process were mandated as compulsory reforms under JnNURM. The objective of Community Participation Law is to institutionalise citizen participation as well as introducing the concept of the Area Sabha in urban areas. The larger objective is to involve citizens in municipal functions, e.g. setting priorities, budgeting provisions, local level monitoring of municipal operations, etc. The enactments must clearly specify functions, duties and powers of each of these tiers, and provide for the appropriate devolution of funds and functionaries as may be decided by the State Government to these levels. All the governments are expected to enact such community participation laws to make the existing ward committee system more effective and decentralised in true sense. The implementation of this reform however seems to be very slow as only the states like Andhra Pradesh, Assam, Bihar, Gujarat, Kerala, Madhya Pradesh, Maharashtra, Rajasthan, Tripura, Uttar Pradesh and West Bengal have enacted the Community Participation Law. It is also observed that the provisions in some of these legislations do not conform to the objectives of the Nagara Raj Bill.[8] The system of ward committee in a few major states is being discussed hereunder.

Andhra Pradesh

The Government of Andhra Pradesh restructured the ward committee system through the community participation law as a mandatory reform. The revamped ward committee is made more representative, purposeful and focused. The ward committee system introduced in Andhra Pradesh is modelled on the lines of the Nagara Raj Bill, 2008. The ward committees in Andhra Pradesh are allocated 20 per cent of the municipal budget. The ward committee can identify works in the wards concerned and get them executed by the Commissioner as per rules. This financial provision is not found in the Nagara Raj Bill. In Andhra Pradesh, it is also provided that fifty per cent of the members of the ward committees must be women. This stipulation is not suggested in the Nagara Raj Bill. Further, functional committees are not provided at the ward level as suggested by the Nagara Raj Bill. Excepting these, all other provisions of the Nagara Raj Bill find place in the Andhra Pradesh Model.

Necessary amendments were made to the Hyderabad Municipal Corporation Act 1955, Andhra Pradesh Municipal Corporations Act and the Andhra Pradesh Municipalities Act 1965 for implementing the new ward committee system in Andhra Pradesh.[9] Rules were issued separately for corporations and municipalities in February, 2010.[10] The new ward committee system provided for constitution of ward committees in every ward in municipalities and corporations.

As per the amended Acts, every ward of the municipalities shall have a ward committee. This committee shall be constituted within three months from the date of notification. The composition of the ward committee is made broad based to make it more representative involving different categories of local people. The ward committee consists of three categories of members viz., municipal councillor; civil society representatives; and the area sabha representatives. The civil society means, any non-governmental organisation or association of persons registered under law and who works for the social welfare of the people. Further, the term includes community based organisations, residents welfare associations, professional institutions and civic, health, education institution, social or cultural body or trade or industrial organisation. It is also provided that 50 per cent of the members shall be women. The process of making ward members is by nomination by the corporation/council. In order to give fair representation to all categories of people, a minimum strength of membership was prescribed for each category.[11]

The ward committee meetings are chaired by the ward councillor. The meeting of the ward committee will be held at least once in two months. The meetings are also open to the general public. In order to assist the ward committee meetings, the ULG appoints an officer as secretary for each ward committee. The secretary prepares the agenda in consultation with the chairperson of the ward committee. Apart from the secretary, the ward level officers, namely, engineers, revenue officers, environment engineers, health officers, town planning officers, etc. shall also attend the meetings of the ward committee. The decisions in the ward committee meetings are made by a majority of votes of the members present at the meeting. The minutes of the ward committee resolutions shall be entered into a book duly signed by the chairperson. Care is taken to make these resolutions accessible to all-officials concerned and the people at large. For instance, the resolution copies are to be kept at the ward committee office, made accessible to taxpayer of the ward, forwarded to the Commissioner as also to the members of the ward committee and affixed on the notice board of the ward committee office for information of general public. The secretary is made custodian of the proceedings and records of the ward committee.

The role of the ward committees is envisaged as a miniature council at the ward level for inclusive governance. The functional responsibilities entrusted include supervision of various works executed and services provided in the respective ward, facilitating collection of taxes and non-taxes, preparation of beneficiary lists for various welfare schemes like pensions, subsidies, etc., and mapping of infrastructure facilities in the ward. Most importantly, the ward committee is also responsible to prepare a ward level development plan for taking up various development activities.[12] The Ward Committee is competent to accord administrative sanction to maintenance works up to an amount of Rs one lakh in case of the corporations and Rs 50,000 in case of the Municipalities in each case subject to the availability of the funds sanctioned every quarter.

West Bengal

In West Bengal[13] ward committee is constituted in every municipality in each ward within one month of the general elections of the municipality. The members of the committee are nominated by the councillor of the ward as well as by the municipality from amongst the citizens of the ward. The elected councillor acts as chairperson of the ward committee.

The Committee, subject to general supervision of the Borough Committee or the Municipality, supervises and monitors: detection of violation of various provisions of the Act such as unlawful construction, encroachments on municipal and public properties, public nuisance, evasion of taxes, unlicensed activities, and the like; arresting wasteful uses of various municipal services like tap-water, street hydrant, street light, parks and playgrounds, community centres and libraries, and the like; collection and removal of garbage; removal of accumulated water on streets, public places due to rain, and other causes; health immunisation services; improvement of slums including its clearance wherever necessary; redressal of public grievances; maintenance of essential statistics; numbering of streets and premises; execution of various obligatory and discretionary functions and functions transferred by the State Government to the Municipality/Notified Area Authority as per provisions of the Act.

The Committee, as per directions of the Borough Committee or the Municipality/Notified Area Authority also monitors and supervises within the ward (a) work of assessment, collection of property tax, other taxes, fees and other sums due to the Municipality/Notified Area Authority; (b) maintenance and repair of roads and drains; (c) execution of public works; (d) maintenance of sanitation and public health; (e) the prevention and control of dangerous diseases; (f) issue of birth and death certificates; (g) all Urban Poverty Alleviation Programmes of the State and the Central Government; (h) tree planting in the municipal properties; (i) maintenance of parks, tanks, and other municipal properties; (j) generation of internal resources; and (k) literacy expansion programmes. The Committee performs other functions as may be entrusted to it by the Municipality or the Borough Committee and supervises implementation of decisions of the Municipality/Notified Area Authority and the other Committees.

The Committee is mandated to consider the opportunities for creation of new sources to augment municipal income by way of (i) using lands vested to the Government or Municipality/Notified Area Authority and any land belonging to any Government department subject to permission of such department for (a) tree plantation on long term and short term basis; or

(b) constructing market complexes, shops and buildings for commercial use; or (c) integrating scattered small scale industries in industrial complexes; or (d) constructing parking spaces for buses, trucks, cars etc; or (e) construction of housing complexes; or (f) any other measure the Committee considers feasible; (ii) using private, public water bodies for promotion of pisciculture or entertainment purposes; (iii) installing unconventional energy sources like solar electricity or bio-gas or any other alternative for commercial distribution; (iv) using private lands for commercial purpose on joint venture of Municipality/Notified Area Authority and the concerned owner; (v) running municipal transport system on roadways and waterways for carrying goods and passengers in such routes which are not covered by adequate public system and are financially viable; and (vi) such other measures which the Committee considers feasible and viable.

Bihar

In Bihar,[14] two types of committees, Wards Committee and Ward Committee are constituted. In every Municipal Corporation having a population of three lakhs or more, a Wards Committee consisting of not less than three wards is constituted. Each Wards Committee consists of councillors representing the constituent wards. The councillors elect one from amongst themselves, who is not being a member of the Empowered Standing Committee, to be its chairperson. The chairperson may resign his office by giving notice in writing to the Mayor and the resignation takes effect from the date of acceptance by the Mayor. A wards committee functions under the general supervision and control of the Empowered Standing Committee. It discharges the functions relating to the provisions of water supply, drainage, sewerage, removal of accumulated water on the streets or public places due to rain or otherwise, collection and removal of solid wastes, disinfection, provision of health immunisation services and slum services, lighting, repair of minor roads, maintenance of parks, drains and gullies, issue of licenses and such other functions as are determined under regulations by the Municipality. The wards committee may conduct public hearings on any major issue of public interest.

The Empowered Standing Committee assigns officers and other employees. One of such officers is designated as Wards Officer.

Each ward of a Municipality has a Ward Committee. It is constituted within six months of the constitution of the municipality. A ward committee consists of the Councillor elected from the ward and all elected representatives of the Area Sabhas falling under the ward. The councillor elected from the ward acts as chairperson of the committee. The term of office of the members is coterminous with the term of the municipality. The Chief Municipal Officer or the Zonal Officer having jurisdiction is entitled to take part in meetings and deliberations. The chairman may request the representatives of other concerned departments as special invitees to participate in the meetings. The Sanitary Inspector, or the Zonal officer, or any other officer, as nominated by the Chief Municipal Officer, acts as secretary of the committee. Secretary calls the meeting on date fixed by the chairman. All minutes are recorded by the secretary and a copy of minutes is forwarded to the municipality. The municipality is responsible for providing administrative and infrastructure support to the ward committee for its proper functioning. The ward committee has to create mechanism for checks and balances over the government personnel in the ward. Ward committee has power to recommend to the competent authority for imposition of penalties in respect of a municipal employee for misconduct and negligence of duties.

A ward committee is obliged to discharge following functions in the ward: provide assistance in solid waste management; supervision of sanitation work; provide assistance for the preparation of the development schemes; encourage harmony and unity among various groups of people; mobilise voluntary labour and donation by way of goods or money for social welfare programmes; provide assistance in the implementation of development schemes; provide assistance for identification of beneficiaries for the implementation of development and welfare schemes; encourage art and cultural activities and activities of sports and games; ensure peoples' participation in the voluntary activities necessary for successful implementation of the developmental activities of the municipality; assist in timely collection of taxes, fees and other sums due to the municipality; maintenance of parks; maintenance of street lighting; assist the municipality in realising the yearly collection targets set by municipality; perform such other functions as may be assigned to it by the municipality.

The chairman and the members of the ward committee have the right to seek information from the Chief Municipal Officer regarding any matter relating to the ward. The ward committee has right to: obtain the full municipal budget to verify, seek clarifications and suggest changes; obtain the requisite financial and administrative support from the municipality in managing accounts; be consulted in the development of land use and zoning regulations within its jurisdiction; obtain full details on all revenue items including taxes and budgetary allocations; get allocation of funds from the municipality for its functioning and submit accounts of funds used; have a proportionate claim on municipal development expenditures, based on the Ward Infrastructure Index of the ward.

Every ward committee has mandatory duty to: prepare the ward plans consistent with the municipal plan; prepare the ward budget in accordance with the ward plans; encourage local-level alternatives for implementation;

ensure optimal collection of all revenue sources; assist in preparing the Ward Infrastructure Index.

The activities of the ward committee are:

(i) ***Preparation of a calendar***—At the first meeting for each budget year, the committee shall decide upon specific obligatory agenda for each monthly meeting.

(ii) ***Preparation and completion of plans***— The ward committee shall prepare the annual ward plan and forward the same to the municipality.

(iii) ***Preparation of ward budget***— A ward level budget calendar is prepared six weeks before the municipal budget. The municipality may suggest changes. The municipality budget shall aggregate all the ward budgets and can have additional account heads for specific receipts and expenditures at the municipal level.

(iv) ***Maintenance of accounts and constitution of ward finance committee***— A ward finance committee consisting of three persons is constituted through nomination by chairman within one month of the constitution of the ward committee. The ward finance committee appoints one of its members as chairperson who is the authorised signatory for maintenance and use of accounts. The committee prepares the annual budget within seven months from the closure of the previous fiscal and places before the ward committee for discussion and approval. The committee also maintains ward level bank account for all the receipts and expenditures on activities of the ward. The committee ensures that all funds transfers from this account are made out to appropriate account heads of the municipality. The committee shall present accounts every quarter at the meeting of the ward committee. The committee prepares a quarterly report of the financial transactions and makes available to every member of the ward committee one week prior to the meeting on discussion on the report. The report is made available for public scrutiny.

(v) ***Penalties***—The chairman of the ward committee can take appropriate action including penalty against the members after giving adequate notice in writing and reasonable opportunity of being heard. The decision of imposing a fine shall come into force only it is approved by a majority vote at the meeting of the ward committee. However, penalty cannot extend to the dismissal of the concerned member.

(vii) ***Perusal of bills***— For every expenditure bills shall be presented. Accounts of ward committee shall be audited annually and report shall be presented in the general body meeting, besides internal audit by municipality and quarterly simple reports.

In every ward a Ward Information and Statistics Committee shall be formed for various developmental and planning works. The committee consists of three members who are not members of the Ward Finance Committee nominated by the chairman of the ward committee. The committee shall appoint one of its members as chairman who presides over its meetings. It is the duty of the committee to compile, maintain and update annually the following on formats prescribed by the municipality: Economic Information including the number and nature of commercial establishments in the ward and employment data in these establishments; Information relating to Land Use including the changes in pattern of land use, data relating to public spaces and civic amenities sites, traffic patterns and public transportation hubs and preservation and restoration of environment, natural resources and heritage sites; and Infrastructure Index including the extent of infrastructural development, current and pending projects, and the infrastructural requirements of the ward.

The ward committee shall participate in all development plans of the city; enforce zoning and land use regulations; participate in the creation and enforcement of new instruments like transferable developments rights etc. The committee is responsible for comprehensive intervention for urban poor by integrating all existing activities undertaken by the municipality; preparing and maintaining beneficiary list for all the programmes and schemes undertaken by the municipality in coordination with the relevant government agencies; and preparing a report on the housing and public distribution system in the ward. The ward committee shall be responsible for decentralised management of primary collection and management of solid wastes; de-silting of drains; maintenance of street lights and parks; de-weeding of ponds; road works including construction, maintenance and restoration; and general beautification of the locality.

Jharkhand

In Jharkhand,[15] ward committee is constituted in every ward of the municipality having a population of one lakh and more. In case of a municipality having a population of less than one lakh, a ward sabha is constituted in every ward. It consists of all electors in the electoral roll relating to the ward. A ward committee consists of the councillor elected from the ward, who acts as chairperson of the committee, the Area Sabha Representatives from the ward and not more than ten persons representing the civil society from the ward nominated by the council. If the population of the ward is not more than five thousand the number of nominated members would be four and thereafter on population of every two thousand an additional member will be added. At least one-third of these members will be from the resident welfare associations and community-based organisations in the ward. About half of the members would be women. The chairperson may invite officials and members of civil society to attend the meetings and participate in deliberations. An official nominated by the Municipal Commissioner or the Executive Officer acts as member-secretary. He records minutes of the meetings and forwards to the Municipal Commissioner or Executive Officer with the approval of the chairperson within ten days of the meeting. The municipality allots twenty per cent of the amount earmarked in the maintenance provision of municipal budget to ward committees for maintenance of services like water supply, sanitation, drains, street lights, parks, markets etc. The ward committee may appoint sub-committees for opinion to enquiry on any matter relating to the functions entrusted to it.

The ward committee is obliged to perform following functions in the ward: to supervise solid waste management, sanitation works, water supply, maintenance of parks, playgrounds and market places, working of street lights and repairs of roads, and implementation of poverty alleviation programmes and development schemes; to monitor the working of schools, dispensaries, health centres etc. under the control of the municipality; to assist in the preparation of development schemes; to encourage harmony and unity among various groups of people; to mobilise voluntary labour and donations by way of goods or money for welfare programmes; to assist in identifying beneficiaries for the implementation of development and welfare schemes; to encourage art and cultural activities and activities of sports and games; to ensure people's participation in voluntary activities necessary for successful implementation of the developmental activities of the municipality; to facilitate collection of taxes, fees and other sums due to the municipality; and to perform such other functions as may be entrusted.

The chairperson and members of the ward committee have right to seek information from the Municipal Commissioner or Executive Officer regarding any matter relating to the ward. The ward committee has right to: obtain information about the master plan and development plans of the municipality; obtain information from the Municipal Commissioner or Executive Officer on any matter relating to the ward; obtain information on municipal budget and details of all revenue items relating to the ward; and be consulted in the development of land and zoning regulations within the ward.

Uttar Pradesh

In Uttar Pradesh,[16] the earlier provisions made in the municipal laws were to constitute wards committee in municipalities having a population of three lakhs and more. These committees were to consist of five wards in municipal councils and ten wards in municipal corporations. In 2009, the municipal laws were amended and it was provided to constitute a ward committee for each ward in a municipality having population of three lakhs or more. The rules were also amended accordingly in 2014. A ward committee consists of the councillor elected from the ward and ten other members nominated by the municipality from amongst persons registered as elector in the ward, office bearers of citizen welfare societies and representatives of non-government organisations, who have special knowledge or experience of municipal administration. About one-third of the members should be women. The councillor representing the ward acts as chairperson of the committee. An employee of the municipality authorised by the municipal commissioner or executive officer is the secretary of the committee.

The ward committee exercises the powers to: ensure that the municipality is fully responsive to the aspirations of the local residents; ensure greater citizen participation in the servicing and development of local areas; serve as a medium for the execution of local works through allocated funds and contributions involving voluntary agencies; review the progress of works and to suggest measures for accelerating the execution of the works included in the budget for the area; identify lacunae and lapses in the maintenance and the municipal services, facilities and amenities and call upon the relevant bodies for action; seek information from the municipal commissioner or executive officer regarding any matter relating to the ward; make periodical reports to the municipality; obtain full information about the municipal plans; have a proportionate claim on municipal development expenditure based on the ward infrastructure index; obtain the full budget to verify, seek clarification and suggest changes; and obtain full details on all revenue items including taxes and budgetary allocations.

The ward committee has to perform following functions: serve as a forum for residents to express their grievances related to municipal functions; ensure implementation of the plans of the municipality for construction, reconstruction, repair, maintenance, renovation of the roads, streets, lanes, drains, sewers, water-connections, street light, sweeping city, cleansing parks, playgrounds and open spaces; ensure watering, scavenging and proper cleansing of roads, streets, lanes, drains, sewers and all the public places, play grounds and the collection and removal of sewage, offensive matters and rubbish, guarding from pollution water used for human consumption, preventing polluted water from being so used; ensure that the property tax and other taxes, charges and fees are being realised as per demands and also to suggest measures necessary for the collection of taxes and fees being made up to the maximum percentage of the demands; ensure that the numbering of houses and the naming and renaming of roads, streets, lanes and the localities and also to conduct time-to-time checking thereof; ensure that the licenses granted by the municipality are not being misused; ensure the upkeep of the fire hydrants; provide assistance in solid waste management; supervise sanitation works; provide assistance for the preparation and encouragement of development schemes; encourage harmony and unity among various groups of people; mobilise voluntary labour and donation by way of goods or money for social welfare programmes; provide assistance for identification of beneficiaries for the implementation of development and welfare schemes; encourage art and cultural activities and activities of sports and games; ensure people participation in voluntary activities necessary for successful implementation of development activities of the municipality; produce ward plans in a manner consistent with the District Plan, City Development Plan and Metropolitan City Development Plan; encourage local-level alternatives and initiatives for implementation of plans; ensure optimum collection of revenue from all sources; map ward infrastructure index; and perform other functions assigned by the municipality.

Madhya Pradesh

In Madhya Pradesh,[17] Ward Committees are constituted in every municipal corporation having population of three lakhs or more. The corporation having less population may constitute Ward Committees. The number of ward committees in a corporation is equal to the population of the corporation divided by one lakh. The number of wards in such wards committees in a corporation should as nearly as possible be equal.

Every elected councillor representing a ward within the territorial area of wards committee and two persons residing within the area as nominated by the Mayor are the members. Only persons eligible for election as councillor and having special knowledge or experience in municipal administration are nominated on the recommendation of the chairman of the wards committee. The nominated members have no voting rights in the meetings. The committee at its first meeting elects one of the elected councillors as chairman who holds office until the duration of the corporation. Where the Speaker is a member he is ex-officio chairman of the committee.

For the territorial area of the wards committee it is responsible for: construction and repairs of new roads and drains; provide water supply and pipe connections; sanitation; necessary recommendation for licences; collection of tax, fee, rates etc.; prepare and execute the schemes of markets, development and beautification; execution of national programmes relating to health, women and child welfare, destitute pensions, handicapped and to run the old age homes; remove encroachments and unauthorised constructions; arrange for lighting of streets; supervision of schools up to primary standards and hospitals up to the primary health centres; development and maintenance of gardens, squares and public places; activities relating to libraries, reading rooms and sports; works relating to control of the food adulteration;

supervision of public distribution system; plantation and betterment of environment; and any other functions allotted by the corporation.

A ward committee is empowered to accord sanction up to Rs fifty thousand for executing its functions; to inspect and supervise the execution of functions; and to administer and control the staff working with the committee. The committee can invite any officer of the corporation to attend the meeting and render his advice. The committee prepares a plan and estimates of work to be done in the ward and submits the same to the corporation for allotment of budget.

Every ward committee has its own separate office. The committee meets every fortnight to conduct its business. The chairman of the committee decides the date of the meeting. The staffs provided by the municipal corporation are responsible for discharge of its functions. The quorum of a meeting is half of the total members. Every decision is taken on the basis of consensus.

Constituting Area Sabhas

The concept of area sabha is new to the municipal governance in India excepting Kerala, Madhya Pradesh and Chhattisgarh. In Madhya Pradesh and Chhattisgarh these are known as Mohalla Committee. The level of people's involvement in urban areas, to day, is negligible. There exists a wide gap between municipal government and urban communities. If municipal governance is to become truly accountable to the people, the level of people's involvement should be enhanced. The institution of area sabha aims at not only improving municipality-citizen interface but also facilitating truly participatory municipal governance penetrating deep at the community level. The community participation reform, therefore, accords top priority to this institution for mobilising the local communities for improving the levels of municipal services and infrastructure. The provisions related to constitution and functions of area sabha in a few major states find mention hereunder.

Andhra Pradesh

The area sabhas are to be formed in all the municipalities in Andhra Pradesh having a population of one lakh and above.[18] Every ward shall be divided into areas based on the population of the ward. In the division of ward into areas, it is stipulated that natural boundaries and geographical contiguity of the area must be considered as far as possible. The existing poling booths also should be taken into consideration while dividing the ward into areas. In municipalities, one area sabha will be constituted for every 1500 population with a variation of 10 per cent more or less and in the corporations, the population of area sabha varies between 2,000 and 5,000.[19]

The area sabhas consist of all the electors in the jurisdiction of that particular area who registered as voters prior to three months from the date of constitution of the ward committee. Each area will be represented by an area representative nominated by the corporation or council. The qualifications and disqualifications as prescribed for the elected members of the corporation/ council shall apply *mutatis mutandis* to the area sabha representatives also. Similar to ward committee members, the term of the area sabha representative is co-terminus with that of the corporation/council. The eligibility for nomination as area sabha representative is that primarily he should be a registered voter of the ward and area concerned. The municipal corporation/municipality concerned will divide the ward into certain number of areas as per the prescribed method.

The area sabha is primarily responsible for identifying local problems in respect of water supply, roads, street lighting, etc., and needs of the people and, send proposals to the ward committee for taking-up necessary remedial measures and development works as needed. Being grass root level institution, it has to identify the beneficiaries, prioritise and send the list to the ward committee for consideration. The system of area sabha introduced in Andhra Pradesh is similar to the model suggested in the Nagara Raj Bill excepting that the area sabha representative is nominated by the corporation or council as against direct election by the voters.

Bihar

In Bihar,[20] Area Sabha is constituted in each ward of every municipality having population of one lakh fifty thousand and above. Area Sabha consists of all persons registered in the electoral rolls pertaining to every polling booth. An Area Sabha Representative is elected from every area. Any registered voter in an area of which he is a resident and who has attained the age of eighteen years may file his nomination for the office of the Area Sabha Representative unless he is disqualified for being elected as such is disqualified for the purpose of elections to the municipality or to the State, Legislature. The elections are conducted under the control, direction and superintendence of the State Election Commission. Not more than fifty per cent of seats are reserved for women. Further seats are reserved for the members of the Scheduled Castes, Scheduled Tribes and Backward Classes in proportion of their population to the total population of the area. Fifty per cent of seats reserved for these categories are reserved for women belonging to these classes. The reservation of seats is done on rotation. The term of Area Sabha Representative is ordinarily coterminous with that of the municipality concerned. He may be recalled by the registered voters. The Area Sabha Representative convenes and conducts the meeting of the Area Sabha and chairs its meetings. Area Sabha Representative is ex-officio member of the ward committee constituted for the ward within which that Area is situated.

An Area Sabha may perform and discharge functions and duties : to generate proposals and determine the priority of schemes and development programmes and forward the same to the ward committee or in its absence for inclusion in the developmental plans; to identify the most eligible persons for beneficiary oriented schemes such as housing, training, economic development etc.; to certify the eligibility of person for various welfare schemes; to suggest the location of street lights, street or community water taps, public wells, public toilets, community halls and other public amenities; to suggest renovation, up-gradation and construction of public parks and their maintenance; to point out the paucity of water supply, road or lighting arrangements and suggest remedial measures; to address and monitor issues of public health, especially of diseases, outbreak of epidemics and incidents of natural calamity and to make arrangements to inform competent authorities; to provide and mobilise voluntary labour and contributions in cash and kind for developmental programmes and to supervise such developmental works; to sensitise the community towards their duty to pay taxes and help in collection of the same; to ensure realisation of yearly tax and non-tax revenue collection targets set by the ward committee; to impart awareness on matters of public interest such as cleanliness, preservation of environment and prevention of pollution; to encourage friendship and social harmony between various communities and organise cultural and sports activities; and to help the ward committee in maintaining cleanliness.

The Area Sabha may exercise following rights and powers: to get information from the officials as to the services they render and the works they propose in succeeding three months after the meeting; to be informed by the ward committee about every decision and the rationale of such decisions; to be informed by the ward committee of the follow up action taken on the decisions; to receive initial grant from the municipality for holding of meetings and maintenance of records and accounts; and to receive funds from the municipality for performance of its duties and fulfillment of its goals.

The municipality shall guide and monitor formation and functioning of the Area Sabhas; budgetary provision; inclusion of suggestions/proposals received from Area Sabha/Ward Committee; selection of suitable NGOs for imparting training and capacity building programmes of the chairpersons and members on priority and help in organisation of such activities.

Jharkhand

In this newly created state,[21] a municipality having population of one lakh and more, each ward is divided into Areas consisting of two or more, but not exceeding five, contiguous polling stations. For each Area, an Area Sabha is constituted by consisting of all persons registered as voters of the area. From each such Area an Area Sabha Representative is nominated by the council. The term of Area Sabha Representative is coterminous with the term of the council. The Area Sabha meets once in two months for conducting its business.

The ward sabha and area sabha are obliged to perform and discharge the following functions and duties: to identify eligible persons for beneficiary-oriented schemes, and to prepare lists of eligible persons in order of priority and forward to the municipality; to verify eligibility of persons getting of welfare assistance such as pension and subsidies; to suggest locations of street lights, public water taps, community/public sanitation units and other public amenities; to identify the deficiencies in the water supply and street lighting arrangements and suggest remedial measures; to assist in activities of urban public health centres, especially in disease prevention, family welfare, and immediately report the incidence of epidemics and natural calamities; to impart awareness on matters of public interest such as cleanliness, preservation of environment and prevention of pollution; to promote harmony and unity among various groups of people; to arrange cultural festivals, sports meets etc; and such other functions and duties as may be assigned by the municipality.

Madhya Pradesh

In every area[22] determined by the government a Mohalla Committee is constituted within three months from the date of notification. The number of committees and number of members are prescribed by the government. A Mohalla Committee consists of a Chairman, a Vice-Chairman, a Treasurer, and the other members determined by the government. The elected councillor of ward concerned is a member patron in all such committees within the ward. For the purpose of nominating Chairman, Vice-Chairman, Treasurer and members, the names are obtained by a committee consisting of members of Lok Sabha registered as electorate in the municipal area, members of State Legislature and Rajya Sabha representing the area, Mayor, in case of Municipal Corporation, and Chairman, in case of Nagar Palika Parishad and Nagar Panchayat. The persons who are chosen for nomination shall be resident of the concerned Mohalla and be eligible for election as councillor. The Collector presents the names suggested to the Minister-in-charge, who accords his approval to the names suggested and thus nominates the members of the Mohalla Committee. While doing so he will give adequate representation to women, Scheduled Castes, Scheduled Tribes and Backward Classes in every Mohalla Committee. The Collector notifies the nominations and sends list to the municipalities and the persons nominated. Copy of the list is put on the notice board in the office of the Collector as well as the respective municipality. The term of the committee shall be co-terminus to that of the municipality. However, the Chairman, Vice-Chairman, Treasure and members shall hold the office on the pleasure of the Minister-in-Charge.

Every Mohalla Committee is responsible for enjoying following powers and discharging functions within its territorial area: monitor matters related to sanitation, road, drain, water supply, street light and solid waste management; and arrange for redressal of grievances related to them; making the public aware for conserving the environment from pollution and discarding the use of polythene; assisting in implementing national programmes and schemes related to health, women and child welfare, pension to destitute, urban poverty alleviation, differently-abled and immunisation; promoting plantation of trees through public participation; maintaining municipal properties and save them from encroachment; expansion of education especially assisting the dropout children in re-admission and other in taking admission in primary and secondary schools; encouraging people's participation in development programmes; functions related to assessment and collection of tax/fees as entrusted by the municipality; campaign against illegal activities like child labour, liquor, baiting, gambling etc; and any other functions as entrusted by municipality or any department of the government.

Every Mohalla Committee has its office manned by a Secretary who is provided by the municipality. Mohalla Committee meets at least every fortnight. The date and time is determined by the Chairman. The Secretary intimates Chairman and every member about meeting at least three days prior to the meeting. The Chairman has a casting vote. The Secretary attends every meeting and maintains minute book and proceedings.

Chhattisgarh

In Chhattisgarh,[23] each ward of every Municipal Corporation, Municipal Council and Nagar Panchayat is treated as territorial area for constituting a Mohalla Committee. It consists of councillor elected from the ward and the nominated councillor residing therein. The other members nominated from amongst the residents of ward are ten, in case of Municipal Corporation, seven, in case of Municipal Council, and five, in case of a Nagar Panchayat. Such persons should be eligible for election of councillor. The names of persons to be nominated are presented by the Collector to the Chairman of the District Planning Committee, who while nominating members ensures that at least half of the members belong to women, SCs, STs, Backward Classes and non-governmental organisations working in the ward. The elected councillor is the chairman of the Mohalla Committee. The Vice-Chairman and Treasurer are nominated by the Chairman of the District Planning Committee out of the residents who are eligible for election to councillor. The term of the Vice-Chairman and Treasure is coterminous to that of the Chairman of the District Planning Committee. The term of the Mohalla Committee is coterminous to that of the council. Mayor, in case of Municipal Corporation, and Chairman, in case of Municipal Council acts as patron to the Mohalla Committee.

Every Mohalla Committee discharges following functions: preparation of proposals or schemes for development of Mohalla; generate finances from the residents and execute development activities; promote people's participation in developmental activities; assisting in maintaining and preserving municipal properties; assessment and collection of tax/fee as assigned by the municipality; and functions entrusted by the municipality or any department of the government.

Every Mohalla Committee has its office manned by a Secretary who is provided by the municipality. Mohalla Committee meets at least every fortnight. The date and time is determined by the Chairman. The Secretary intimates Chairman and every member about meeting at least three days prior to the meeting. The Chairman has a casting vote. The Secretary attends every meeting and is responsible for maintaining minute book and proceedings.

The Greater Hyderabad Municipal Corporation (GHMC)-A Case Study

The revamped system of ward committees was instituted in Andhra Pradesh by an Act of 2008 in the Corporations. The formation of ward committees and area sabhas was taken up after the elections. The ward committees were constituted in all the 150 electoral wards in GHMC. The ward committee consists of the Corporator representing the ward, nominated members and the area sabha representatives as ex-officio members. The strength of nominated members is restricted to 10 as per the Act. The actual strength of each ward was calculated in GHMC based on the population of that ward as per 2001 population. Minimum four members are provided for wards having a population of ten thousand and an additional member added for every four thousand population. Similarly, area sabhas were demarcated as per the Act. The area sabhas are to be constituted in the corporations for every 5000 population. The members of the ward committees and area sabha representatives were nominated by the Corporation by inviting applications as per the prescribed procedure. In GHMC, a total of 1077 area sabhas were constituted based on the above criteria. The number of area sabhas in each ward ranged from five in Jubilee Hills and Ameerpet Wards to 11 in Banjara Hills Ward based on the population of the ward concerned.

The GHMC has done very good preparatory work to enable ward committees to function effectively. The Commissioner has circulated special guidelines for better understanding of the ward committee system. Instructions were given to important officials like the Deputy Commissioner at circle level and the Zonal Commissioner at Zone level and the other officials concerned to provide necessary assistance to ward committees and area sabhas and implement the decisions taken by ward committees. Technical officers mostly engineers were appointed as the secretaries of ward committees and area sabhas to assist in convening the meetings and liaisoning with other officials for implementing decisions and resolving problems identified by members of the ward committees. The area sabhas and ward committees will identify a list of works which are called Citizen Priority Request. The Zonal Commissioners are directed to examine the proposals and options and, execute the works as per the rules. It is suggested that the ward committees can explore involvement of the private sector on public-private partnership model to fund the city projects. Further, the officials of GHMC should share the information pertaining to aspects like DCB of property tax, trade licences, rents, etc. with the ward committee for getting necessary help from the ward committees. The ward committee has to monitor and review various maintenance works in respect of water supply, street lighting, road repairs, markets, parks, playgrounds, etc. Apart from these, some schemes need to be monitored by the ward committee, especially poverty alleviation programmes. Necessary instructions have been issued to different functionaries within the GHMC. For instance, through a standing committee resolution, the ward committees were involved in the activities of the GHMC sports complexes.[24] The GHMC maintains certain sports complexes in Hyderabad. These sports complexes are managed by a management committee. It is instructed that the managing committee has to furnish information regarding all sports activities within the ward to the ward committee concerned. The sport management committee should represent in the ward committee meetings and give feedback. Similarly, standing orders were issued by the Commissioner for involving ward committees in the maintenance of municipal market complexes. The GHMC maintains 697 shops/units in its jurisdiction. These complexes are located in different circles. In the standing instructions to the market committees, it is directed that market committee shall furnish information regarding activities of municipal markets to the ward committee and get the feedback. Specific guidelines were provided to different ward level functionaries to attend the meetings of the ward committees and render necessary support. In the guidelines, it is mentioned that the officers viz., bill collector of taxation section; community organiser or project officer of UCD section; section officer of town planning section; AE of engineering section; AE of street lighting section; sanitary jawan/sanitary supervisor of sanitary section; supervisor of entomology section; and supervisor of horticulture section must attend the ward committee meetings and provide necessary help.

In Hyderabad, the most important civic service, i.e., water supply and sewerage is the responsibility of the Hyderabad Metropolitan Water Supply and Sewerage Board (HMWS&SB). On the request of the Commissioner, GHMC, the Managing Director, HMWS&SB had issued a Memo instructing the ward level officers to involve in the ward committee meetings and provide necessary assistance.[25] The water supply activities which come under the co-supervision of ward committees were also listed. The details of these activities are as below:

(i) Fixing of manhole covers and fixing of safety grills to manhole covers;
(ii) De-silting by air-tech machines, pull through and bucket cleaning methods, etc;
(iii) Water supply timings regulation;
(iv) Pollution related water quality issues;
(v) Chlorination of water;
(vi) Sewerage blockages;
(vii) Checking illegal connections;
(viii) Road restoration in respect of works;
(ix) Billing and collection of water and sewerage cess;
(x) Identification of water leakages;
(xi) Citizen awareness campaigns and community participation.

The HMWS&SB provided detailed guidelines to its officers about the ward committee and their role in its working. The ward committee can inspect the works taken up by the HMWS&SB in that ward and bring its observations to the notice of the General Manager concerned. Another interesting feature has been that the HMWS&SB has made financial allocations to take up water supply and sewerage maintenance works for each ward. In case of wards where both, water supply and sewerage functions are vested with HMWS&SB, a sum of Rs 1.00 lakh is allocated and in wards where only water supply is being looked after by the HMWS&SB, Rs 50,000 per ward has been allocated. In order to facilitate participation of the HMWS&SB officers in the ward committee meetings, ward level officers of the rank of Manager (Eng.) were designated for each ward. One hundred fifty officers were thus designated as ward level officers to attend the meetings of the ward committees and to act as liaison officers between GHMC and HMWSSB at the ward level. These ward level officers will attend the meetings of the ward committees in respective wards and furnish ATRs to the ward committee within 15 days. It is specifically mentioned that the general managers shall give priority to the list of works identified by the ward committee while sanctioning the works. The general managers also should indicate the resolution of the ward committee in the sanction proceedings issued by him. In respect of quality control, the general manager will rectify the mistakes if any within seven days and submit an ATR to the ward committee meeting. It is also mentioned that the quality control wing will inspect these corrections and give report to the Chief General Manager and Director concerned. The works taken up on the suggestions of the ward committee shall be executed as per the rules and procedures of the HMWS&SB. The general managers are directed to maintain separate registers for the works taken up through ward committee advice and they should ensure that there won't be any duplication or overlapping of the works. The guidelines also specified some details of different statistical data which has to be shared with the ward committees, viz., IT support, billing information, equipment given to sanitary workers, details of outsourcing workers, slum level activities, fogging and spraying schedules, sanctioned work lists for the ward, list of public toilets, master plan and development plan, parking slots, disaster response plans, etc. The involvement of the ward committee in these activities is also sought.

This apart, the activities like implementation of hawkers' policy, development of master plans/zonal plans, road and nala widening, greenery plans, door-to-door collections, etc. are to be shared with the ward committees for getting their cooperation. Every year the ward committee has to prepare an annual report by June end while giving details of fund allocation, functions taken up, coordination and cooperation obtained from the municipal functionaries.

The ward committee meetings are chaired by the ward corporator. He has a greater role for effective working of the committee followed by the interest and commitment of the ward committee members. The ward committee meetings are generally held once in two months. In order to facilitate attendance of officials, staggered dates for every month have been suggested for each of the six zones giving a range of five days for each circle to convene the meetings. Ward committees have to discuss about problems and issues being encountered in the respective wards. The minutes of the meetings are monitored by the Zonal Commissioners. The area sabha shall generate capital and maintenance works proposals and determine the priorities of development activities and forward them to the ward committee. The proposals generated by the area sabhas are finalised in the special meetings of the ward committee. The Additional Commissioner (Finance) is authorised to release the required money to take up the works including 20 per cent earmarked in the annual budget for every quarter to the Zonal Commissioners. Most common issues and problems discussed in the ward committee meetings relate to provision of civic services like water supply, drainage, street lighting, sanitation, solid waste management, roads and some aspects relating to construction of community halls, identification of beneficiaries, etc.

Resolutions in the ward committee should be passed by majority votes. But, in practice, formal voting is not taking place. Chairperson signs the resolution document and it is kept under the custody of secretary. In this system, some members feel that their proposals are not given due importance. This informal method will work as long as the relations between members are cordial. Though the members may not express freely in the meetings, discussions with the members reveal that some of their proposals are not considered and they feel sidelined.

The members, in general, are aligned on party lines. Though the ward committee rules do not mention about the role of political parties in the activities and functioning, actual working reflected such polarisation among the members. The role of political parities can not be ruled out in a democracy, but this trend must be arrested to make the ward committee system truly participatory. Another trend noticed was that the ward committee members who are nominated by GHMC feel that they are different from the area sabha representatives who are also nominated by GHMC on similar lines. In the ward committee meetings also they are asked to sit separately. This virtual division between the members may adversely affect functioning of the ward committees as ward committee is a unit and all the members are of equal importance though area representatives may focus on the area level problems. This trend must be arrested for smoother working of the ward committees where the chairman could play a proactive role.

The secretaries of ward committees have to play critical role for effective functioning of ward committees and area sabhas. They have to prepare the agenda in consultation with the Chairman, provide all administrative support needed for conduct of the meetings and take follow-up action on the resolutions of ward committee. They should also act as liaison officer between the ward committee and the other officers from external organisations. The coordinated action depends on the ability of the secretary. The secretary is also custodian of all the official correspondence and documents of the wards committee. The secretary also performs similar functions in respect of the area sabhas too. They have to attend to these functions apart from performing their routine ward level functions. In this background, majority of the secretaries feel that they are overburdened. They, therefore, suggest an exclusive officer as secretary of the ward committee to attend to all functions effectively.

Ward committee system in the new format was introduced by GHMC for the first time based on the Nagara Raj Bill, as part of the Community Participation Law. Very good ground work was done before forming the ward committees. Circle level awareness camps were also organised. In spite of this, several ward committee members expressed that they lacked adequate knowledge about purpose, functions and roles of officials and the members. Some even reported that basic information was also not supplied to all the members. Some of the area sabha representatives are illiterate; hence, need proper orientation to understand the system. It is interesting to note that some of the area sabha representatives are not aware that they should conduct meetings in their respective areas. There is, therefore, need for conducting capacity building programmes to all the ward members including the area sabha representatives quite frequently by higher officials like the zonal commissioners and the deputy commissioners. Some of the ward members suggested that the deputy commissioner and the zonal commissioner should attend some ward committee meetings to provide necessary support and build confidence among members. They feel that presence of higher officials will certainly help in solving their problems quickly. The capacity building is also needed for the secretaries and other related officials to improve their understanding of the ward committee system.

Decentralised and people centred urban planning are the objectives of the ward level planning exercise. In fact, the Constitution 74th Amendment Act stipulates that municipalities should be entrusted with urban planning. The ward committee provides excellent platform for active municipal government role in urban planning. It is in this background, the ward committee is entrusted with the responsibility of preparing the Ward Development Plan in Andhra Pradesh. The Ward Development Plan (WDP) has to provide the vision and development strategy for future development of the ward concerned. In urban planning, WDP is the basic unit of planning at local level. The city/town level plan has to be prepared by consolidating and integrating WDPs with the city-wide vision and needs. The effectiveness of city level planning, therefore, rests on the ward level planning process. The WDP primarily focuses on ward level issues and concerns in terms of infrastructure development, social development and service delivery. In order to make the planning process more participatory and people centred, the area sabhas are given important role in preparation of the WDP. Ward level planning process as envisaged in GHMC is outlined below:

(a) First, area sabha will identify the issues, problems and needs in the area sabha meetings convened by the area sabha representative in respective area. The local communities may have several problems and needs in respect of water supply, sewerage, storm water drains, solid waste management, poverty alleviation, power and street lighting, etc. The identified local needs must be prioritized based on the expenditure requirements by the area sabhas.

(b) In the second stage, the problems and needs identified by area sabhas will have to be presented by the respective area sabha representative and discussed in the ward committee meeting specially convened to discuss the WDP. The ward committee members and, all the circle level, ward level and line department officers must attend this meeting to provide the necessary inputs.

(c) Proposals are finally vetted by the concerned officials. The officials later work out details of repairs and capital works and estimate the costs involved for executing the proposals. The officials finally prepare a plan of action giving details like the agency responsible for execution and schedule of executing the works and funding responsibilities.

A study[26] brought-out some interesting facts about priority works identified. The number of projects varied from ward to ward depending on the local needs. Of these, water supply and sewerage appear to be the most common area of concern. Another important observation was that the ward development plans were prepared without taking into consideration the budgetary allocations made for each ward committee. In the absence of financial commitments, the Plans ultimately turned-out to be to do lists. It was in fact observed in some wards; the members feel that their proposals are not given importance by the authorities. The success of this exercise ultimately depends on the seriousness of the officials concerned in giving priority to execute the works proposed by the ward committees. If the officials ignore these proposals, the ward committee members and the local people will lose faith in the ward committee mechanism itself. There should be proper follow-up and monitoring mechanism to ensure their implementation.

Improving the Ward Committee System

The ward committees and area sabhas are institutional mechanisms designed to promote participatory and decentralised governance in urban areas. All municipalities in the country should constitute these institutions. It is unfortunate that in majority of states these committees are still waiting to see the light of the day. In many states the related provisions are confined to the respective laws. It is in this context, the experiences like GHMC become relevant and beneficial. The municipal authorities in many states are successful in constituting ward committees and area sabhas but, they are facing certain operational problems impacting their working on desired lines. These problems broadly relate to proper understanding, coordination and support system.

At the outset, there exists considerable confusion among the members of ward committees and officials about their roles, functions and powers. Adequate training and awareness camps are essential to the ward committee members and area sabha representatives. The responsibilities of ward level officers and external agencies also need to be explained properly. The capacity building is important to bring better clarity, sensitise the officials and members of ward committees including area sabha representatives about the underlying philosophy and operational mechanisms of ward committees and area sabhas.

The past experiences with ward committees in India reveal that the state governments as well as ULGs give very less importance to the working of the ward committees. Due to lack of commitment and effective monitoring, even the meetings are not conducted regularly. Regular conduct of the ward committee and area sabha meetings is very important to ensure effective peoples' participation. The state governments should therefore give top priority in getting these committees constituted and the ULGs should pay attention on proper awareness building and sensitisation exercises and, monitoring the working of the ward committees. Many a time, non-availability of adequate space is taken as reason for not conducting ward committee meetings. The municipalities should provide necessary place to facilitate regular conduct of ward committee meeting. The ward level offices should be strengthened with proper building, meeting place, infrastructure and staff which in turn would facilitate effective functioning of the ward committees.

Urban communities comprise of diverse groups having the tendency of polarisation. They show reluctance, apathy and isolation as far as local level cooperation is concerned. The area sabhas will have to break these self imposed barriers and encourage community level cooperation and participation in solving local problems. It is observed that the local communities are not aware about ward committees and area sabhas. The ULGs should do extensive publicity about ward committees and area sabhas through various media and their local offices. The officers should monitor and guide wards committees and area sabhas regularly so that members become active and confident to take the message of participatory governance to common people. It is, no doubt, a slow process but essential to make area sabhas successful.

Execution of the proposals made by the ward committees is crucial for the success of the local level planning exercise. The crux of the problem appears to be lack of seriousness among the executing officials about the proposals of the ward committees. As the execution tasks are taking long time, the ward committee members feel that their proposals are not given due priority. The ULGs should evolve a planning framework wherein the local proposals are integrated with the city development plan. Unfortunately, the city development plans are biased toward the city wide projects. At present, there is no proper system to plan for the local level works. The ward committee system provides excellent opportunity in this regard. The success of the Kerala ward committee planning experience should inspire other states to make the ward committee system centre of the local level planning process.

The ward level planning exercise is basically micro level infrastructure development planning in the municipal services sector. The municipalities are implementing Service Level Benchmarks (SLB) scheme to improve the level of municipal services in India. Ideally, ward is the appropriate administrative unit for operationalising the SLBs by municipalities. Presently, SLB implementation is suffering from lack of proper data and information on identified parameters of urban services. The SLB exercise may turn out to be ritualistic if enough care is not taken on this front. The ward committees will be of immense benefit in this respect. They can collect the required information ward-wise and the ward-wise information reports could be consolidated at the municipality level. As the main objective of both, the SLB scheme and ward committees is improving the municipal services, the integrated mechanism will reinforce both the reforms. In the greater cities where the ward populations are on par with many municipalities, ward appears to be the right choice for decentralisation and it is high time that the local administration must become ward centred.

The reform of community participation has not received required attention either from the state governments or from the ULGs. Many a time the judiciary or civil society organisations had to compel the governments to move forward on the reform. The problems basically relate to lack of proper understanding of the system by the officials as well as the ward committee members, lack of necessary place and support system for the ward committees and the area sabhas, political influences and lack of enough budget allocations for execution of works identified by the ward committees or area sabhas. If their priorities are not given importance by the municipal authorities, the members will lose faith in the entire reform. In the light of these constraints, the following measures are necessary to make the reform purposeful and realise the goal of involving people in the local governance:

(i) The state authorities in general have not shown interest in the community participation reform. The utmost need therefore, is commitment of the state authorities for the participatory urban governance;

(ii) Educating the ward members and the area representatives about their roles, responsibilities and powers should be done by capacity building programmes and proper guidance from the higher level officers;

(iii) Ideally, the city or town level development plan must be initiated from below by the ward committees and enough care should be taken to integrate the ward level plans with the city level plans through proper budgetary linkages;

(iv) There is need for giving total attention to the working of the ward committees. Full time, secretaries must exclusively be appointed for this purpose;

(v) The ward committees should be made responsible to prepare ward level SLB status data and regularly monitor their implementation;

(vi) The participatory urban governance cannot be achieved unless the ward level systems are strengthened and streamlined to support the institutional mechanisms of ward committees and area sabhas.

The ward level offices should be strengthened accordingly in terms of space, infrastructure and staff; and

(vii) Finally, the ward committee and area sabha system must be incentivised to improve their working.[27]

References

1. UNHABITAT (2000):, Global Campaign for Urban Good Governance.
2. The Constitution 74th Amendment Act, 1992, Section 243 (S).
3. The Constitution Seventy-Fourth Amendment Act 1992, 243(S).
4. The Energy Resources Institute (2009): Enhancing public participation through effective functioning of Area Sabhas, p. 9.
5. Planning Commission, Report of the Working Group on Urban Governance for 12th Plan, 10.
6. *Ibid.*, p. 15.
7. India's Community Participation Law: The Model Nagara Raj Bill, 2008 (http://www.criticaltwenties.in/lawthejudiciary/india%E2%80%99s-community-participation-law-the-model-nagara-raj-bill-2008). retrieved on February 24, 2011
8. The Energy Resources Institute (2009): op.cit, pp. 17-18.
9. Hyderabad Municipal Corporation Act, 1955, Section 8-A and B and A.P. Municipalities Act, 1965, Section 5-B, C and D.
10. Municipal Administration and Urban Development (UBS) Department, Government of Andhra Pradesh, G.O. Ms. No. 57 and 58, both dated 10th February, 2010.
11. G.O. Ms. No. 57 and 58, Municipal Administration and Urban Development (UBS) Department, Government of Andhra Pradesh, Hyderabad, both dated 10th February, 2010.
12. Hyderabad Municipal Corporation Act, op. cit., Sub-Section (6) of Section 5-B of the Act.
13. The West Bengal Municipal (Ward Committee) Rules, 2001.
14. Bihar Urban Local Bodies (Community Participation) Rules, 2013.
15. Jharkhand Municipal Act, 2011.
16. U.P. Municipal Corporation Act,1959; U.P. Municipality Act,1916; U.P. Municipal Corporation (Ward Committee) Rules, 2014 ; and U.P. Municipality (Ward Committee) Rules, 2014.
17. Madhya Pradesh Nagar Palik Nigam Adhiniyam,1956; and Madhya Pradesh Nagar Palik Nigam (functions, authority and conduct of business of Ward Committees) Rules, 1995.
18. Municipal Administration and Urban (UBS) Development, Government of Andhra Pradesh, GO Ms. No. 58, dated 10th February, 2010.
19. GO Ms No. 57, Municipal Administration and Urban Development (UBS) Department, Government of Andhra Pradesh, dated 10th February, 2010.

20. Bihar Municipal Act, 2007; and Bihar Urban Local Bodies (Community Participation) Rules, 2013.
21. Jharkhand Municipal Act, 2011.
22. Madhya Pradesh Nagar Palika Mohalla Committee (Constitution, Functions, Duties and Conduct of Business Procedure) Rules, 2001.
23. Chhattisgarh Nagar Palika Mohalla Committee (Constitution, Functions, Duties and Conduct of Business Procedure) Rules, 2005.
24. Greater Hyderabad Municipal Corporation, Standing Committee Resolution No. 284, dated 29-09-2010.
25. Hyderabad Metropolitan Water Supply and Sewerage Board, Managing Director, Circular Memo No. HMWSSB-CGM(P&C)/CCM-2010-1081/1832, dated 14-07-2010.
26. Gnaneshwar, V. (2011): Working of the Ward Committee and Area Sabha System in GHMC, pp. 52-53.
27. Working Group on Urban Governance for the Twelfth Plan, op. cit.

Chapter - 5

Property Tax Reforms

Introduction

Cities across the world are considered major territorial nodes for driving the economic prosperity of countries. But, the cities in developing countries are caught in a low equilibrium trap, a vicious cycle of low civic services in terms of quantity as well as quality, low tax compliance, low revenue and again low services. Unless this syndrome is broken, the cities and their prosperity cannot be improved and sustained. It is pertinent to mention the comment made by the High Power Committee (HPEC) (2011) of the Ministry of Urban Development[1] that the urban local governments in India are among the weakest in the world both, in terms of capacity to raise resources and low revenue buoyancy.

Globally, property tax is considered most suitable and appropriate for the urban local governments because of certain advantages viz., immobile nature, visibility and simplicity. Some thinkers consider it a benefit tax as it is linked to the benefits of various civic services provided by the municipalities for local residents: "the 'benefit view' of the property tax is the older of the two conceptions....the residential property tax is seen to be the key source of revenue for local governments, and strict zoning is added to incorporate homeowners' pressure on local officials to make land policy decisions that protect the value of their principal asset—their home. Furthermore, because "capitalisation is everywhere," the property tax serves to assign tax liabilities in line with the public service benefits received, and thus it becomes a "benefit tax." Another view is local property tax is, in effect, an excise tax on capital. As a tax on capital it discourages building relative to other capital investment and thus distorts local housing markets and local fiscal decisions.[2] Further, in respect of the non-residential properties, it is considered regressive in terms of its negative impact on the local businesses and the possibility of shifting the burden on the consumers.[3] It may look paradoxical as most of the municipalities in India give top priority to non-residential properties which can yield better revenues.

Property tax is being levied by the urban local governments in India by virtue of its constitutional status as Entry 49 in the State List of the Seventh Schedule.[4] The property tax constitutes the single largest source of revenue to the municipalities in India in spite of several loopholes in the levy, assessment and collection. The composition of the revenues shows that property tax is the mainstay of the municipalities. According to a study conducted in 36 largest cities of India for the 13th Finance Commission, property tax constituted 23 per cent of the total municipal revenues and 28.5 per cent of own source revenues.

The average revenue yield was Rs 486 per capita with wide variations across the cities ranging from Rs 1334 in Bombay Municipal Corporation to Rs 25 in Patna Municipal Corporation.[5] The share of assessed properties to the total properties was about 56 per cent. Further, it was estimated on this basis that the total property tax collected in the country in all the municipalities would be in the range of Rs 6274 crore and Rs 9424 crore constituting between 0.16 and 0.24 per cent of the country's GDP as against 0.6 per cent in the developing countries.[6] In the absence of reliable information on property tax collection in the country, these figures should be taken only as broad indicators of the macro picture.

It is evident from the above that there is considerable potential for improvement of revenues from the property tax in India. It is estimated that the property tax revenues could increase to Rs 22,000-32,000 crore by increasing coverage and collection efficiency to 85 per cent in all cities and towns in India.[7] There is a general criticism that the property tax is not exploited to the full potential in spite of rapid urban expansion and consequential economic growth.

The yield from property tax is not encouraging primarily due to structural deficiencies like dominance of informal property market, lack of reliable property information system, inelastic tax base, defective assessment, inefficient tax administration and liberal exemptions and concessions. Article 285 of the Constitution exempts property tax on all Central Government properties. The municipalities, however, are permitted to levy service charges on the properties belonging to the Government of India but the cooperation from the central government organisations is not forthcoming making them helpless. The most concerning aspect is that because of its local nature property tax is very unpopular tax as citizens find no commensurate benefits. Assessment, the most critical aspect of property administration, is riddled with several problems, viz., unscientific method of assessment, lack of transparency, lack of proper property recording, official discretion and tax-payer-official collision—all leading to revenue loss to the municipalities.

The property tax also suffers from following administrative deficiencies:

- ***Poor coverage***: Widely prevalent un-assessed and under-assessed properties.
- ***Misuse of exemptions***: The exemptions granted by the state governments for various categories of properties viz., owner occupied houses, properties of educational, religious and charitable institutions, ex-service men, concessions under age of the building, etc., are widely misused.
- ***Inability to tax vacant lands***: Due to several administrative problems, the municipalities are not able to levy tax on vacant plots despite legal provisions.
- ***Inefficient collection***: Poor collection efficiency due to constraints like lack of adequate staff, lack of commitment among the staff and taxpayer apathy.
- ***Non-revision of property taxes at regular intervals***: All municipal laws provide for revision of assessment once in four to five years. But in practice, revision is hardly done regularly. Whenever revision is made after long gap, the tax naturally increases steeply and people raise voice against the tax increase. Often, the state governments cap the increase bowing to popular pressures.
- ***Rent control constraints***: The municipalities are constrained to assess the buildings coming under rent control legislations at low rates fixed decades back leading to loss of income from such properties.
- ***Failure to collect taxes from government properties***: The municipalities find it difficult to collect property tax from state government properties and other public sector institutions located in cities and towns. The government support also is lacking in this area.

Reforming Property Tax Administration

If the problems confronting the administration of property tax are effectively resolved, the full potential could be realised. Minimising the discretion to avoid arbitrary assessment by the tax officials; delinking the property tax from the Rent Control Act provisions; making the process of levy, assessment and collection rational, simple and transparent; ensuring equity across the tax payers; and introducing self-assessment by property owners are some measures proposed to improve the system.

Building integrated property information system is essential to realise full potential and improve interface between the tax payers and the municipalities. The reform initiatives in the country are aimed at achieving these goals.

Efforts are made in India since 1980's to reform the property tax. If the earlier reform initiatives focused on standardisation of the assessment methods, new dimensions like GIS, computerised information system and web based interface are being added currently to take advantage of the happening IT revolution. The major areas of reform have been standardisation of assessment methodology, self-assessment system, integrated property information system, improving collection efficiency and amending rent control legislations (Fig. 5.1).

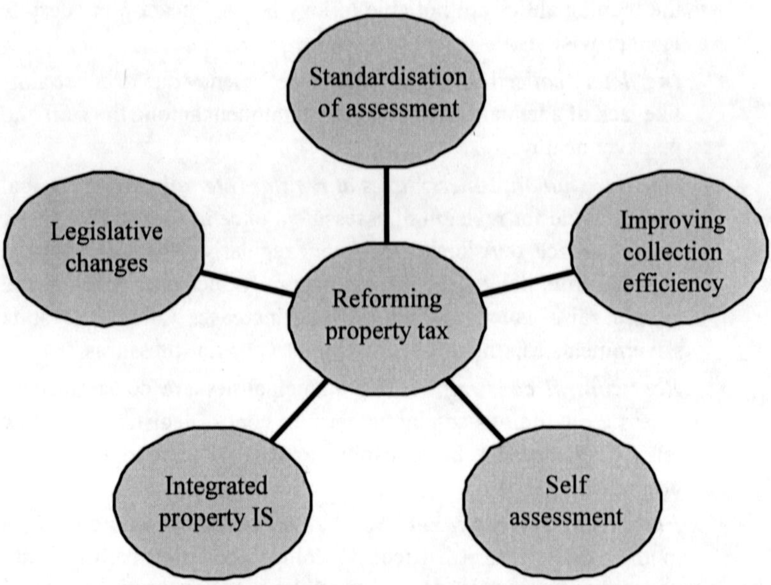

Fig. 5.1: Components of Property Tax Reform.

Standardisation of Assessment

The assessment of lands and buildings for property tax purposes is done either based on Annual Rental Value (ARV) or the Capital Value (CV). The CV method is mostly used in respect of the vacant lands, non-residential properties and some special buildings which are not ordinarily let out. The capital value is arrived at based on market value of properties. The ARV method of assessment is widely applied in respect of the residential buildings. Main problem with the ARV method is suppression of actual rents by various methods like payment of premium, multiple leases, separate charging for fixtures, etc. In the absence of transparent rental market, the ARV is mostly based on notional rental values of similar properties. The notional rent is often misleading due to subjective element involved, providing wide scope for exercise of discretion by the assessors. The net result is collision between the tax assessor and tax payer to the disadvantage of the municipalities. Both, ARV and CV systems suffer from systemic problems arising out of dominance of informal property market in India and therefore, are based on presumptive notional rents or under quoted market values. The option left is to standardize the assessment methodology by making the system rational, measurable and transparent.

Standardisation makes the assessment system scientific, simple and transparent. Standardisation aims to rationalise the system by way of setting up certain measurable norms in the assessment procedure irrespective of the basis of assessment, i.e., rental value or capital value. In past, most of the cases of property tax assessment were struck down by courts on the grounds of equity and arbitrariness. Major challenge for the standardisation process, therefore, is to evolve a system that can withstand the rigours of law and ensure smooth implementation.

The area-based approach provides a way out to these problems. The state and the local governments are experimenting with various models of area-based taxation. Area-based valuation, in simple terms; means assessing the properties with reference to tangible attributes viz., plinth area/ carpet area, location, type of building and use of building.

Factors of Standardisation

The values of properties – either residential or non-residential, primarily depend on three tangible factors viz., location, type and use of the building. Standardisation is achieved by fixing unit value and proper classification of these factors (Fig. 5.2).

- ***Location Factor:*** Location value is captured either by zoning or type of the road on which the property is located. Ideally, zoning coupled with type of road can be considered. In zoning, the entire city/town will be divided into certain zones based on homogeneity, availability of the infrastructure and civic facilities and administrative viability. Location advantage can also be captured by the type of the road on which the property is located viz., highways, main arterial roads, sub-arterial roads, bye-lanes, etc.
- ***Building Type Factor:*** Buildings are generally classified into categories like RCC- posh, RCC, tiled, thatched, etc., based on type of construction.

- ***Use Factor:*** Use of the buildings is generally classified into residential, commercial, industrial, etc.

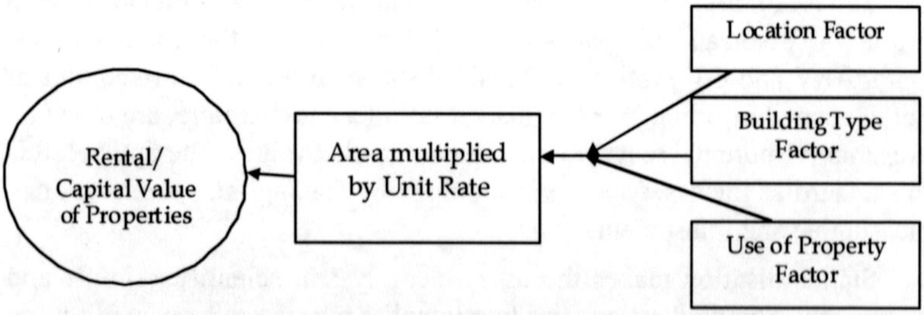

Fig. 5.2: Method of Standardisation of Assessment.

Plinth Area Based Property Tax — Andhra Pradesh Model

Property tax in Andhra Pradesh is a composite tax comprising of general purpose tax, water and drainage tax, lighting tax and scavenging tax.[8] The property tax assessment was based on two methods: Fair Rent as fixed under the Rent Control Act (RCA); and Annual Rent received or receivable on a hypothetical basis for the buildings not covered under RCA. There was considerable discretion exercised by the assessment officers which often led to discrepancies in the rental values fixed for similar properties under similar use in the same location. Equity, the basic tenet of taxation has been compromised due to the discretion enjoyed by the tax authorities.

The Government of Andhra Pradesh initiated state-wide reform of the property tax as back as 1993 quite successfully. Plinth area based tax was introduced in 1993 which made the assessment process rational and objective. The reform was initiated in 1989 and it was made effective from October, 1993 onwards in all the municipalities in the state except in Greater Hyderabad Municipal Corporation. The objectives of reform are to: evolve a scientific method in the assessment and levy of property tax incorporating the principles of equity, objectivity, fairness and simplicity; de-link the operation of Rent Control Act from the assessment of property tax; make tax administration efficient, transparent and effective in the levy of property tax; and improve buoyancy in the property tax revenues.

Features of the Reformed Model

The concept of plinth area has been under discussion for quite some time for reforming property tax in India. It is a simple and easily measurable criterion. It was introduced in Andhra Pradesh for arriving at the ARV of constructed buildings. The basis of assessment of property tax remained ARV but certain parameters viz., plinth area, location and type of building based on construction, age of the building and use were introduced to capture the unique characters of properties that determine the rental values in an urban setting.[9] It is also provided in the Act that in case of buildings which are not ordinarily let and the authorities cannot estimate the annual rent, viz., government buildings, railway buildings and others, the annual value of such buildings or lands will be the nine per cent of the land value and the cost of construction less depreciation for building at the maximum of nine per cent.[10]

The zoning was introduced to assess the location advantages of the properties. Under this method, the entire municipal area is divided into certain territorial zones with reference to expected rental values based on availability of civic amenities e.g., water supply, street lighting, roads, proximity to markets, shopping centres, educational and medical institutions, banks, offices, factories, etc. As rental values of properties differ based on the type of road on which they are located, sub-zoning is done with reference to type of roads.

The classification of buildings is done on two parameters: type of construction and the nature of use. The buildings located in each zone are classified into: RCC posh; RCC ordinary; Madras or Jack Arch or stone slabs or slate roofed buildings; Mangalore tile or Asbestos or G.I. roofed buildings; Country tiled buildings and Huts. Further, it is stipulated that the difference in rental value between the RCC posh and RCC ordinary buildings should not be more than 20 per cent. The buildings are further classified based on the nature of use into: residential; shops/shopping complexes; public use (office complexes, public and private offices, banks and ATMs, hospitals and nursing homes; educational institutions); commercial use (hotels, lodges and restaurants, and godowns, marriage halls, etc.); industrial use (factories, mills, workshops, etc.); and cinema theatres or places of public entertainment.

Assessment Methodology

All buildings located in a zone are classified based on the type of construction and nature of use. A maximum of 60 attributes of buildings can be identified in each zone based on the above criteria. These attributes are selected to capture more realistic rental values of properties. Based on a sample survey of 20 per cent properties, information relating to the prevailing rental values of various categories of buildings in various zones is gathered, and average monthly rental values per square meter of plinth area are fixed. In respect of multi-storied buildings, rental values are fixed separately for the cellar, ground and first floor and other floors. These unit rates are basically presumptive in nature although based on sample surveys. The unit rates are published for wider circulation across the local people inviting objections. After duly disposing of the objections, the unit rates are notified officially. The tax authorities simply apply these rates for assessing the properties.

The deductions are allowed from the Annual Rental Value (ARV) attributable to the building in lieu of all allowances for repairs while assessing the properties. The deductions vary between 10-30 per cent based on the age of the building. Further, rebate of 40 per cent of the ARV is also allowed in respect of the residential buildings occupied by the owner inclusive of the deduction permissible towards the age of the building.[11] Further, owner occupied residential properties are exempt from tax, if the ARV is less than Rs 900 in municipal corporations, Rs 450 in selection and special grade municipalities and Rs 300 in case of other municipalities. In respect of houses constructed through the agencies of State Government under Weaker Section Housing Programmes, the municipality has to collect one rupee per every half-year towards property tax. An additional deduction of five per cent of ARV attributable to the building is allowed in respect of municipalities and municipal corporations located on the sea-shore. The rates of tax are 25 per cent for residential and 33 per cent for non-residential properties. In respect of vacant lands, the rate of tax is 0.20 per cent of the market value of lands in municipalities and 0.50 per cent in case of the municipal corporations.

Another important reform introduced in Andhra Pradesh is the provision of penalty. The penalties are of two types. In case of the unauthorised buildings, the penalty is 25 per cent till they are regularised. In respect of the late payments, the penalty is two per cent per month for not paying tax within 15 days from receiving the demand notice.

Impact and Sustainability of the Reform

The property tax reform introduced in Andhra Pradesh could be considered a success from the point of standardisation and tax payer acceptability. It developed tax payer's confidence due to the rational and transparent method adopted and also provided guidelines for tax assessing officials. A respectable relationship emerged between the tax-payers and the municipal tax authorities. There was four to five fold increases in the property tax which led to tax payer resentment. The government while yielding to the pressures of tax payers restricted the increase in tax after revision to 100 per cent in case of properties whose tax was revised prior to 1-10-1983 and to 75 per cent in case of properties revised after 1-10-1983.

The general revision of property tax took place in 2002 after a gap of about 10 years. In this revision also there was higher increase due to longer gap and the government resorted to capping the tax increase, a regular practice. In case of the residential properties, the maximum increase was limited to 75 per cent of previous tax and for the non-residential properties; the limit was 100 per cent for 25 years old buildings and 150 per cent for more than 25 years old buildings. Again the general revision was done in the state in October, 2007 but it was confined to the non-residential properties due to protests from the people. In every revision, the plinth area-based assessment was conducted and there was increase in the revenues due to rising rental values.

The reformed model also received the backing of the judiciary. The High Court while upholding the provisions of the Rent Control legislation wherever applicable, concluded that the area based taxation as being implemented in Andhra Pradesh is right in so far as it maintains objectivity and rationality[12]. Later, the Supreme Court also upheld the new assessment procedure[13] and observed that the Act and Rules provided a complete code for assessment of the property tax to be levied on buildings and lands and the provisions of the Rent Control Act should not hinder the fixation of property tax. The model adopted in Andhra Pradesh primarily focuses on administrative process. The system became more transparent and equitable. There was also some immediate revenue improvement. But, the model has limited success with respect of capturing of real market values of individual properties. The system is based on averages arrived by limited surveys. In the absence of regular revisions, the impact of the reform has dried-up from buoyancy point of view. In spite of these limitations, the reform has sustained for more than two decades mainly because of the institutionalisation process.

Plinth Area Based Property Tax—Tamil Nadu Model

The municipalities in Tamil Nadu have been following the Annual Value Method of property taxation. The annual value is deemed to be the gross annual rent at which they may reasonably be expected from month to month or from year to year less a deduction from the cost of building of 10 per cent for maintenance. In case of buildings not ordinarily let, the annual value shall be deemed to be of the total estimated value of land and the estimated present cost of erecting the building less 10 per cent depreciation.[14] The main problem has been the subjectivity and discretion available with the assessing officers under these provisions. To overcome this, and to ensure objectivity in the levy and assessment of property tax, the state government issued guidelines, which were adopted by the executive authorities of municipal councils in the general revision conducted by the municipalities with effect from 01.10.1993. The guidelines in simple adopt area-based assessment similar to Andhra Pradesh. The major features of the reform are:[15]

- Division of the town/city into viable zones- A, B and D
- Fixation of basic values for different zones for purpose of fixation of ARV of buildings and lands. The basic value is probable rental yield per sq. ft. per month of residential properties. The basic values were fixed based on the rental values obtained for newly constructed RCC buildings measuring 1000 sq. ft., and converting these into rental values per sq. ft.

- Next stage was to allow discount based on the age of the building. The discount is nil for the buildings of five years age; for buildings between 5-15 years, it is 10 per cent; for buildings between 15-20 years, it is 15 per cent; and for the buildings above 25 years old, it is 20 per cent.
- A discount of 30 per cent is also allowed in respect of self-occupied buildings.
- As the basic values are fixed in case of RCC buildings as a standard, a discount is allowed in case of the other buildings. A 25 per cent discount is allowed for tiled, AC sheet and GI sheet roof buildings and 50 per cent discount in case of thatched roof buildings.
- On the same analogy, appreciated values are applied in case of non-residential buildings. In case of buildings used for industrial purpose, the basic value is double that of residential buildings and in case of commercial uses like nursing homes, Kalyana Mantapams, etc., the basic value will be thrice that of residential buildings.

The general revision took place again in 1998 in Tamil Nadu. The guidelines, which were issued in 1993, were actually operationalised during the 1998 revision. A model format of assessment sheet was circulated to help the assessee in providing the necessary information properly. The format was prepared to facilitate calculation of property tax by them and also to make the whole process computerised. A ceiling on enhancement of tax was imposed in 1998. The ceiling on increase due to revision was 50 per cent in case of owner occupied residential buildings, 100 per cent for rented buildings, 150 per cent for commercial buildings and industrial buildings.[16] The ceilings imposed on the rise though benefited the existing buildings; it led to wide gap between the old and new assessments. The solution to this problem lies in the conduct of general revision regularly as postponement always leads to high tax incidence due to revision and the consequential taxpayer protests.

The second State Finance Commission (SFC) recommended several measures to overcome these problems and improve yield from the property tax. Important suggestions include: ensuring that all properties are assessed through regular surveys and inspections; issue of assessee cards for all tax payers to maintain a record of tax payment; computerising the entire tax administration; introduction of self-declaration to make tax administration simple, transparent and assessee friendly; constitution of a guideline committee at the municipality level to rationalise the assessment and zoning process and zonal rates; levy of two per cent fine per month for not filing or wrong filing of returns; bringing a parity between the old and new assessments in the next general revisions; and there should be no time limit for proceeding with legal charges for recovery of tax (at present the statutory limitation is six years) except the provisions under the Limitations Act that is, 12 years.[17]

Capital Value Based Property Tax—Karnataka Model

In Karnataka, the system of property tax assessment was based on ARV till 2000. The tax was being assessed based on the expected ARV and it was calculated as per the discretion of the assessing officer as the actual rental value was difficult to get in case of self-occupied properties and special properties like hotels, nursing homes, etc. Often there was criticism and allegations of corruption against the assessing officers as the system of arriving at the rental value was mostly invisible, notional and subjective. In view of these problems, a major shift was made in the levy of property tax in Karnataka. The Capital Value based Self-Assessment Scheme was introduced in both, municipalities and municipal corporations with effect from 19.11.2001. The Karnataka Municipalities Act and the Karnataka Municipal Corporations Act were amended thereby, giving a legal backing to the reform. The existing system of assessing properties based on ARV was replaced by the Capital Value System. There are two aspects to this reform—one relates to the method of assessment and the other self-assessment.

Assessment Methodology

Area-based approach was adopted to assess the properties based on capital values. The property tax is levied on a percentage of the taxable value of the property calculated with reference to location, area of land, plinth area, type of building, age of the building and nature of use. The capital value of the building includes built-up area as well as the site. The property on which building is not constructed is called a vacant land. Property tax is levied on constructed properties as well as vacant lands.

Vacant Lands: Imposition of tax on land and building comes in the purview of the state government. The state government had taken responsibly to levy tax on urban land as municipalities failed to levy the property tax on vacant land effectively. In order to levy tax on urban vacant land, the state government enacted the Karnataka Urban Land Tax Act. Provision was made in this Act to collect the tax at one per cent (Rs 1000 per lakh rupees) of the actual market value of urban vacant land. The urban land tax officer could levy the tax after conducting the spot examination.

After the 74th Amendment to the Constitution, with a view to give more taxation powers to municipalities, the Karnataka Urban Land Tax Act was repealed in 2000 and municipalities were delegated with the power of levying tax on vacant land. Currently, the capital value of the vacant land is based on the estimated market value of the land as published by the committee constituted under Sec. 45B of the Karnataka Stamp Act. The capital value is simply arrived at by multiplying area of the land with the value of the land in sq. ft./sq.mt.

Buildings: The taxable capital value of the building is determined including the land occupied by the building as per the following formula:

CVB = (CC X PLA) – (CC X PLA X DQA) + LV X PLA

Where, CVB = Capital value of the building including land

CC = Cost of construction per sq.ft. / meter as published

DQA = Depreciation Quotient approved by PWD

PLA = Plinth area of the building/area of the land occupied by the building

LV = Estimated Land Value per sq. ft. or sq. mtr.

The capital value of the building is calculated by subtracting the depreciation out of the cost of construction of building and by adding the capital value of the land on which the building is erected. The land appurtenant to the building shall be exempt from levy of property tax. A rebate of 50 per cent of the property tax is allowed for the owner occupied residential properties. Further, a rebate of five per cent is allowed if the owner or occupier files his returns and also pays the tax due within one month from the date of commencement of the financial year.[18] The following rates of tax are applied in the municipalities and municipal corporations in Karnataka.

Commercial Buildings: Between 0.5 per cent (Rs five per thousand) and two per cent of (Rs twenty per thousand) taxable capital value. The maximum limit of property tax in case of commercial building situated in a municipal council whose population is below one lakh shall be 0.9 per cent.

Residential Buildings: Between 0.3 per cent (Rs three per thousand) and one per cent (Rs ten per thousand) of taxable capital value. The maximum limit of property tax in case of a residential building situated in a municipal council whose population does not exceed one lakh shall be 0.6 per cent.

Vacant Land:

- For the first 10,760 sq.ft.- 0.1 per cent of taxable capital value.
- From 10,760 sq. ft. to 43,040 sq. ft – Rs 25 for every one lakh of taxable capital value.
- Above 43.040 sq. ft. –Rs 10 for every one lakh of taxable capital value.
- Further, it is provided that no vacant land tax shall be levied on the vacant lands situated within the municipal council whose population does not exceed one lakh.[19]

The municipality determines the rate of tax and publishes the same. The taxpayer will have to assess his property and file tax returns under the newly introduced Self Assessment Scheme (SAS) of Property Tax. There was a sharp rise in the property tax due to the introduction of the new system. Expectedly, the state government provided some concession for the first three years. Accordingly, the new tax should not exceed two times the property tax including all cesses. A penalty of two per cent of the property tax per month shall be levied for the payment made after the due date.[20] In spite of many concessions, there was over hundred crore revenue increase in respect of municipal corporations in the state. But, the increase was negligible (thirty lakhs) in respect of the other municipalities. Another important feature of the new property tax system relates to periodic revision.[21] It was stipulated in the Act that the property tax would be automatically enhanced by 15 per cent once in three years, provided there is no addition or alteration or variation to the building. The municipal council/corporation can enhance the rate upto 30 per cent also if it decides so.

Area-based Model—Bihar[22]

For the purposes of levying Holding Tax (Property Tax in Bihar is termed as Holding Tax), the holdings in the municipal area are classified on the basis of:

(a) Situation of Holding
- (i) Holdings on the Principal Main Road,
- (ii) Holdings on the Main Road,
- (iii) Holdings on roads other than (i) and (ii)

(b) Use of the Holding
- (i) Purely residential,
- (ii) Purely commercial or industrial,
- (iii) Partly residential and partly commercial/residential
- (iv) All holdings other than (i), (ii) and (iii)

(c) Type of Construction
- (i) Pucca building with RCC roof,
- (ii) Pucca building with asbestos/corrugated sheet roof,
- (iii) All other buildings not covered under (i) and (ii)

(d) Type of Occupancy
- (i) Self-occupied
- (ii) Tenant occupied

(e) Type of non-residential use of Holdings
- (i) Hotels, restaurants, clubs, cinema houses, guest houses, marriage halls and all places of entertainment

(ii) Shops, showrooms
(iii) Commercial offices, banks, hospitals and nursing homes, dispensaries, laboratories
(iv) Government offices and institutions
(v) Industries, workshops
(vi) Schools, colleges and other educational institutions, research institutions
(vii) Educational and social institutions run by charitable trust on no-profit no-loss basis for the benefit of poor, physically challenged, social security of women and children
(viii) Religious places
(ix) Any other holding not covered under (i) to (viii).

For the purpose of calculation of Annual Rental Value (ARV) of a holding measurement of total built-up area is calculated as under:

(i) Individual Residential Property— 70 per cent of total built-up area
(ii) Individual Non-Residential Property— 80 per cent of total built-up area
(iii) Individual Residential Multi-storey buildings/apartments— 70 per cent of total built-up area
(iv) Individual Non-Residential Multi-storey buildings/ apartments— 80 per cent of total built-up area

The rate of rental value per sq.ft. is fixed by the municipality with the prior approval of the state government having regard to the situation, use, type of construction, occupancy of the buildings and type of non-residential use of buildings. The rental value of the built-up area is increased by minimum 15 per cent every five years when fresh assessment takes place. The municipality may also increase the rental value and rates any time during the five year period with the prior approval of the government. The Annual Rental Value shall increase 1.5 times in case of buildings on rent. The state government may provide differential methods of calculation of annual rental value of certain categories of non-residential holdings. Those portions of the buildings which are places, centres and institutions of spiritual and religious nature are used for commercial purposes, office buildings, boarding and/ or lodging facilities whether free of charge or whether fee or donation is charged shall be charged property tax. However, all hutments or dwelling units situated in slums having a total plinth area of less than 250 sq.ft. are exempt from payment of property tax. Any property that has adopted the technique and structure of rainwater harvesting may be given relief to a determined percentage as the government prescribes. Annual Rental Value of non-residential holdings is calculated by using multiplying factors given in Table 5.1.

Table 5.1: Multiplying Factors for Non-Residential Buildings

Sl No.	*Type of non-residential holdings*	*MultiplyingFactors*
i.	Hotels, bars, clubs, health club, marriage halls	3
ii.	Shops (with less than 250 sq.ft. area)	1
iii.	Shops (other than (ii)), showroom, shopping malls, cinema houses, multiplexes, dispensaries, laboratories, restaurants, guest houses	1.5
iv.	Commercial offices, financial institutions, banks, insurance offices, private hospitals and nursing homes	3
v.	Industries, workshops, storage, godowns, warehouses	2
vi.	Commercial establishments and undertakings of State and Central government	2
vii.	Coaching classes, guidance and training centres and their hostels	1.5
viii.	State and Central government offices Other than their commercial establishments and undertakings	1
ix.	Private schools, private colleges, private research institutes, other private educational institutions and their hostels	1.5
x.	Places, centres and institutions of spiritual and religious nature	0
xi.	Educational and social institutions run by the charitable trusts on no-profit no-loss basis for benefit of poor, physically challenged, social security of women and children	1
xii	Any other holdings not covered under (i) to (xi)	1.5

Property Tax is levied by the municipality at the rate of a minimum of nine per cent and a maximum of 15 per cent of the annual rental value. Property tax on vacant land can be levied annually as per rate given hereunder depending on the situation. However no such tax shall be levied within two years from the date of acquisition of land by the owner. The rates are furnished in Table 5.2.

Table 5.2: Rate of Property Tax

Type of Municipality	*Rate per sq.metre (Rs.)*		
	Principal Main Road	*Main road*	*Other Roads*
Municipal Corporation	5	4	3
Municipal Council	4	3	2
Nagar Panchayat	3	2	1

Area-based System—Uttar Pradesh[23]

The U.P. Municipal Corporation Act, 1959 and the Uttar Pradesh Municipalities Act, 1916 provide for levying Property Taxes as a major tax by the municipalities in the state. The valuation of properties for tax purposes is done keeping in view the location, usage and type of construction of a building. The unit rates are determined by the municipalities every two years. For assessing the tax liability, maximum rates have been prescribed by the state government. The property taxes consist of: General Tax; Water Tax; Drainage Tax leviable in areas provided with sewer system; and Conservancy Tax in areas in which the municipality undertakes the collection, removal and disposal of excrementitious and polluted matter from privies, urinals and cess-pools.

The system of deriving Annual Rental Value (ARV) has been switched over from letting value to area based method (based on Patna Model). The valuation of a property is made on the basis of location, construction type and use. The rates for the constituent components are 10 to 15 per cent, 7.5 to 12.5 per cent, 2.5 to 5 per cent and not more than 2 per cent respectively, subject to total tax ranging between 22 to 32 per cent of Annual Rental Value (ARV) of a building. The non-residential properties were earlier taxed on the basis of capital value, but it has now been changed and the ARV of industrial and commercial properties is multiple to that of the residential buildings. The tax-payers have been provided the option of self-assessment as well. A tax-payer may file his tax return on the prescribed format after valuating his property and assessing the tax and pay tax directly. The response is quite satisfactory and the municipalities aim to cover hundred per cent under self-assessment scheme. A number of exemptions based on life of building and usage are granted to the assessees.

Plinth Area-based Model—Jharkhand[24]

For the purpose of levying the property tax, the holdings in municipal area are classified on the following criteria:

(a) *Location of the holding*—
 (i) holding on the Principal Main Road
 (ii) holding on the Main Road
 (iii) holding other than above;

(b) *Use of the holding*—
 (i) purely residential
 (ii) purely commercial
 (iii) purely industrial
 (iv) holding other than the above;

(c) Type of construction—

(i) pucca buildings with RCC or R.B. roof
(ii) pucca buildings with asbestos or corrugated sheet roof
(iii) buildings other than the above.

For the purpose of calculation of Annual Rental Value of a holding, the rental value per sq.ft. of carpet area for different classes of holdings is determined by the municipality with the prior approval of the state government. Measurement of carpet area is calculated as under:

(a) rooms—full measurement
(b) covered verandah— full measurement
(c) balcony/corridor, kitchen and store— fifty per cent measurement
(d) garage—one-fourth measurement
(e) area covered by bathroom, latrines, portico and staircase is exempted

The rate of tax is revised by the municipality every five years or earlier with the prior approval of the state government.

Built-up Area Basis— Madhya Pradesh[25]

Property tax is levied on lands and buildings at the rate not less than six per cent and not more than ten per cent of the annual letting value. The annual letting value of any building or land is determined by the municipality on the basis of per square meter of the built up area of a building or land taking into consideration the area in which property is situated, its location, situation, purpose for which it is used, its capacity for profitable user, quality of construction of the building and other relevant factors.

The property tax is not levied on buildings and lands owned by or vesting in the Union Government, State Government, the municipality; buildings and lands the annual value of which does not exceed six thousand rupees in case of the municipality having population of one lakh or above and four thousand eight hundred rupees in case of municipality having population less than one lakh; buildings and lands or portions thereof used exclusively for educational purposes or are owned by the educational institutions or have been placed at the disposal of educational institutions without any rent; public parks and play grounds which are open to the public and building and land attached thereto; buildings and lands or portions used exclusively for public worship or public charity and public burial grounds and other places of disposal of dead; buildings and lands owned by widows or minors or differently-abled persons where the main source of livelihood is the rent derived from such building and land; buildings and lands owned by freedom fighters, retired members of Defence Services and their widows during their life time if they are exempted from income tax; buildings and lands owned by blind persons, abandoned women and mentally incapacitated persons if the main source of their livelihood is the rent derived from such buildings and lands; buildings and lands in occupation of the owner for his residence are exempted to the extent of fifty per cent; the electric pole; and the property owned by recognised political party in the state.

Self Assessment System

Emergence of self assessment models is an important phenomenon which makes taxpayers part of the reform process. The self assessment system is expected to build better taxpayer-municipality interface. The taxpayers are also expected to become more tax compliant and the municipalities will get relieved of certain duties. The provisions of standardisation of assessment are inbuilt into the self-assessment systems to guide the taxpayers who are supposed to submit tax returns on their own. If Karnataka has introduced compulsory self assessment scheme with legal provision, others are implementing it as an alternative option. The models implemented in the Karnataka state and, Kanpur, Hyderabad, and Bangaluru Municipal Corporations are discussed to examine the efficacy of the system.

Self-assessment Scheme in Karnataka

Under the Self-Assessment Scheme (SAS), the owner or occupier is liable to submit the returns every financial year. He has to pay full amount of the self-assessed tax payable in advance and submit the tax payment receipt along with the returns before 30th June of every financial year.[26] The taxpayers should calculate the tax for their properties and file the returns. The municipality issued guidelines and provided the information to facilitate self-assessment, viz., the rate of tax fixed and published by the municipality; estimated market value of land published in each ward or locality or street as per the Stamp Act; types of building and their estimated cost of construction; depreciation quotient based on the age of the building; rate of tax with reference to nature of use of the premises; property tax return form and the method of submission; name of the bank receiving tax payment and the challan form; and procedure for giving rebate to the tax-payers paying tax within the prescribed time.

In order to make the self-assessment system effective, penalty provisions are also made in the Act itself. If the owner or occupier fails to submit the return or pay advance tax or submits false information, the Municipal Commissioner or the Chief Officer shall assess the property. While making such an assessment, the authorised officer may also impose a penalty as below:[27]

- In case of failure to submit returns and pay advance tax: two per cent of the property tax due

- In case of submitting false information in the returns: not exceeding two times of the difference between tax assessed and the tax paid along with the returns.

The self-assessment scheme is now operational in all the municipalities in Karnataka. The legal backing has facilitated the process. The Directorate of Municipal Administration is monitoring the implementation of the scheme and the progress made periodically. The reformed model of property tax in Karnataka is being implemented at slow pace since about a decade as the municipalities are gradually adapting to the new system. The burden on the municipality is a bit reduced due to introduction of self-assessment. The decade long experience shows that tax payer compliance is not total. The allegations are that the officials are taking very long time for disposal of the Self Assessment System forms.

Apart from these teething problems, the new system has improved transparency and equity across the similarly placed tax payers. The assessment system has become more objective and rational. The GIS-based survey was also done to identify un-assessed properties to support the SAS system. Both these initiatives have had a good impact on improving the coverage of assessments, and revenue yield while making the system more transparent and equitable. However, the administrative aspects still need streamlining for reaping the benefits totally. The buoyancy also improved to some extent due to tax compliance and better coverage.

Limitations of Self Assessment System

Self Assessment, no doubt, has certain advantages for the taxpayers as well as the municipalities. But, the important challenge for implementing this system is taxpayer compliance. Further, the assessment machinery at the municipality level has to be cooperative and efficient in completing the process of assessment in time. The common complaint has been that the assessment process is slow consuming lot of time which tests the patience of the tax-payers losing their interest in the system. The Karnataka model of self assessment through legal backing is preferred as it will have some built-in features for smoother implementation. The experience shows that the municipal bureaucracy appears to be very slow in disposing of the self-assessment returns. The self assessment returns needs to be scrutinised and cross-verified by the officials for accuracy of the information. If these bureaucratic issues are tackled, the system can work well. In the Indian situation, the self assessment system can only supplement the existing system but cannot replace it as is being experienced in Uttar Pradesh.

The reforms of property tax in the states reviewed here have a common feature that is, resolving official discretion and equity issues which often lead to legal litigations. A number of building attributes are identified with reference to type of construction and use of the buildings to capture realistic market values in standardised way. Another advantage has been gaining peoples' acceptability through a process of standardisation. Though there was improvement in the revenues especially during the initial periods, the buoyancy aspect was secondary.

Reforming Property Taxation in Kanpur Municipal Corporation[28]

Property tax reforms in Kanpur Municipal Corporation (KMC) in Uttar Pradesh established the system of scientific, transparent and self-assessed system of tax collection, based on rational parameters which are now acceptable by the citizens. This reform has worked in way that the whole system became responsive to citizens as well as to the service provider.

To streamline the whole process, KMC took the initiative of property tax reforms in its jurisdiction. To start the process, KMC required the latest and accurate database so that they have the exact number of properties. The previous data available was ten years old and additional properties had been constructed. These new constructions were mostly not paying any tax and most of them were illegal and therefore, the municipal authorities were not able to collect property tax from them which resulted in tremendous loss of revenues for KMC.

KMC involved the Indian Institute of Technology (IIT), Kanpur in this reform process. After a series of meetings with IIT Kanpur, it was decided that GIS mapping of the property can sort out this issue for KMC. Funds available under Urban Reforms Incentive Funds (URIF) were used to kick start the reform.

Goal of the Reform Initiative

The goal of this reform was the improvement of the performance of property tax system, tapping the potential, and making it an important source of revenue for financing and maintaining municipal services and infrastructure.

This was attempted to achieve by fulfillment of some of following objectives:

- Modernisation of property tax system by linking GIS and property data base.
- Reassessment of all properties on unit area basis.
- Improving rentals of municipal properties by amendment of Rent Control Act.

Implementation Strategies

No proper map of the Kanpur city or the areas of tax collection were available. Nobody actually knew how many taxable property units were there in Kanpur metropolitan area because of constant construction work. The tax collectors (Revenue Inspectors) had minimum training to use any computer assisted methods and the existing assessment method was extremely tedious resulting in a growing number of un-assessed property. To implement the reform and to get the revenue collection on a systematic track, it was necessary to agree on the following:

- A new base map was needed to make it possible to identify all taxpaying units (houses).
- A method of identification of the taxpaying units needed to be developed.
- A supporting Geographical Information System (GIS) Base was needed to be developed.
- The revenue collection needed to be reorganised to utilise the GIS application.
- The staff needs to be trained to utilise the new system.

KMC issued notice and tender regarding this reform, conventional method of preparing a base map was adopted and survey was started with enthusiasm to know the existing situation of the city. Once the survey was in its initial stage, there were protests and agitations from the citizens which took place because a message went to citizens that the government was trying to set up a new mechanism wherein the tax would increase manifold. This wrong impression resulted in snatching of instruments from surveyors and other forms of aggression and non-cooperation from the people. Though mass awareness programmes were taken up through the mass media, the survey had to be stopped resulting in loss of instruments and injuries to the surveyors.

Even the elected representatives opposed the tax reforms and survey done by KMC, but KMC tried to convince them with logic and broaden their vision for future profit and thus, the survey restarted. This time with police protection and with the help of mass media, the surveyors reached every nook and corner of the city. Extensive media campaign was done which helped in eliminating the communication gap between the KMC and citizens, and finally citizens were assured that the reform was for their benefits.

As a part of the e-governance initiatives, KMC envisaged development of an online web and GIS based property tax calculation information and payment system as a pioneering effort in the country. The scope of this work covered the follwing area for the major activities:

(i) Creation of a GIS based spatial property database involving:

  - Procurement and supply of Quick bird Image (0.6 m resolution) for the entire area under the KMC jurisdiction.

- Geo-rectification of the satellite image form the GCPs collected from DGPS survey.
- Updation of digital GIS maps of KMC with the help of geo-corrected Quick bird data and preparation of data for assessment list.

(ii) Conduct a property level survey for every property unit.

(iii) Development of Informative and Interactive web GIS system for online property tax calculation and payment.

(iv) Development of desktop application for property database repository and tax assessment.

(v) Supply and Installation of software, hardware and networking with server and additional peripherals.

(vi) Training of KMC Personnel.

(vii) Re-numbering of properties with unique premises no. and fixing of house number plates.

Outcome of the Reform

The outcome of the reform was very productive as the number of properties covered under the tax net increased manifold and added to the revenues of the KMC. The increased revenue was used for infrastructural development.

- Information Availability
  - (a) An overview of the City
  - (b) On-line property tax calculation
  - (c) Hassle free online property tax payment facilities
  - (d) A facility to access different maps based on Zones, Wards and Property
  - (e) GIS-based spatial searching and identification of properties
  - (f) Monthly per sq. feet rates used for calculation of property tax
  - (g) Circle rates of property used for the calculation of property tax for non-residential properties
  - (h) Ward-wise/Property-wise maps with civic infrastructure details like roads, lanes, by lanes, parks, electric poles, telephone poles, railway lines, hotels, hospitals, police stations, schools, cinema hall etc.
- Process Re-Engineering
- Information Collection system changes as follows:-
  - (a) Kanpur City Level information collection system

(b) Kanpur City level information especially the basics of the city and landmark are available on-line

(c) Digital Maps of the City is available

(d) The city map which was previously on tracing paper and ammonia print is available in 1:1000 scale in digital format with details up to the property level

- Property Tax Data

(a) Property tax data now is globally available through website

(b) Property Tax liabilities

(c) Existing status of paid taxes

(d) Annual taxes due to be paid

- Tax Calculation System

(a) Previously it was manually calculated

(b) Now software based calculation on the Unit Area Method and On-line tax payment is available through KMC web-site

(c) One can calculate and view stepwise calculation

- On-Line Birth and Death registration

(a) Birth and Death registration is now available On-Line

- Different type of reports based on above mentioned facilities is always available on server or on-line
- Citizen Centricity and User Convenience
- The functionalities in the web-site provide following services/ conveniences to stakeholders:

(a) On-line property tax calculation and view steps of calculation

(b) On-line property tax payment facilities

(c) A facility to access different maps based on Zones, Wards and Property

(d) GIS based spatial searching and identification of properties

(e) Ward-wise/Property-wise maps with civic infrastructure details like roads, lanes, by lanes, parks, electric poles, telephone poles, railway lines, hotels, hospitals, police stations, schools, cinema hall etc. The information is being used for R&D, planning of infrastructure and other projects.

- Transparency in Government Process

(a) Almost 100 per cent property tax coverage in the KMC area

(b) Transparency and speed in working

(c) On-line tax payment and status verification facility removes intermediaries

(d) Better revenues for investments into civic facilities

(e) Property details (attribute data) and GIS database (spatial data) available

- Time Saving/ Efficiency improvement initiatives

  (a) Reduction of time in surveying the area and mapping

  (b) Time saving in tax calculation

  (c) Better understanding of property tax calculation

  (d) User friendly application for day-to-day tax management

  (e) Web application for hassle free Tax calculation and payments

  (f) Creation of GIS-based spatial property database

  (g) Map-based tracking of property tax defaulters

  (h) Revenue generation efficiency

  (i) Renumbering of properties with unique premises no. and fixing of plates.

Base data generated during this reform has been used by the KMC at different fronts which includes citizen centric initiatives to make government more responsive for development purposes.

Achievements of Reform

This reform became a huge success in Kanpur to create an umbrella effect for covering large number of properties under it. The major achievements are as under:

(i) Increase in number of Properties

It is found that in every zone number of properties and also that of tax assessees have increased tremendously. In one zone number has just doubled. Surprisingly in one zone number has decreased. It needs further probing. However, overall the number of properties increased from 2.74 lakhs to 4.76 lakhs showing more than 70 per cent increase. The increase in number of assessees has been about 50 per cent. This is bound to make an impact on revenue income of the Kanpur Municipal Corporation.

(ii) Increase in Demand of Tax

Increase in number of properties has obviously resulted into corresponding increase in demand of property tax. Interestingly, tax demand in post-GIS era was found increased manifolds. The demand nearly doubled even in the zone recording decrease in number of properties. This has caused because of unprecedented increase in number of properties identified through GIS activity and assessed for tax purposes. The tax network has increased tremendously and coverage widened. This has led to increase in tax demand even without any increase in tax rates.

(iii) Increased Efficiency in Collection

The inclusion of all properties under tax net-work and adoption of area-based system of valuating properties has eliminated the element of discretion and it has resulted into efficient collection of property taxes. The tax-payers are coming forward to self-assessment of tax liability on their properties and paying tax in time. It has almost eliminated the tax-related litigations.

Greater Hyderabad Municipal Corporation (GHMC) –A Case

The system of property tax in Greater Hyderabad Municipal Corporation (GHMC) is similar to other municipalities in Andhra Pradesh. It is a composite tax comprising of general tax, water tax, drainage tax, lighting tax and conservancy tax. The tax can be levied even on unauthorised buildings but at the penalty of 20 per cent till they are regularised. The annual rental value of the buildings shall be deemed to be gross annual rent at which they may be reasonably expected to let from month to month or from year to year with reference to its location, type of construction, plinth area, age of the building and nature of use. A deduction of 10 per cent for buildings aged below 25 years and 20 per cent for buildings of above 25 years is also allowed while assessing the property tax. Further, a rebate of 40 per cent for owner occupied buildings. The rates of taxation for residential properties are fixed progressively from 17 per cent to 30 per cent based on ARV slab and for non-residential properties; the tax rate is flatly fixed at 30 per cent of the ARV. The Act however, prescribes 15 per cent minimum and 30 per cent maximum for residential properties.[29]

The Greater Hyderabad Municipal Corporation (GHMC) suffered heavily on financial front from two complex problems viz., a spate of legal litigations prolonging years together and non-conduct of the general revision of the property tax for decades together. The tax revision was pending for about 17 to 23 years in different circles. The tax-payers continued to pay the taxes fixed decades back leading to huge revenue loss to the GHMC. Further, there was rampant under-assessment and non-assessment of properties. It was estimated that the number of assessments were only 4.1 lakhs in 1998-99 with a demand of only Rs 49 crores as against the estimated number of properties of over 6 lakhs.

The plinth-area system, which was introduced in the other municipalities in the state in 1993, could not be implemented in the GHMC due to legal litigations for a period of over seven years.[30] In view of these problems and the continued loss of revenues to the GHMC, it was decided to take a positive stance and gain the confidence of the tax-payers by introducing a method of self-assessment of property tax. This method was conceived as an alternative and a way out to overcome the litigations. Interestingly, this was done within the existing legal provisions[31] which provide that a Municipal Commissioner may call for information like name of the owner/ occupier, dimensions of the building, etc. or returns from owner or occupier and the taxpayers are bound to provide the information / returns within the specified time.

The objectives of self-assessment were to: ensure complete transparency and openness in the levy and collection of property tax and to enable citizens/ tax-payers to understand the basis of taxation so as to calculate the tax by themselves; build a computerised property tax database with each property in the twin cities being assigned a unique Property Tax Identification Number (PTIN); promote equity in tax assessment; link services with tax payment so that tax-payers get value for money; minimise prolonged disputes between taxpayers and GHMC running over years; and establish a healthy relationship between GHMC and tax-payers.

All the property owners, occupiers, lessees, etc., are required to file the self-assessment returns in the prescribed form. The notification was published in newspapers along with the rationale for introduction of self-assessment, form and procedure to be followed for filing the returns.[32] Wide publicity was given to self-assessment scheme covering aspects like legal provisions, existence of low property tax in Hyderabad city compared to other cities like Ahmedabad, Bengaluru and Chennai, development needs of the city, assurance to the property owners that the Municipal Corporation would spend to the tune of two to three times of the tax paid in their areas on various services, especially in the low income areas, etc. Major property holders including bureaucrats, politicians, etc., were targeted in the initial stages. Wide publicity was given to model taxpayers as against publishing of the names of the professional litigants and habitual evaders in the newspapers.

Several teams of officials were formed to assist the taxpayers and clear their doubts and clarifications. Information was communicated to various categories of taxpayers – individual and groups like welfare associations. The GHMC made it clear that it will not harass the taxpayers if they provide correct self-assessment information and tax is paid as per the benchmark rates. All the assessment returns along with the unique PTIN were computerised and the rates adopted in these returns were given wide publicity to motivate others to follow the system. They were also put on website.

Though the Municipal Corporation had not prescribed any fixed formula for assessing the rental values by the taxpayers, certain guidelines were issued. The guidelines indicated that the properties were supposed to be assessed based on plinth area with reference to type of the building, age of the building and use of the building. The taxpayers were themselves to fix their existing market rental values in the neighbourhood for similar properties. No fixed rates were given unlike the methodology adopted in municipalities in the state. Under the Self Assessment Scheme, no stipulation was made by the Municipal Corporation as to at what rental values taxpayers should file returns. The resident welfare associations were educated regarding the legal provisions. The Municipal Corporation facilitated the process of formation of groups of resident welfare associations into federations. One of the resident welfare associations came with suggestions regarding benchmarks for acceptance of tax returns during 1999, 2000. The Association resolved that properties could broadly be divided into three categories: (a) high rent areas, (b) middle rent areas and (c) low rent areas.

The benchmark Monthly Rental Values (MRV) for acceptance of tax returns for the above categories were adopted by the association at one rupee, 0.60 paise and 0.40 paise respectively. These rates were accepted by many other resident welfare associations. The information on rates adopted by resident welfare associations were published in print and electronic media and newspapers. Once the MRVs were published, other resident welfare associations adopted the rates and filed self-assessment returns.

This voluntary method was adopted from practical or feasibility point of view, especially when the taxpayers have been paying at so low rates and tax-payers were taking recourse to law suits so frequently. The objective was to ensure that reforms take place in stages and that initial efforts do not bounce back. Accordingly, civic authorities were content with the position that at least minimum taxes would accrue. The honest tax-payers tempted to file returns without taking the risk of being caught and being subject to regular process of valuation with the risk of paying higher tax. The GHMC did not accept the system of slab rates adopted by other Municipal Corporations like Patna, Kanpur, Bengaluru and Chennai as slab rates are averages and averages are affected by extremes of urban properties especially in the cities like Hyderabad where areas especially commercial properties, are heterogeneous. It is argued that averages are representative only when the data is homogeneous and it may work in case of the residential properties to some extent but not in case of heterogeneous non-residential properties, which may turn out to be regressive.

The self-assessment scheme was only a beginning for reforming the property tax administration in Hyderabad city. The GHMC is now in the process of restructuring the assessment system based on the zoning and plinth area concept. The entire city is divided into 47 zones and 140 sub-zones for this purpose. The assessment methodology is similar to that of the system operating in the other municipalities. The reform is basically to formalise informally working system and further standardise the assessment method based on the experiences on the other metro cities. But, the self-assessment system was slowly discontinued due to operational problems and lack of commitment of the GHMC for the reform in the subsequent periods. After a long gap, the voluntary disclosure scheme was again introduced in July, 2013. But, a local media reported that the response has been very poor.[33] It was reported that less than 10 per cent of property owners opted for the scheme. The reasoning given was that approximately 1000 building permissions were given in a month in GHMC whereas the new assessments were just 100 only per month. Lack of proper publicity to the scheme and diversion of the tax department staff to other activities are considered the main reasons for poor response.

The GHMC experience shows that the informal methods cannot sustain unless proper systems are put in place while institutionalising the reforms. Though there was some improvement in the revenue yield immediately, the reform as such did not pick-up and sustain. The GIS survey is now attempted in some areas. The reform approach has been ad hoc and piece-meal lacking any comprehensive perspective. It appears that the authorities are content with the incremental increase of revenues by mere administrative efforts rather than attempting a comprehensive reform of property tax in GHMC.

Brihat Bengaluru Mahanagara Palike (BBMP) System

The method of assessment in Bengaluru city was based on the Annual Rental Value (ARV). The ARV of a property was the gross annual rent at which the building or land may reasonably be expected to let from month to month or from year to year.[34] As there were no standard guidelines to arrive at the annual rental value of properties, official discretion, citizen's dissatisfaction and revenue leakage were common in the assessment. It is said that this system neither helped the citizens nor the Municipal Corporation. Added to this, there was no general revision of property tax since 1972 despite the legal provision to conduct revision every five years. All this contributed for very low yield from the property tax.

The BBMP migrated to the self-assessment system in 2000 even before it was introduced in other municipalities in the state. But, the self-assessment was not a mandatory provision like the other municipalities. Further, self-assessment returns were called-for only once. Thereafter, the tax-payers were paying the tax and obtaining receipts. The officials feel that there is no need for filing returns every year as it only adds to paper work.

Main features of the restructured property tax assessment are zoning and classification of buildings. The entire city is divided into six zones (A to F) based on the valuations done by the surveys of its revenue department. The cost of construction at the current market rate and other factors are taken into consideration for classification of properties. In order to facilitate the taxpayers, the BBMP prepared a handbook incorporating the main features of the new scheme which came into effect from 2000-01 and the method of assessing their properties in simple terms. The basis for assessment is built-up area, the unit being sq. ft. Unit rates were fixed based on the zone with reference to category of the building, use of the property and occupation that is, tenanted or owner occupied. The properties are categorised as below:

Residential: RCC/Madras Terrace with more than Rs 250 per sq. ft; RCC/ Madras in the range of Rs 150-250 per sq. ft; RCC/Madras with less than Rs 150 per sq. ft; Tiled with less than Rs.150 per sq.ft.; and Thatched house/ hut estimated to be of Rs 40 per sq ft. value.

***Non-Residential-Group A*:** Commercial complexes, offices, banks, shops, etc.; hotels, pubs, restaurants, lodging homes, etc.; clinics, polyclinics, diagnostic centres, laboratories, etc.; sports centres, gyms, etc.; student hostels, educational institutions, etc.; and other non-residential uses not covered in above categories.

Non-Residential-Group B: Star hotels; cinema theatres—again sub-classified into four categories based on quality of construction and facilities like AC; touring theaters as separate category; kalyana mantapams, community halls, etc.; industrial buildings sub-classified as large scale, medium scale, small scale and tiny industry; and non-government hospitals and nursing homes.

The total built-up area had to be multiplied with monthly rates fixed to arrive at the monthly ratable values and this monthly ratable values were multiplied by ten to arrive at the annual ratable value. If the vacant land exceeded three times of the built-up area, such land had to be assessed at 30 per cent of the rate fixed for built-up area of such property. Unit rates for owner occupied properties were fixed at 50 per cent less than the rented properties both, in respect of residential and non-residential properties. Depreciation was allowed based on the age of the buildings. The rates ranged between ten per cent for five years buildings; with five per cent increase for every five years upto 55 years; and 70 per cent for the buildings exceeding 55 years of age. The rate of tax was 20 per cent for residential use and 25 per cent for non-residential use. On property tax, 34 per cent cess was added— education cess (10 per cent), health cess (15 per cent), beggary cess (3 per cent) and library cess (6 per cent).

In 2002, the KMC Act was amended to bring CV method on par with other municipalities. But the taxpayers resented it. Therefore, it was abandoned. Later, a new system was introduced. The Karnataka Municipal Corporation Act was amended and a new section called 108A was inserted. The new system of assessment called Unit Area Value (UAV) was introduced on 10th February, 2008 (from 2008-09 onwards). The UAV is the average rate of expected returns from the property per sq. ft. per month on the basis of the average market rate determined through mass appraisal method or real estate market information or any other reliable source or combination of these sources.

The unit rates are determined with reference to location, type of construction of building or parking area in non-residential buildings, etc. Zoning system is continued like earlier system.[35] Two factors are taken into consideration to arrive at UAV viz., pre-determined value per sq. ft. of built-up area depending on location of the property and, cost of construction of building depending on roof type or its usage. The residential buildings are classified into five categories based on roof type viz., RCC or Madras terraced; Tiled or sheets; RCC/or tiled with red oxide floor; houses built for the poor; and special category for the newly added villages. The cap fixed in 2000 Self Assessment System was removed.[36]

The Bengaluru model is a hybrid form with a combination of features of rental value and capital value systems. The objective is to capture correct market values of residential and non-residential properties. This strategy in a way has helped the BBMP to shift part of the property tax burden on the non-residential properties.[37] The reform has succeeded in improving tax coverage. Nearly 1.50 lakhs unauthorised buildings were brought into tax-net and the revenue increased from Rs 430 crore in 2007-8 to Rs 780 crore by 2009 after the reform.[38] However, state audit department contradicts it.[39]

Integrated Property Tax Information System—Karnataka Initiative

The potential of GIS in urban planning and administration are being appreciated in recent years. Geographic Information System (GIS) deals with capturing, storing and analysing geographic data. It is observed that since a city's administration and planning are closely tied to its geography, GIS systems are of great help for governing a city efficiently. Several experiments are being made in cities and towns to apply GIS for property tax since the success of Mirzapur (Uttar Pradesh) experiment. The Karnataka Government initiated a state wide GIS-based property survey in 49 municipalities under Nirmalagara Project with ADB assistance in 2004 and later extended to 169 municipalities with the World Bank assistance in 2008. It partnered with the E-Governance Foundation, Survey of India, Karnataka Remote Sensing and Application Centre (KRSAC), Secon Surveys—a private surveying company and ESRI California-based GIS Software Company in this reform.

The objectives of the GIS reform are to: computerise revenue department of municipality for efficient tax management; put in place integrated property tax information system; improve record keeping of properties; increase tax compliance rate; enable municipalities to take informed decisions; maximise the coverage; rationalise property tax; bring transparency in levy and assessment; and improve revenues of municipalities.

A comprehensive survey of all taxable land and buildings within the municipalities was conducted. A pilot study was done in Byatarayanpura before implementing it in other municipalities. The tape survey was adopted on feasibility and cost factors. The Survey of India supplied ward-wise blank maps and the municipal councils hired engineers to survey the wards and collect the basic data as per the prescribed format along with the sketches of the buildings identified in the form of polygons. The maps were not to the scale but were adequate enough to give all information pertaining to property tax. The data format provides comprehensive information about the properties including the details like location, type of the building, area, use, ownership, amenities available, property tax details, etc. Buildings are being numbered on sequential order method with left of road odd numbers and right even numbers. This order is being followed to the extent possible. Both, vacant plots and buildings are being numbered similarly. The house numbers are being given with ward number, block number, street name and house number in the order. The GIS data was digitized. The computerisation of MIS data was done simultaneously.

The human resources were mobilised from different sources. Each Municipality was provided with IT engineer support. Several specialists like programmers, data entry operators, surveyors, etc were hired on contract basis. Assistance of some of the retired municipal staff was also taken. Data validation and close monitoring was done at local, regional and state levels. The data was collected by two streams that is, MIS and GIS format and it was duly integrated. The web applications made the system speedy and citizen friendly.

The GIS based property tax reform in Karnataka resulted in several benefits to the municipalities. Significant of these are:[40]

- *Computing automatic property tax demand*
  - (i) Property tax demand calculation
  - (ii) DCB reports
  - (iii) DCB report at different municipality boundary levels
  - (iv) Demand adjustment due to property modifications
  - (v) Demand adjustments due to any write offs
  - (vi) Auto rebate and cess calculation
  - (vii) Auto calculation of penalties

- *Automatic generation of demand notices and seizure notices*
  - (i) CAL notices are generated on 1st July every year for tax default properties;
  - (ii) Demand Notice is generated if tax is not paid within 30 days from the time of CAL notice served;
  - (iii) Seizure notice is generated if tax is not paid within 30 days from the time of Demand notice served.
- *Provision of online property tax calculator*
  - (i) Online tax calculator facility has been provided to help citizens to calculate their property tax;
  - (ii) Citizen has to furnish their property location and property details and system will calculate the property tax and shows the tax details.

Impact of the GIS Mapping

The major advantage of GIS mapping in Karnataka has improved coverage bringing un-assessed properties into the tax net. The improvement of coverage of the property tax between 55 to 294 per cent after GIS survey.[41] The municipalities are content with the initial benefits, the reform cannot be sustained and the advantages gained will be lost gradually. This requires strengthening of municipalities in terms of expertise and regular monitoring and, the municipalities should own the reform and continue it.

Improving Collection Efficiency

The ultimate indicator of the success of property tax reform is revenue buoyancy. Unless collection efficiency improves reflecting in the increased tax yield, tax reform cannot be considered a success (collection efficiency = demand/amount collected). The collection efficiency of the property tax has been given top priority in the urban financial reforms sector. The erstwhile Jawaharlal Nehru National Urban Renewal Mission (JnNURM) accorded top priority to the internal revenue mobilisation efforts of the municipalities in its financing scheme. Especially, it was stipulated that 85 per cent of property tax collection was a must for every municipality. The status of average collection efficiency in the municipalities across the country was observed to be mere 37 per cent of the tax demand, a discouraging situation. The better efficiency of municipalities ranged between 40-50 per cent reported from Karnataka, Tamil Nadu, Andhra Pradesh and Kerala. Low collection rates were observed in Maharashtra, Gujarat and Delhi.[42] It is therefore the newly launched Atal Mission for Rejuvenation and Urban Transformation (AMRUT) proposed at least 90 per cent of collection.[43]

Effective Collection Management

In most of the municipalities, proper Demand-Collection-Balance (DCB) statements are not maintained. These registers exist initially but, they are not regularly up-dated during the collection periods. The collection targets should be fixed ward-wise and consolidated at the municipality level. The collection efficiency should be measured against the updated DCB statements.

Another common problem encountered in municipalities is last minute pressure of collection especially, at the end of the year. Collection should be evenly scheduled throughout the year to avoid last minute pressures on the collection machinery. Monthly targets should be fixed not only to avoid end of the year pressure but also to raise revenues constantly to meet the regular expenditure commitments of the municipalities.

The collection machinery of municipalities normally suffers from problems like inadequate staff, diversion of collection machinery to other tasks under pressure, tax payer-collection staff nexus and undue delays in disposing of the self-assessment returns. Unless these bureaucratic deficiencies are rectified, the reform benefits cannot be optimised. The personnel engaged in the tax collection must be properly educated and motivated to perform better. The strategy of introducing incentive-penalty provisions could bring improvements in this area. The collection responsibility cannot be just left to the bill collectors or other ground level staff. Every officer beginning with the Municipal Commissioner should be involved and committed. Another important aspect is involving the municipal councillors including the mayor or chairperson.

The Constitution 74th Amendment has provided for the system of ward committees for municipal decentralisation at the local level. The ward committees are constituted in every ward and all the ward level officers are associated with these committees. The local elite groups and social workers are also represented in these committees. In Andhra Pradesh, these ward committees are also entrusted with the task of supervising the property tax collection in the respective wards. If the ward committee institution is strengthened, the tax collection could be improved with local level support.

Measures to Improve Collection Efficiency

The collection efficiency is the product of proper collection planning, scheduling, efficient collection machinery, improved compliance by the tax-payers and adoption of an incentive-penalty package. The following measures are essential to make tax collection efficient (Fig. 5.3).

(i) Tax mapping by GIS, aerial and manual surveys to increase tax base.

(ii) Proper maintenance of DCB statements.

(iii) Introduction of penal and incentive provisions in the municipal Acts.

(iv) Fixing collection target monthly by a system of benchmarking, milestone setting and regular monitoring.

(v) Wide publicity programmes to educate the tax-payers about the services being rendered in each ward, the costs involved, taxes being collected and the need for prompt payment of taxes.

(vi) Prioritising assessments with top importance given to habitual defaulters and major tax-payers.

(vii) Providing multi-channel access to the tax-payers, viz., on-line submission of tax returns and payments, convenient collection centres- e-seva (AP model), banks, post offices, etc.

(viii) Introducing the system of rewarding prompt tax-payers by way of cash awards. This practice has been adopted in GHMC and the State Bank of Hyderabad sponsor the gift.

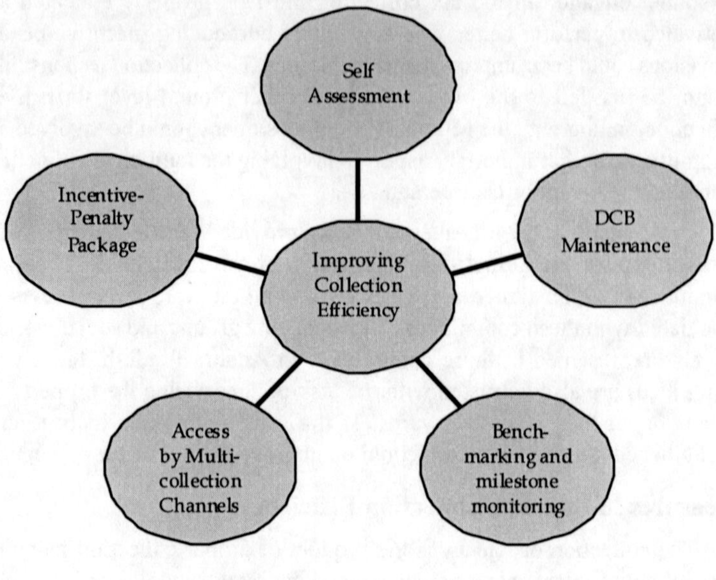

Fig. 5.3: Methodology of Improving Collection.

Amending the Rent Control Acts – The Case of Karnataka

Assessing the properties coming under the ambit of the Rent Control laws has been difficult. The rents fixed long back continued same over the years in the absence of any provision for revision in the laws. As the municipalities had to assess the properties only according to the fair rent, which became static, they were losing considerable revenues. The reform initiative attempted in Karnataka aims to overcome this problem through a new rent control law.

In the Karnataka Rent Control Act, 1961, there was no provision for revising the fair rent once fixed. In order to overcome this constraint, the Government of India and the Apex Courts have been suggesting the state governments to amend the rent control laws appropriately. Responding to this demand, the Government of Karnataka enacted the Karnataka Rent Act, 1999 which came to effect on 31-12-2001. Following are some of the features of the Karnataka Rent Act:[44]

(i) Its application is restricted to the residential building the Standard Rent of which exceeds rupees 3,500 per month in the areas covered by Karnataka Municipal Corporation Act, 1976 and rupees 2,000 per month in other areas and a commercial building having plinth area of not exceeding 14 square meters. These buildings should be more than 15 years old.

(ii) The Rent Deed is required to be in writing and registered.

(iii) Tenancy is made inheritable to a limited extent.

(iv) The standard rent is to be fixed by Rent Controllers in relation to the investment on property. The cost of construction shall be the actual amount spent on construction, and in case where such amount cannot be ascertained, such cost shall be determined as per the scheduled rates of the State Public Works Department for cost of construction for similar construction for the year in which the premises was constructed. Further, the market price of the land shall be the price for which the land was bought as determined from the deed of sale registered under the Registration Act, 1908 (Central Act 16 of 1908), if construction commenced in the year of registration or the land rates notified by the State Government or a local authority for the year in which construction was commenced, whichever is higher (Section, 7(2)).

(v) The most significant feature of this legislation is the provision made for enhancement of rent annually. Further, the enhancement in the case of a tenancy entered into before the commencement of this Act shall be effected gradually in five equal yearly installments and in case the landlord is a widow, a handicapped or a person of the age of sixty-five years or more, the enhancement of rent shall not be spread over a period of five years but shall come into force with immediate effect (Third Schedule).

(vi) The middlemen and estate agents should be registered.

(vii) Adjudication of eviction application by Rent Courts will be only Right of Revision, but no appeal.

(viii) Provision is made for immediate eviction of tenants of State or Central Government Employees, members of Armed Forces, widows, handicapped persons and persons above the age of 65 years under certain circumstances.

(ix) Special Procedure for trial of cases before the controllers and also the Courts so as to achieve quick disposals and negotiated settlement.

In the new legislation, provision has been made to revise the fair rent every year as below while restricting the maximum enhancement to 25 per cent to 100 per cent:

(a) Rents may be raised by two per cent per annum for buildings constructed before 1949;

(b) Rents may be raised by four per cent per annum for buildings constructed during 1950-1960;

(c) Rents may be raised by six per cent per annum for buildings constructed during 1960-1970;

(d) Rents may be raised by eight per cent per annum for buildings constructed during 1970-1994; and

(e) After 1994, 75 per cent of annual inflation rate based on whole sale price index.

This amendment facilitated increasing the property tax annually. But, it is provided that 10 per cent of the capital value will be taken as the rental values of the properties and tax can be levied on rental values so worked out. As per the rental value calculations, the tax rate would have been 16.6 per cent of rental value, which is considered exorbitant. In order to overcome this problem, the rate of tax was reduced to 0.3 per cent in the newly introduced capital value system.

Looking Forward

Property tax continues to be the main source of revenue to the Municipalities in India. Decades of efforts seem to be giving some benefits in recent years. The Government of India has been playing significant role in binding the states to reform the property tax. But, there is a strong feeling that in spite of being the key revenue source, the property tax is relatively under-used compared to its potential. The ability to finance growing local government expenditures *via* property taxes is severely constrained by administrative, regulatory, and technical shortfalls. In particular, weak administration and strong political interests limit the extent to which local government can tap on an expanding tax base and enforce compliance with taxes.[45]

The focus of reforms has been on standardisation and making the system taxpayer friendly so that the reform can sustain. The controversy of capital value vs. rental value has become irrelevant today, as the focus is on standardisation. Both the systems have certain advantages and limitations. It is observed that the concept of rent has an edge in respect of special properties like kalyana mandapams, cinema theaters, hotels, nursing homes, etc. The Bengaluru implements a hybrid model incorporating features of rental value system and capital value systems. The practice of notifying the rental or capital values have made the system more transparent and tax-payer friendly.

The initiative to link the property tax with GIS mapping is considered innovative and cost effective worth emulating by other states for improving transparency, coverage and thereby revenues. Two main tangible benefits accrued in the entire reforming process: one is people's acceptability of the standardisation based on unit area concept and the other is increased coverage and revenue yield to the municipalities.

One common problem noticed is reluctance to conduct the general revision regularly in spite of the legal provisions. Whenever revision was done after a long gap, the incidence of tax was naturally very high inviting the tax-payers' resentment and the consequential reduction of the increased tax bowing to pressures. It is in this context, the Thirteenth Finance Commission condition has come handy to remove such constraining provisions. Unless revision is done regularly, the fruits of standardisation of assessment cannot be realised fully in spite of the reform initiatives. The Karnataka initiative of legislative provisions for automatic tax increase irrespective of revisions in the property tax is worth adopting in all states. This will assure a minimum increase in revenues regularly.

The property tax reforms so far reviewed here resulted in certain improvements in the assessment methodology. But, the property tax reforms should be broad based focusing on critical structural and systemic aspects like political interference, prevalence of populism in local taxation, administrative efficiency and regular revision. A World Bank Report[46] observes that India faces a major structural problem with its property tax systems, resulting from the failure to resolve conflicts between assessing the true market value of property with rent control ordinances, and other limitations such as the FSI. Moreover, government officials have generally been unwilling to issue new valuation rolls, in some cases for many years. Much of the recent property tax reform in India has entailed stop-gap measures to overcome these problems, rather than engaging in comprehensive reform. Meanwhile, the growth in property tax revenues has remained anaemic. Unless these structural issues in properly valuing of property are resolved, improved administration will do little to make the property tax a viable revenue source for local governments, and the gap between local expenditures and revenues is likely to grow over time.

The objectives of property tax reform are ultimately increased revenue yield and tax compliance. This demands broad-basing the property tax reforms targeting increased revenue yield, administrative efficiency and tax compliance. The area-based taxation has yielded initial benefits wherever it is implemented effectively. But, the presumptive average unit rates are only the beginning and should be considered as a transition phase of the reform as some thinkers consider.[47] The reform has to be taken toward the logical end wherein the property tax valuation truly reflects the existing market on a sustainable basis. In spite of the tangible variables introduced like plinth area, use of buildings, location and construction quality, these attributes still cannot totally reflect the actual market values until the property market is highly organised and developed. Further, the experience shows that due to absence of continuous revaluation of properties, the presumptive unit rates have become stagnant. In the first instance, lower rates were fixed to gain tax-payer acceptability, and lack of updation has negated the reform impact. It is suggested that a method of property price indexation could be done at national and state levels covering all the cities and towns.[48] This exercise can be taken by the State Property Tax Boards. At the national level also, similar arrangements should be done.

Another important area of attention should be the ownership of the reform by municipal tax administration department.[49] The municipal officials not only should understand the reform but also initiate measures to convince the tax payers. Integrated urban property information system should be developed by convergence of all the stakeholders like the revenue department, registration department, town planning department and the municipality concerned. The following steps in brief, are necessary to realise the objectives of property tax reforms:

- Bringing parity between assessed values with the existing market values by a system of indexation of property prices;
- Rationalising exemption policies;
- Freeing property tax from various state government cess;
- Regular revision of property values either every five years or annually with reconciliation at the end of the fifth year;
- Periodic re-valuation of property values to capture market changes;
- Freeing property tax from rent control constraints;
- Property mapping to identify un-assessed and under-assessed properties on sustainable basis;

- Improving collection performance to over 90 per cent;
- Transparency by publication of ratable values of properties and lands;
- Improving service delivery for better tax compliance.

The Government of India, under AMRUT has made 90 per cent collection efficiency as an important reform to be achieved by all the municipalities in the country. The Thirteenth Finance Commission made two conditions relating to property tax, viz., removing all constraints for the levy of property tax and constitution of a state level Property Tax Board. The Property Tax Board apart from standardizing assessment should also be entrusted with the responsibility of setting guidance values and subsequent up dating. Most of the states are taking steps to implement these conditions. The state level Property Tax Board if made effective (not haven for favourable retired civil servants) can yield better results primarily from the point of setting realistic guidance value and re-valuation of properties. This strategy appears to be more beneficial in the context of widely prevalent resistance at the local levels for tax increases and the helplessness of the municipalities. It could be a way out for such systemic problems.

This apart, the Thirteenth Finance Commission also suggested broadening the Tax-base by instituting a geographic information system for mapping properties in all cities and introducing a specific conditionality in JnNURM (may be replaced by AMRUT) aimed at reducing the gap between the assessed and market values of properties.[50] To conclude, though certain reforms are being implemented to standardise the assessment process, irregular revisions, collection inefficiency and structural issues are still the important concerns to be addressed on priority basis by the state governments and municipalities to make the property tax reform purposeful and sustainable.

References

1. High Power Committee on Infrastructure and Services (HPEC) (2011): Ministry of Urban Development, Government of India, p. xxvii.
2. Krane, Dale, Ebdon, Carol and John Bartle (2004): Devolution, Fiscal Federalism and Changing Patterns of Municipal Revenues The Mismatch Between Theory and Reality, Journal of Public Administration Research and Theory, 14(4), p. 523.
3. Rao, M. Govinda (2013): Property Tax System in India : Problems and Prospects of Reform, Working Paper No. 2013-114, National Institute of Public Finance and Policy, New Delhi, pp. 4-5.
4. Thakur, Debulal and Rajadhyaksha, Nilesh with assistance from Roy Bahl (2009): Urban Property Tax Potential in India, National Institute of Public Finance and Policy, New Delhi, p. x.
5. *Ibid.*, p. viii.

6. *Ibid.*, pp. ix-xi.
7. Om Prakash Mathur, p. vii.
8. The Andhra Pradesh Municipalities Act, 1965, Section, 85.
9. *Ibid.*, Section, 87.
10. *Ibid.*
11. *Ibid,* Section, 88.
12. AP High Court Order, dated 29th September, 1994.
13. CA No. 4052 of 1996, dated 2nd May, 2001, Supreme Court of India.
14. Tamil Nadu District Municipalities Act, 1920, Section 82.
15. Commissioner of Municipal Administration, Government of Tamil Nadu, Property Tax Assessment Sheet-Tamil Nadu ULBs.
16. ROC No. 60572/98/R1, Office of the Commissioner of Municipal Administration, Government of Tamil Nadu, dated 23-10-19989.
17. 2nd SFC Report, Government of Tamil Nadu, para 2.13, pp. 19-21.
18. Karnataka Municipalities Act, 1964, Section 103.
19. Karnataka Municipalities Act, 1964 (as amended in 2005), Section 101.
20. Circular No. UDD14TCT 2004, Urban Development Department, Government of Karnataka, 22nd February, 2005.
21. Karnataka Municipalities Act, 1964, *Ibid.*
22. Bihar Municipal Act, 2007; Bihar Municipal Property Tax (Assessment, Collection, Recovery) Rules, 2013.
23. U.P. Municipal Corporation Act, 1959; U.P. Municipality Act, 1916; and U.P. Property Tax Rules, 2000 with amendments.
24. Jharkhand Municipal Act, 2011.
25. M.P. Municipal Corporation Act,1956; M.P. Municipality Act, 1961.
26. Sec 105, KM Act, 1964 and KM Taxation (Amendment) Rules, 2002.
27. *Ibid.*
28. Singh, U.B.(2013): *Property Tax Reforms in Kanpur: An Impact Study, Nagarlok,* XLV (1), pp. 42-62.
29. Greater Hyderabad Municipal Corporation Act, 1955, Section, 199 and 212.
30. The legal provisions were made in the GHMC Act, 1955 by amending the section 212 in 1990 to introduce plinth area basis similar to other ULBs in the state.
31. Greater Hyderabad Municipal Corporation Act 1955, Section 213.
32. GHMC, Property Tax Guidelines for Calculating Self Assessment Tax.
33. The Times of India, 20th July, 2013, Hyderabad.
34. Karnataka Municipal Corporation Act, 1976, Section 109.
35. Property Taxation-Bruhat Bengaluru Mahanagara Palike, Implementation of Best Practices, Vol. 2, Urban Reforms, National Institute of Urban Affairs, New Delhi, (www.IndiaUrban Portal.in), pp. 32-39.
36. *Ibid.*
37. *Ibid.*
38. Rao, M. (2013): *op.cit.,* p. 10.

32. *Ibid.*
39. State Audit Department, Audit Report (Local Bodies) for the year ended 31st March, 2010, Unit area value based self assessment scheme of property tax in Bruhat Bangalore Mahanagara Palike, performance Review. (www.sailindia.gov.in/nglish/home/owner_production/audit/chap_4), pp. 89-93.
40. Municipal Reforms Cell, Directorate of Municipal Administration, Government Karnataka, Bangalore.
41. Municipal Reforms Cell, *Ibid.*
42. Om Prakash Mathur, op.cit., p. x.
43. Government of India: AMRUT Guidelines.
44. The Karnataka Rent Act, 1999.
45. Lall, Somik V. and Uwe Deichmann (2006): Fiscal and Distributional Implications of Property Tax Reform in Indian Cities, Working Paper, No. 39, National Institute of Public Finance, New Delhi.
46. The World Bank (2004): India-Urban Finance and Governance Review, Vol. I (Executive Summary and Main Report), December.
47. Rao, M. Govinda, *op.cit.,* p. 16.
48. *Ibid.*
49. *Ibid.,* p. 17.
50. Thirteenth Finance Commission (2010-15), Vol. I, Report, Government of India, December, 2009, pp. 163-164.

Chapter - 6

Municipal Accounting Reforms

Introduction

A vibrant municipal financial management system is essential to meet the needs, demands and challenges of urban development in India. The financial management critically depends on effective accounting practices of municipalities. The accounting system should reflect the financial transactions as well as performance of the municipalities in terms of achieving developmental goals, efficiency in raising and use of financial resources and, improving income, assets and liabilities. The accrual double entry accounting system can provide such holistic financial information of an organisation. The good governance demands that this information should be shared with all the stakeholders, viz., elected representatives, citizens, auditors, concerned authorities, lenders, etc., from the accountability and transparency perspective.

The accounting system practiced in municipalities was introduced during the British regime in India which is suitable for predominantly regulatory environment. In spite of incremental changes and superimposed modifications after Independence, the general criticism has been that it remains rudimentary, transaction oriented, dated, hence, inadequate to meet the present financial management needs and accountability demands of the municipalities. The over emphasis on legal and procedural aspects and cash based single entry method of accounting served to control the flow of funds as per the budgetary provisions and adherence to legal sanctions to some extent but, it has failed to support the municipalities from the financial management and performance perspective in the changing urban context. The municipal accounting reforms in India stem from these inadequacies. The international funding agencies like the World Bank, USAID (FIRE-D) and ADB were instrumental for earlier accounting reform initiatives which began in the 1980's. But, the increased flow of funds in the urban development sector in recent periods and the dated accounting practices of the municipalities necessitated the national level interventions and finally, accounting reforms became part of the national reform agenda under erstwhile JnNURM and presently under AMRUT schemes. The key objective of accounting reform is to "improve municipal accounting system based on double entry and accrual principles leading to better financial management, transparency and self reliance."[1] A perceptive conceptual overview is prefaced in this chapter followed by a critical discussion on the accounting reform initiatives in India.

The Accounting System—Conceptual Overview

An accounting system measures and records financial data and converts these data to information that is then analysed, interpreted and reported to various groups both, within and outside the organisation.[2] The purpose of accounting is three-fold: reporting on the financial status of the organisation to the stakeholders; providing accurate quantitative financial information for the internal use of the organisation on routine operations; and assisting the organisation in formulating policies and long range plans. In short, accounting is a system of recognising, calculating, classifying, posting, analysing, summarising and reporting financial transactions.

Funds: as Accounting Entities

In public organisations, funds are established as accounting entities. The funds are independent accounting and fiscal entities (mostly legal) to which resources are assigned along with related liabilities, obligations, reserves and equities.[3] Financial reporting statements are made for each of these funds. In fund accounting, revenues are controlled through an appropriation process and the proposed expenditure by a line-item budget. Revenue is classified by a fund/source while the expenditures are classified in a detailed manner including organisational unit, function, programme, etc. Fund accounting has a built-in control through budget where expenditures are monitored by various appropriations.

Though the municipalities maintain funds—single or more, accounts are maintained according to the budget heads, not according to the funds, which in turn are broadly related to the functions of the municipality concerned. Under each head e.g., water supply, health, etc., income and expenditure details are maintained. Generally, municipalities maintain limited number of funds, many a time a single general fund only depending on the need and operation. In this situation, it becomes difficult to track the sources of funds-grants, loans, or internal sources, as the accounts don't reveal about the sources of funds. Though the existing systems permit separate accounts based on the sources especially, in case of loans, the repayments are mostly from the general funds. Hence, the self-sustaining nature of an investment cannot be ascertained. This becomes significant in the context of increased investments on urban infrastructure through loans, capital market, etc.

As against this, in a truly fund-based system accounts are maintained according to the funds- government funds, proprietary funds, and fiduciary funds. Government funds are general in nature not specifically linked to a particular source viz., taxes, non-taxes, grants, transfers, etc. Generally, municipalities constitute these funds to maintain their administration, pay salaries and incur expenditure on maintenance of amenities and services. Proprietary funds are established to run some remunerative enterprises or services which could be run on self-sustaining basis viz., water supply, solid waste disposal, etc., through cost recovery. The fiduciary funds are created to manage external funds. For instance, pension funds, which are maintained by the municipalities. It is considered a good practice to maintain these funds separately not mixing with the general municipal funds.

Account Book-keeping—Single Entry and Double Entry

Every financial transaction has two main implications—either receipt of a benefit in the shape of cash (by taxes, non-taxes, and other sources), goods and services, or giving away a benefit in the form of incurring expenditures for payment of salaries, purchases, advertising, etc. In the account book keeping parlance, there are mainly two methods of recording a financial transaction viz., Single Entry and Double Entry. In the Double Entry method both, receiving and giving of a benefit, in other words, credit and debit—the two possible sides of a financial transaction—are simultaneously recorded while in the Single Entry only one side gets reflected and the other is ignored. In simple terms, for every debit, there should be a corresponding credit. The Double-Entry method of accounting has certain advantages over the Single Entry method. The Double Entry method reflects both, personal and impersonal accounts while the Single Entry method covers only personal accounts. As the Double Entry method records the two sides of a financial transaction, it is possible to prepare a trial balance more accurately. Further, it is possible to prepare income and expenditure account which in turn would facilitate preparation of the balance sheet that gives true picture of the financial status of the organisation. As Double-Entry method has built-in crosschecks, it is claimed, the fraud can easily be detected.

Basis of Accounting—Cash Basis and Accrual Basis

The basis of accounting provides as to when and how the revenues and expenditures are to be recorded. There are two bases in vogue in modern government accounting on revenue side viz., cash basis and accrued basis while there are four bases viz., cash, obligations, accrued expenditure and accrued cost on the expenditure side.[4] Under purely cash basis, revenues are recorded only when they are received and expenditures are recorded only when payments are made in the Cash Book without any distinction of revenue and capital receipts or payments. It starts with a cash balance and the difference between the debit and credit become the cash balance at the end of the financial year. It captures flow of funds into and out of a budgetary account but it may not give complete picture of funds availability for expenditure at a point of time as the cash-based accounting system does not permit a clear understanding of actual position of debits and credits and hence, it conceals transactional flaws in the system. For instance, receivables under property tax are not accounted and certain obligations like purchase orders and payables are not accounted. Further, purely cash basis does not support the assessment of assets and liabilities at any point of time, reconciliation of bank account is not done and monitoring of the fiscal transactions becomes difficult. Under cash basis, the reporting system generally include a statement of receipts and payments along with the Demand, Collection and Balance statement, statements regarding any loans and schemes. Under cash basis:

(i) A statement of account only could be prepared based on actual cash receipts and cash payments, recorded in the Cash Book of the year;

(ii) This statement viz. Receipts and Payments statement will commence with the opening balance - both cash on hand and cash at bank;

(iii) It need not necessarily be accompanied by a statement of Assets and Liabilities; and

(iv) The difference between the two sides - debit and credit - will indicate the cash balance at the end of the period.

Accrual System of Accounting

Accrual system of accounting means a method of recording financial transactions based on accrual, i.e., on occurrence of claims and obligations in respect of incomes or expenditures, assets or liabilities based on happening of any event, passage of time, rendering of services, fulfillment (partially or fully) of contracts, diminution in values, etc., even though actual receipts or payments of money may not have taken place. Under a purely accrual basis of accounting incomes are accounted when the right to receive them arises, irrespective of actual realisation or collection e.g., when they are levied, billed or earned. On the same analogy, expenditures are accounted when liability or obligation to pay arises irrespective of the fact when they are paid. In simple terms, any cost incurred even if it is not yet paid gets reflected in the accounting. It will not commence with a balance and treats revenue and capital receipts and payments separately. The accounts reflect the figures of the current year only and do not include items of past or future years. The deficit or surplus position can be ascertained by comparing the income and expenditure status any time and therefore, it is possible to prepare the income and expenditure statements monthly or periodically based on necessity. It provides for a total asset and balance statement including not only current assets and liabilities such as cash or payables but also capital assets and long-term liabilities such as buildings, bridges, roads, general obligation, debt, etc. Accrual accounting system, therefore, provides comprehensive financial information including information like whether the services received were paid from this year or the costs are shifted to the future years and the extent investments are made on capital assets including infrastructure which will be useful to the officials to focus on short-term and long-term planning.

In this system, there is a change in accounting for transactions and reporting the financial results so as to provide the municipalities and the government with the Financial Reports, in the form of two important financial statements for the purposes noted against each:

Statement	***Purpose***
Income and Expenditure Statement	To determine the financial performance of the municipalities
Balance Sheet	To assess the financial status of the municipalities

Benefits of Accrual System of Accounting

The main benefits of accrual based accounting system are enumerated below:

(i) Revenue is recognised as it is earned and thus "Income" constitutes both revenue received and receivable. The accrual basis not only records the actual income but also highlights the level and efficacy of revenue collection, thereby assisting decision-makers in taking financial decisions.

(ii) Expenditure is recognised as and when the liability for payment arises and thus it constitutes both amount paid and payable. In accrual basis of accounting, expenditure incurred on repairs and maintenance shall be recognised as expense of the period in which they are incurred and, if not paid for during the year, shall be treated as a liability (payable) and be disclosed as such in the Balance Sheet.

(iii) Expenses are matched with the income earned in that year. Thus, it provides a very effective basis to understand the true performance of the organisation for the operations that is conducted in that year.

(iv) A distinct difference is maintained between items of revenue nature and capital nature. This helps in correct presentation of financial statements, viz., the Income and Expenditure Statement and the Balance Sheet.

(v) Costs which are not charged to Income and Expenditure Account are carried forward and kept under continuous review. Any cost that appears to have lost its utility or its power to generate future revenue is written-off.

(vi) The surplus or deficit as shown at the year-end represents the correct financial position of the organisation arising out of the various transactions during that year.

(vii) It assists in effective follow-up of receivables by the municipal body and proper ascertainment of payables by the municipal body.

(viii) It captures "full" cost of servicing and helps in identifying financial viability of rendering services.

(ix) It helps in providing timely and right quality information for planning, decision-making and control at each level of management.

(x) One of the distinct advantages of adopting accrual accounting system is ease in financial appraisals by the financial institutions. It also facilitates credit rating through approved Credit Rating Agencies, which is a pre-requisite for mobilising funds in the financial markets through debt instruments. Credit rating of municipalities is one of the important reforms.

(xi) As it presents a true picture of the financial position of an organisation, it facilitats proper financial analysis, reporting and therby, better financial management.

The accrual basis of accounting results in recording of transactions and events on the basis of their substance, rather than merely when cash is received or disbursed, and thus, enhances their relevance, neutrality, timeliness, completeness and comparability.

Municipal Accounting System in India—Need for Reform

The existing system of accounting in the municipalities is cash-based and for a variety of reasons the system continues. The system simply follows the budget code providing information budget item-wise receipts and payments and serves only legal purpose of preparation, authorisation and control. The municipal accounting system has been distorted over a period of time due to superimposition of a kind of peripheral accounting with certain modern features without changing or strengthening the core structure. Municipal asset position is not correctly reported due to absence of the capitalisation concept. Increased transactions, lack of internal controls, long-term perspective and dearth of qualified staff-all contributed for deteriorating situation. The current reporting system comprising of the annual budget, annual administration report and the audit reports fails to provide a comprehensive financial picture of the municipality as they focus more on legal and administrative aspects. They are prepared in dated formats and in most cases; they are not seriously taken by the users as also by the controlling authorities. The solution does not rest with simply conversion of the cash-based accounting to the double entry accrual based system, as the municipal functional and administrative environment is complex which cannot be compared to a single department or a corporate body. The accounting reform initiatives in India in recent periods have amply demonstrated the structural and operational limitations. There is hence, need to devise appropriate customised models based on the size of the municipality, nature of transactions, political environment and availability of qualified and trained manpower. The need to improve the accounting and budgeting system arises primarily from the following factors:

(i) The municipalities are required to take a clear view of their financial position and plan accordingly in the context of alternate funding requirements for urban services and infrastructure development viz., accessing capital market, international and domestic borrowing, privatisation and public private partnerships, etc.

(ii) The municipalities should draw clear-cut strategies for future investments and locate areas of immediate and long-term interventions.

(iii) Actual asset and liability positioning of the municipalities would facilitate their credit rating. This, in turn will enable them to raise funds from financial institutions and market.

(iv) The correct assessment of the financial status would enable identification of weak sectors and the sectors which should be financed through innovative financing methods.

The worsening financial status of the municipalities coupled with inability of the central and state governments to fund the urban infrastructure on fast expanding scale, necessitated search for alternate funding in India. In this context, governments are forced to go for international borrowing from institutions like the World Bank and ADB and other institutional funding. The JnNURM scheme with ambitious investments plans in cities by a combination of funding options including grants, borrowing and internal mobilisation demanded that these investments were properly accounted for and managed efficiently. This has been re-emphasised under newly launched AMRUT scheme. All these factors have led to certain compulsions to streamline financial management at the municipal level especially, their accounting and financial management practices. The good governance, which is a catchword at the national and international levels to-day, demand participatory, transparent and accountable processes reflected in the municipal financial management and accounting. Entry of the private sector and civil society organisations into the urban governance necessitates that the municipal accounts also reflect their transactions implemented through various forms of contracts. The demand for municipal accounting to be transparent, comparable, accessible and user-friendly to all the stakeholders has compellingly necessitated municipal accounting reform initiatives in India.

Municipal Accounting Reforms in India

The municipal accounting reforms which initially began during 1980-90 picked-up momentum in the subsequent decades. Interestingly, the accounting reform was initiated at the instance of the World Bank and the purpose was to shift towards accrual based double entry accounting to fit into the commercial and corporate urban governance environment. During the decade of 1980-90, the metro cities of Mumbai and Chennai introduced the accrual based accounting. If the former confined this system to water supply and sanitation accounts, the latter applied this to all the civic operations. The accounting reforms in municipalities in India generally started with some financial problems[5] viz. inadequate and unreliable data; lacunae in the single-entry accounting system, improper distinction of receipts and payments under different heads such as revenue, capital and extraordinary; problems in tallying and cross-checking accounts; and inordinate delays in finalising the accounts. But, in Gujarat, the reforms started with inability of the Ahmedabad Municipal Corporation to raise finances because of the lack of transparency in its financial accounts and the transition to an accrual-based accounting system became a device to enhance transparency around a loan. It was one of seven municipalities to pursue reforms at the behest of the World Bank through Gujarat Urban Development Project during 1990-95. This was followed by the Tamil Nadu Urban Development Fund in 1998. It was a statewide initiative in Tamil Nadu at the instance of the World Bank. The Government of Maharashtra initiated accounting reform in the municipalities under the USAID assistance. Further, if the Bangalore Corporation started as part of the complete revamping of the public expenditure management system, the transition was part of the computerisation projects in Indore and Hyderabad Municipal Corporations.

The municipal accounting reforms assumed national dimension with the recommendations of the Eleventh Finance Commission (2001). The Commission suggested[6] that the Comptroller & Accountant General (C&AG) should be entrusted with the responsibility of providing technical guidance and supervision (TGS) over proper maintenance of accounts and audit of all tiers of Panchayati Raj Institutions (PRIs) and municipalities; C&AG should prescribe auditing standards for PRIs and municipalities, guidelines for certification audit of the accounts of PRIs, budget and accounts formats for PRIs and municipalities and list of codes for programmes, functions and activities for PRIs; and C&AG should conduct comprehensive training programmes to upgrade the skills of the staff of local fund audit department and PRIs. The Supreme Court of India opined that double entry accounting system be introduced in municipalities in view of increased flow of funds in the urban development sector. In the light of these suggestions, the Government of India constituted a task-force to look into the issue of municipal accounts under C&AG which submitted the report in 2002. The Task Force of Comptroller & Accountant General in its report recommended: uniform formats for financial statements and budget; formats for determining the cost of important utilities and services like water supply and showing this information by way of disclosure to the accounts; significant accounting policies to be followed by municipalities as per model accounting policies need to be disclosed in a separate schedule forming part of the accounts and initiate steps for smooth switch over to the double entry accrual system of accounting.[7] In order to motivate the state governments and municipalities to take-up urban reforms on priority basis, the Government of India started a scheme called the Urban Reforms Incentive Fund (URIF) in 2002-03 to provide special financial assistance to municipalities. This fund had a provision of 10 per cent of total budget for introduction of Double Entry System of Accounting in municipalities.

All this led to the preparation of the National Municipal Accounts Manual (NMAM) by the Ministry of Urban Development in collaboration with the C&AG, National Institute of Urban Affairs and the INDO-USAID (Fire-D) project in 2005. The objective was to prepare a generic framework of National Municipal Accounting with comprehensive coverage of accounting policies, procedures, and guidelines designed to ensure correct, complete and timely reporting of municipal transactions and produce accurate and relevant financial reports. The NMAM has been deliberated widely at national and state level platforms. Later, a Committee on Accounting Standards for Local Bodies was constituted by the Council of the Institute of Chartered Accountants of India (ICFAI) in 2005 to formulate accounting standards for local bodies to facilitate preparation of accounts on accrual basis. In order to guide the capacity building programmes, a National Municipal Accounting Training Manual was also prepared in 2007 by the Ministry of Urban Development. The implementation of accrual based double accounting reforms by the municipalities across the country was initiated through the urban reform agenda under the erstwhile JnNURM. This move has been given more focused emphasis under AMRUT.

The JnNURM scheme envisaged a broader perspective for accounting reforms in municipalities as listed below:[8]

(i) Increased access to resources for urban services

(ii) Improved financial management, accountability, transparency of management and improved governance

(iii) Accurate costing for all urban civic services

(iv) Timely and better MIS for decision-making

(v) Accurate reporting of subsidies for better management and targeting

(vi) Better control and utilisation of assets

(vii) Publishing of audited financial statements on a timely basis

(viii) Better management of resources and risks

(ix) Responsible civil society that is prepared to partner with local government.

Under AMRUT, the augmentation of implementation of Double Entry Accounting process has been given importance and for the purpose milestones with time-lines have been purported. These milestones are as under:[9]

(i) Complete migration to double entry accounting system and obtaining an audit certificate to the effect from FY2012-13 onwards;

(ii) Appointment of internal auditor; and

(iii) Publication of annual financial statement on website.

The financial assistance is provided especially for activities such as: budgeting, accounting, internal controls and auditing; reengineering business processes to align with accrual-based accounting system; integrate financial management systems with financial accounting system; and staff training and building financial management capacity. The state governments are mandated to develop state-level accounting manual; manuals on expenditure management and performance-based budgets; policies and guidelines on training and handholding; and accounting software integrating financial management systems with the financial accounting system. Under JnNURM initiative the municipalities and state governments were required to take following steps:[10]

(i) Establish a committee to implement improved financial management reforms

(ii) Appoint chartered accountants for providing handholding support

(iii) Implement training programmes

(iv) Prepare inventory of assets and liabilities

(v) Value all assets and liabilities, finalise opening balance sheet

(vi) Re-engineer business processes to align with accrual-based accounting system

(vii) Adopt and implement an accrual-based accounting system

(viii) Appoint an external auditor

(ix) Publish financial statements along with audit reports

(x) Carry out parallel runs of both existing and new system of accounting till the latter gets stabilised

(xi) Implement transparent, multi-year budget system with the priorities set by stakeholders

(xii) Institute internal control procedures and appoint an internal auditor

(xiii) Undertake credit rating

(xiv) Install the selected accounting software, and integrate financial management systems with the financial accounting system.

A fixed time line and milestones were prescribed in the JnNURM reform agenda to implement the accounting reforms by the states and the municipalities. However, the scenario in many states has been far from satisfactory. Keeping this in mind, the AMRUT has prescribed to achieve the focused milestones within the stipulated time-lines. This attracts incentive as well.

The accounting reforms which have been initiated in the states and the municipalities across the country are customised forms of double entry accrual accounting systems adopted to suit the local municipal financial environment. If the Tamil Nadu introduced double entry accrual accounting system in all the municipalities with fair amount of success, fund based accounting system is being implemented in Karnataka. The BBMP introduced fund-based accounting along with double entry and partial accrual accounting features in 2001-02. The GHMC introduced the double entry accrual accounting system on ERP format in 2006. We shall discuss these models in the following pages.

Integrated Accrual Based Double-entry Accounting and Financial Management System in Greater Hyderabad Municipal Corporation

If introduction of accrual accounting reforms the municipal accounts to reflect the true financial status of the municipalities while making them competitive and market friendly, computerisation facilitates efficient, user friendly and transparent financial transactions. Computerisation also promotes efficient decision-making build on sound financial information system. The National Municipal Accounts Manual has prescribed standard codes for various accounts and these are being digitalised. Unless accounting operations are reformed in an integrated manner, the financial management cannot be improved. Accrual accounting, computerisation and financial management information system are integral parts of this integrated system. The accounting reforms project in Greater Hyderabad Municipal Corporation (GHMC) is envisioned in this perspective.

The GHMC was adopting the traditional cash-based government accounting with single entry system. Transactions were recorded only when they were received or payment of money as a singly entry in a cash book. Subsidiary ledgers were maintained for aggregation of receipts and payments under the respective budget heads. All payments and receipts were treated same way whether capital or revenue except for classification purpose and the primary control was budgetary framework without any facility for self-sufficiency measure. In the post 74th Constitution Amendment era, the municipalities are expected to be vibrant and financially self-sustaining. In tune with this national objective, the GHMC has embarked on to introduce certain budgetary and accounting reforms.

The budget heads were classified with the objective of enforcing accountability on the part of the various functionaries for the functions assigned to them and for facilitating the introduction of accrual based commercial accounting system. Receipts and expenditures were classified into certain major heads. Receipt heads include revenue receipts, capital receipts, revenue advances (recoveries), capital advances (recoveries) and deposits. The expenditure heads include revenue expenditure, capital expenditure, loan repayment, revenue advances (outgoings), capital advances (outgoings) and deposits (outgoings). Further, revenue receipts include own taxes, fees and user charges, sale and hire charges, revenue grants and reimbursements, assigned revenues and other revenues. Revenue expenditure cover personnel costs including terminal benefits, operational costs, repairs and maintenance, programme expenses, finance expenses and administrative expenses. The GHMC adopted the codes prescribed in the AP Municipal Accounts Manual which again is based on NMAM.

The double-entry accrual system of accounting was initiated in April, 2002. M/s F. Ferguson & Co., Chennai assisted the Corporation but the M/s Bloom Solutions Pvt. Ltd. developed the present model and is also maintaining the system. The double-entry accrual system of accounting was designed incorporating both, double entry and accrual concepts. Certain re-engineering and operational changes were made to make the system functional and IT-based. The accrual-based accounting system introduced in GHMC is customised to suit the requirements and operational feasibilities. The key features of this model are outlined below.

(i) Combining commercial with government accounting procedures while retaining the budgetary control.

(ii) Compliance with existing laws, rules and regulations.

(iii) Needs of stakeholders like revenue providers, benefactors and facilitators.

(iv) Evaluation of efficiency and effectiveness.

(v) The accounts operated are as:

(a) Income accounting - as and when a demand is made;

(b) Expenses accounting - as and when an expenditure is incurred and bill is admitted;

(c) Assets accounting-by carrying over the assets to its value as against charging it as expenditure which was the practice earlier;

(d) Liability accounting-money received with a condition of repayment is recognised with the liability to the respective partners, for instance, loans.

The GHMC is preparing the income and expenditure statements on weekly basis and the Balance Sheet every year. The MIS reports viz. final accounts, performance reports by function, cash flow statement, daily collection report, investment report, scheme utilisation, budget utilisation and cheques pending for issue are also generated. An accounting manual was prepared in 2002 to guide the accounting functionaries. First phase of new accounting manual has been operationalised since 2002-03. The reformed accounting model has thus facilitated considerable improvements to the financial transactions.[11]

E-governance Solution for Integrated Accrual Based Double Entry Accounting

The e-governance accounting reform project commenced on 5th Aug. 2005 and a road map was prepared to implement Oracle Application Software for all the financial activities of GHMC with the support of Centre for Good Governance, Hyderabad. Oracle Financial was officially launched on 20th April 2006 with the target to cover all the DDO locations by end of October, 2006. Oracle Financials-Enterprise Resource Planning (ERP) application is an end to end integrated software solution, popular in public and private organisations across the globe. The application suite provides *integrated* solutions for entire spectrum of activities of an organisation. It has several advantages. The key features are:[12] It aims integration between the functions; ensures quality and completeness in data capture and validation; the system has in-built internal controls and audit trail; it has capacity to handle huge data and can generate flexible reports; empower users with the cutting edge technology and enhancement of security; information flow is allowed in controlled manner; though it aims for paperless office yet works effectively with the paper; and finally, automation of end-to-end transaction process could be achieved.

The software is being implemented in GHMC for the first time in a municipality in India. The high end Oracle Financials was customised in line with the National Municipal Accounts Manual. The Oracle e-Business Suite, Works Module and Grievance Redressal Module are made online in GHMC. Solution Designed for GHMC is being implemented in 5 Zones, 18 Circles, Pre-Audit and Finance and Accounts departments DDO locations (across Circles, Zonal Office and Head Office). The ERP software is operationalised in following departments and operations:

(i) *DDO Locations:* Receipt entry; monitoring of receipts; bills entry; submission of bills to Pre-Audit for approval; and bills monitoring.

(ii) *Pre-Audit:* On-line receipt verification; on-line verification of bills and management of audit approvals.

(iii) *Finance and Accounts:* DDOs, contractors, suppliers creation and maintenance; budget preparation, maintenance and monitoring; generation of cheques for audit approved bills; cheques delivery remittances of deductions; asset accounting; treasury; vouchers maintenance; reconciliations; and financial reports.

The Budget has been designed and implemented in line with State Municipal Accounting Manual and, uniform codification structure is used for budgeting and accounting. Oracle Hyperion Planning Solution has been implemented in GHMC and budgets from the financial year 2009-10 are generated from this system. All the advantages attributable to ERP systems can be derived from Oracle Financials system also. Some of the advantages are mentioned below:

(i) World best practices are embedded in the system followed by diverse organisation in public and private sector- flexible to make the solutions designed to meet the exact needs of the GHMC.

(ii) Solution for every functional area of an organisation with a facility of integration with the each other-IT solution can be provided diverse functional areas of GHMC and seamlessly integrate with all the functional areas resulting in avoiding data redundancy and improvement in quality of information.

(iii) Solution is divided into modules based on the functional areas such as financials, projects, HRM, Inventory, CRM, Business Intelligence, Portal, etc.

(iv) Solution is open to interface with the external systems- facilitates integration with the well functioning existing IT initiatives of the organisation.

(v) Resources are readily available for implementation and maintaining the system-moving from proprietary developed system to universal system shall help sustainability to maintain the system in long run.

(vi) Continuous support provided shall enhance the reliability of the system – problems or bugs can be reported to support system online and necessary guidance or patches shall be made available.

Limitations

The ERP model introduced in GHMC has certain limitations. At the outset, it is generic in nature and most of the features provided in the solution may not be relevant or useful to the organisation. The other important limitation is that it is expensive system to acquire as well as maintain. As the system is based on central server concept, good back-up systems are needed to provide continuous access in case of main server down time due to any contingency. The critical problem is that it requires high skilled technical personnel to maintain the system properly and lack of internal competency to handle the high end functional activates results in outsourcing of these activities to external agencies. The IT-based accounting reform being implemented in GHMC raises the crucial requirement of developing internal personal competencies to sustain.

Accrual Based Accounting System in Vijayawada Municipal Corporation

The Vijayawada Municipal Corporation (VMC) implemented accrual-based accounting in partnership with the Centre for Good Governance, Hyderabad which front ended the project and the e-Governments Foundation, Bangalore has developed and customised the software for application in the corporation. The system came to live operation on 1st April, 2007. The system was launched after imparting training to the functionaries concerned at the Corporation. The VMC prepared the income and expenditure statement and balance sheet under accrual system of accounting for the FY 2007-08.

The e-Gov Financials which was originally developed by the e-Governments Foundation, Bangalore was customised to suit the municipalities in Andhra Pradesh. It is considered a fully integrated financial accounting and budgeting system. The main features are:

(i) Full compliance with the National Municipal Accounts Manual, approved by GOI-Ministry of Urban Development.

(ii) Double-Entry Accounting System with the ability to classify expenditure by capital and revenue; and receipts by fund, function, functionary, field and account codes.

(iii) Internet-based, which makes the rollout of the system extremely cost-effective.

(iv) Built mostly on open-source tools (except the RDBMS) which significantly reduces the total cost of deployment of the solution, making it perhaps the only enterprise class system in the market today which is built on open source technologies.

(v) It has the following sets:

(a) Transactions
- Receipts
- Bills Accounting
- Payments
- Journal Entry
- Contra Entry

(b) Reports

(c) Master Data and Configuration[13]

Accounting Reforms in Chennai Municipal Corporation

It is interesting to note that Chennai Municipal Corporation introduced the accrual based double-entry accounting system as back as 1982. The experimentation in Chennai, in fact, acted as a training ground and helped the Government of Tamil Nadu to extend this system to all the municipalities and corporations in the state under the World Bank assisted TNUDF. The accounts are being maintained on historical cost and on accrual basis. The Corporation maintains three major funds viz., Revenue Funds, Capital Funds and Elementary Education Fund and a separate Balance Sheet is prepared for each of the fund. The Corporation prepares two main financial reports namely, Income and Expenditure Statement and Balance Sheet. The treatment of income, expenditure, assets, advances, etc shows that the accrual system is not being followed totally and a customised version adopted is in vogue. Some of the variations are explained below.

Treatment of Income

(i) Property tax is accounted on accrual basis while the other receipts are accounted on cash basis. Further, property tax recoverable for the latest six half years and property tax recoverable prior to six half years are shown separately.

(ii) Elementary education collection charges, project appropriation interest are accounted notionally.

(iii) Other receipts include adjustment entries relating to prior period.

(iv) EPF accumulation as on 1-8-84 includes the amount representing management contribution to be treated as revenue.

Treatment of Expenditure

(i) Retirement benefits are accounted on payment basis.

(ii) Nutritious Noon Scheme expenditure is accounted for as and when the amount is reimbursed by the government and includes prior period expenditure also.

(iii) Interest on loan is accounted on accrual basis.

(iv) Miscellaneous expenditure includes prior period expenditure also.

(v) All other expenses are accounted on payment basis with a provision for bills received at the end of the year.

(vi) Depreciation is provided as per Companies Act 1956 (since revised in 2013) on written down value on the opening balance and it is calculated on addition during the year.

(vii) Capital projects include final payment made but completion report not received.

(viii) Election expenditure recoverable represents election expenses incurred on behalf of the State Government but not reimbursed.

Fixed Assets

(i) Fixed assets are shown as a Gross Block and are accounted only on payment of final bill on receipt of completion report.

(ii) Stock is valued on weighted average method.

Others

(i) Advances include amount utilised for services and/or materials supplied but not adjusted.

(ii) Bank account maintained by the drawing officers is not reflected in the Book of Accounts.

(iii) Cash and bank balances include bank accounts closed with the banks and are not reconciled in the books.

(iv) Investments represent fund created for training Corporation employees and it is shown as cost.

(v) Contribution represents grants received and utilised.

Fund-based Accounting in Brihat Bangalore Mahanagara Palike

The Brihat Bangalore Mahanagar Palike's (BBMP) financial transactions increased with growing investments on urban facilities and infrastructure since 1990's. But, the civic authorities were not having a clear picture of the resources generated, use of funds, assets created, liability status, etc. which are essential for financial decision making to plan, appraise performance and monitor the implementation process. In order to overcome these inadequacies, the BBMP embarked on a holistic accounting reform called Fund-Based Accounting System or FBAS. Implementation of the FBAS came into effect from 01-04-2001. The accounting reform was synchronised with the Management Information and Reporting System of the Municipal Corporation. The entire accounting data was computerised and an information structure was created to enter and retrieve the data in appropriate form by customising the software. It took over a year to complete the entire process and the BBMP could prepare budget in the new FBAS form with computerisation for the year 2002-03.

The BBMP switched over to Fund Based Accounting System based on Generally Accepted Accounting Principles and Double Entry Accounting System. Three funds viz., Governmental Fund, Enterprise Fund and Fiduciary Fund were created for this purpose. If the Governmental Fund covers most of the maintenance functions of the BBMP, the Enterprise Fund covers all the commercial activities like shopping complexes, remunerative enterprises, etc. The Fiduciary Fund deals with the externally aided projects. All the transactions are classified Fund-wise. All the accounts are being maintained in double entry method irrespective of accrual or cash basis. Regarding accrual accounting, a modified system is being followed. Accrual is 100 per cent in case of the Enterprise Fund transactions and it is modified in case of the Government Fund. Incomes like taxes, non-taxes and grants are being maintained on accrual basis while the loans are being maintained on cash basis. All expenditure is on cash basis. In respect of loans, interest part is maintained on accrual basis and principal payment on cash basis.

Accounts are totally computerised. Mainly, three reports viz., Annual Revenue and Expenditure Statement, Balance Sheet and Cash Flow Statement are being generated apart from other specific purpose reports to serve each department. The entire accounting was restructured and the BBMP (Accounts) Regulations were adopted by the Municipal Corporation in 2001. The accounting formats prepared are reported to be well appreciated by the officials of the C&AG and other government departments. These formats facilitated maintenance of the municipal accounts more efficiently. The coding of payment bills for works also improved financial control. Another innovative reform introduced in the BBMP has been the practice of publishing the annual accounts along with the Balance Sheet in the national newspapers since 2002-03.

The most significant impact of the accounting reform has been on the flow of information across various departments and levels within the department. It is observed that the flow of information is made quickly.

For instance, duration was reduced from 48 days to 48 hours in the finance department.[14] The other advantage was a single data source was created. The reform helped producing daily reports on various activities which are very helpful in monitoring and controlling. Further, the accounts have become more transparent, accurate and reliable. The budget information was disseminated to the citizens through quarterly public meetings organised by the Public Record of Operational Finance (PROOF). It is brought-out that the accurate recording of data has brought down the perceived corruption index from 22 per cent to 9 per cent in three years as per the report card published by the Public Affairs Centre, Bangalore in an independent survey.[15]

State Level Initiatives

The state level initiatives are essential to make the reform initiatives cost-effective sustainable and comprehensive to cover all categories of the municipalities. The state governments should mainly provide the required state level policy support, guidance, legal backing and institutional support in cost-effective manner. The JnNURM accounting reform agenda entrusted following responsibilities to the state governments.[16]

(i) Constitute a state-level steering committee with the authority to promulgate accounting standards, formats for budgeting and costing, financial statements and audit reports.

(ii) Pass a resolution expressing commitment to establish modern accounting system.

(iii) Appoint a consultant for development of a state-level municipal financial accounting manual.

(iv) Establish a state-level project management unit under the guidance of the steering committee to oversee implementation and conduct capacity building and training for municipal officials.

(v) Review the legislative framework to identify the changes necessary to implement the improved financial management system. Implementation of the improved financial management system in all municipalities and urban civic service providers may require amendments to relevant legislation, byelaws, rules/regulations, etc.

(vi) Provide guidelines on expenditure management and budgeting to better control expenditures, prepare outcome-based budgets, ensure accountability, and involve stakeholders in setting budget priorities.

(vii) Develop a state-level accrual-based accounting manual detailing the accounting system, policies and procedures, the records and documents to be maintained, and the form, contents and periodicity of various accounting reports.

(viii) Provide guidance on required manpower, training and implementation to switch over to the improved financial management system. Formulate appropriate policies for necessary handholding assistance and internal audit from chartered accounting firms. Based on the National Municipal Accounts Training Manual, state governments shall prepare a state-level training manual. Further, states shall evolve a suitable recruitment and transfer policy for municipal officials to ensure reasonable stability of tenure of finance personnel.

(ix) Develop/customise accounting software to provide meaningful MIS for managerial decision-making through the new financial management system.

Under AMRUT the state governments and the municipalities are required to augment the process of implementing Double-Entry Accounting system initiated earlier under JnNURM.

Tamil Nadu Accounting Reform - an Experience

Tamil Nadu is considered progressive in respect of the municipal accounting reforms. Accrual system of accounting was first introduced in the Chennai Municipal Corporation as back as 1982. The system was improved over a period of time with computerisation. It is noteworthy that Tamil Nadu is the first state in India to introduce the Double Entry Accrual Accounting System in all the municipalities and corporations. Initially, this system was introduced in ten municipalities on 01-04-1999 and it was extended to all other municipalities from 01-04-2000. The World Bank assisted Tamil Nadu Urban Development Programme played a key role in devising the accrual system of accounting for the municipalities, training and introducing the new system there. In fact, this accounting reform could be considered a World Bank driven initiative. The accrual system is customized to cover limited transactions like revenues. The Municipal Administration and Water Supply Department, Government of Tamil Nadu prepared the state level Accounting Manual for Urban Local Bodies in Tamil Nadu to guide the accounting personnel at the municipal level and other state officials. Training was imparted to the functionaries looking after the municipal accounts for implementing the new accounts system.

Prior to the introduction of the accrual system of accounting, the accounts were maintained on cash basis wherein the receipts actually received and payments actually made under various heads of account were accounted in the form of a Receipts and Payments Statement. The annual accounts prepared, therefore, reflected only a cash flow that is, money actually received and spent during a financial year without taking into account the accrued income and outstanding liabilities. This apart, a Demand, Collection and Balance

(DCB) statement for tax and non-tax revenues was also annexed to the Annual Account. The Statement of Assets and Liabilities was also prepared but not in the form of a Balance Sheet to represent the true financial status of the municipality concerned. Because of the advantages, the old system of cash-based single entry accounting system was replaced with the accrual double entry system of accounting.

The new system introduced in Tamil Nadu produces two main financial reports, viz., Income and Expenditure Statement and the Balance Sheet. If the former facilitates determining the financial performance, the latter helps in assessing the financial strength of the municipality concerned. These reports are prepared on accrual basis so that they conform to the Generally Accepted Accounting Principles (GAAP). It presumes that the financial transactions and events are governed by the substance and not by mere legal form. Main features of the new system are:[17]

- It continues the existing Funds Concept under broad categories viz., General Fund, Water Supply and Drainage Fund and Elementary Education Fund. Transactions are recorded for each fund separately and separate financial statements are prepared.
- The Revenue and Capital Fund are distinguished based on the nature of transactions.
- Assets are properly valued and capitalised by providing for depreciation every year at the prescribed rates.
- It provides for creating a Sinking Fund wherever necessary for repayment of government loans, loans obtained from financial institutions and to invest such funds.
- It provides for doubtful revenue items of taxes, fees, etc.
- Subsidiary ledgers are maintained to analyse certain accounts in detail.
- The account books are closed on monthly basis by preparing the prescribed primary records like receipt book, payment book and general ledger.
- A monthly Trial Balance is prepared to verify the correctness of the posting in the ledgers. Based on trial balance, monthly accounts are prepared.
- Income and Expenditure Account is prepared every year.
- The Balance Sheet is prepared every year, which includes all the assets and liabilities of the municipality with proper valuation. It is observed that the valuation of the buildings is done by providing for depreciation at the prescribed rates. But in case of the lands, the initial values are taken and they are not revised every year.

- The DCB Statements are also prepared but as per the existing system.
- In the year of introduction of the new system, Opening Balance Sheet was prepared which included 24 schedules. Again in the subsequent years, Opening Balance Sheet is not required to be prepared as the Balance Sheet prepared at the end of the financial year covers all the information.
- Codification of Accounts was done with Alpha Codes for various departments within the municipality for identification.
- The accounts are maintained in double entry system of book-keeping, which covers both, personal and impersonal accounts. Personal accounts represent money due from a person or group of persons. The impersonal accounts are of two types - real or asset accounts and nominal or fictitious accounts that relate to income and expenditure accounts. In the double entry system, both, the debit and credit are treated and recorded. That is, two-fold effect of a financial transaction is recorded to make the transaction meaningful and full-fledged.

Further, a User Manual is also prepared by the Tamil Nadu Urban Development Project-II which explains the accounting flow, documentation, record maintenance and report generation under the accrual accounting system using the double entry system of book-keeping. The user manual is prepared to facilitate computerised accounting in the municipalities. For this, account heads are classified into four major heads each series starting with 1000, 2000, 3000 and 4000.

The following illustrations[18] bring-out the differences between the cash basis and the accrual basis with reference to accounting of revenues and expenditure in municipalities in Tamil Nadu. Under the cash basis, the Revenue Department sends the demand notices to assess, but the Accounts Department has nothing to do with it except, when money is received there. On receipt of money by the Revenue Assistants, it is remitted into the Municipal Treasury. The Treasury itself may also receive money directly. The Shroff (Cashier) remits the money into the bank by challan and sends the counterfoil with an abstract of receipts for the day and the Accounts Department classifies revenue according to the heads of account and enter the same in the Receipt Cash Book. If assessees remit money directly into the bank, the scroll is collected and such receipts accounted in the Receipt Cash Book. Assessee-wise details are furnished to the Revenue Dept. to prepare DCB (Demand, Collection and Balance) statements for follow-up of arrears. Though it is ascertainable at the end of the year whether all demands raised are paid or not, such data is not made use of for any accounting purpose as no cognizance is taken in accounts of arrears of revenue. The receipts during a particular period may relate to the current period or a previous period, but that does not make any difference under cash accounting.

As against this, in case of the accrual system, as soon as the Revenue Department raises the periodic demand notice, sends an abstract thereof to the Accounts Department. It treats the entire amount of money is due whether received immediately or not. A journal entry is passed to take credit of the revenue accrued for the particular quarter or half year as the case may be under different heads of accounts like property tax, water supply and drainage tax, education tax, etc. and the sum total of the due is debited to a "Recoverable" account. As and when money is received, the Shroff prepares a Bank Receipt Voucher (following the same procedure as in the other case) and the Accounts Department verifies the same and the amount is set off against the recoverable account. By this, the revenue due for a particular period gets credited fully and what is remaining uncollected is shown as due from assessees and is displayed as an asset in the Balance Sheet. When money is received against these overdue accounts in the succeeding year, the recoverable account only gets credited instead of the concerned revenue heads. Thus, each year (period) account gets its due credit for the entire tax due and receipt for different periods do not get mixed up. The process of preparing DCB statements goes on simultaneously as in the case of cash basis accounting.

Regarding the expenditure under cash basis accounting, as and when the telephone bill is paid, the amount is booked under the head of account concerned. It does not make any difference whether the bill for the month of January/February 2000 due to be paid by 7.4.2000 is received on 20.3.2000 but not paid before 31.3.2000, except that the payment is monitored through a separate bill register called "Register of Contingent Charges". In April 2000, when the payment is made, the amount is booked. Under accrual accounting, the accounting period being year ending 31.3.2000, the bill for charges up to 31.3.2000 has to be necessarily accounted in that year itself. If the bill has already been received but not paid, a journal entry is passed to create a liability for payment (Debit- Telephone Charges A/c and Credit- Telephone Charges Payable A/c) for the exact amount. If the bill has not been received, a provisional amount is booked. When the payment is made in April 2000, the liability account "Telephone Charges Payable" is debited and bank account credited, thus squaring off the payable account.

Fund-based Accounting System—Karnataka Initiative

In Karnataka, the Karnataka Municipalities (Accounting) Rules, 1965 (KMAR) prescribe the accounting formats and requirements for municipalities while the corporations can formulate their own regulations and book keeping formats along with other rules and prescriptions by the government. Accordingly, BBMP had adopted separate Accounting Regulations in March, 2001 while paving way for introduction of the Fund Based Accounting.

The municipalities in Karnataka were following the cash-based single entry system. This system is based on maintenance of registers on a single entry system. Several registers and forms (over 90) are maintained to record the financial information. Cross-verification and interlinking of information across the registers to get a comprehensive financial status is missing in this system. Another weakness of this system is limited distinction between the revenue and capital account transactions. In view of weak segregation of revenue and capital financial transactions and lack of proper asset maintenance, the municipalities are not in a position to prepare a Balance Sheet indicating financial position more accurately.

Rationale and Features of Fund Based Accounting

In recent years, the functional and developmental role of the municipalities has been expanding. Funds are being raised from different sources viz., loans, capital market, specific purpose grants, etc. unlike in the past wherein the municipalities depended mostly on government grants. This demands that the municipalities need to treat themselves not as a single entity in financial terms but a combination of several entities deriving income from different sources and being responsible differently for different funding agencies. Presently, the scheme funds are generally kept in separate accounts in banks as per government instructions. But, it is desirable from accounting point of view that they need to be accounted separately along with the assets and liabilities created under each fund. This will facilitate control and monitoring of performance of each scheme as each scheme has different focus and objective. The fund-based accounting provides for such facility. The fund-based accounting, in simple terms, has the following features:

(i) It is funding centered.

(ii) Treats each fund as a distinct accounting entity with separate book of accounts.

(iii) It has a self-balancing set of accounts recording cash and other financial resources, together with all related liabilities and residual balances and changes therein.

(iv) It accounts for specific set of activities with special laws and regulations of internal requirements.

(v) It facilitates better decision-making.[19]

In order to facilitate the change over to fund-based accounting, The Karnataka Municipalities (Accounting and Budgeting) Rules, 2006

(KMABR) were formulated and these Rules were initially implemented in 57 municipalities and 6 municipal corporations in 2006. The KMABR are made primarily to meet the requirements of the double entry accrual accounting system. The key features of the KMABR are:

(i) Every Municipal Council shall maintain its books of account using the double entry system of book-keeping following accrual accounting principles.

(ii) Municipalities are required to maintain all their accounts in three funds initially viz., General Fund, Water Supply and Sewerage Fund and Enterprise Fund. The water supply and sewerage fund relates to all activities like acquisition or purchase of water, treatment and distribution and underground drainage and sewerage services. The commercial activities of the municipality like shops/markets and all other activities which are intended to make profit or at least cost recovery are covered in the enterprise fund. The general fund is like a residual fund covering all other activities which don't fall into the above two categories.

(iii) Fixed Asset Register has to be maintained for land, immovable assets and fixed assets separately while applying the depreciation rates prescribed.

(iv) The budget has to be prepared as per new format keeping in view the fund-based accounting.

(v) A financial statement audit has to be conducted.

(vi) The DCB for property tax has to be prepared in a new format which is designed to suit the self-assessment scheme.

(vii) A Balance Sheet has to be prepared.

Each fund shall have a separate book of account. Each financial transaction like vouchers, receipts and payments need to be identified and recorded into this fund. Each fund should have a separate bank account. For each inter-fund transfers, the money should be physically transferred from bank account to another through proper recording into the fund accounts. At the end of each month, the amount appearing as 'Receivables from other Funds' should be reconciled with the 'Payable to other Funds' amounts in other funds and at the consolidated level, the inter-fund accounts display nil balance which will make the system perfect.

The account codes are designed in Karnataka on the lines of the National Municipal Accounts Manual for major heads (two digits) and minor heads (one digit), The sub-minor heads (one digit) are to be prescribed by the state government and the detailed account codes (two digits) could be added at the municipality level. Further, Fund Codes, Source of Financing Codes and Function Codes are also prescribed which are to be followed in the state. They are:

(A) Function Codes

• General Fund	1-01
• Water Supply and Sewerage Fund	1-02
• Enterprise Fund	2-00

(B) Source of Financing Codes

• Own Sources	01
• Specific Grants	10-49
• Loans and Bonds	50-89
• Earmarked Funds	90-99

Function Codes are prescribed group-wise and function-wise. The group codes are in single digit ranging from 0 to 9. The function codes are in double digits. The major function groups identified are: general administration, planning and regulation, public works, health, sanitation and solid waste management, civic amenities, urban forestry, urban poverty and social welfare, other services and revenues. Each function of the municipality comes under one of these functional groups.

Implementation of Fund-based Accounting in Municipalities

Successful implementation of the Fund-Based Accounting System with Double Entry Features in BBMP and Tumkur City Municipal Council had motivated the Government of Karnataka to extend this to 57 more municipalities under the Nirmala Nagar Project taken-up with ADB assistance and to the remaining 163 municipalities under the Karnataka Municipal Reforms Project assisted by the World Bank. The Director of Municipal Administration is the nodal agency for these accounting reforms introduced in the municipalities. A FBAS Coordinator is appointed to assist the Director of Municipal Administration. A Technical Committee was constituted to review, comment and approve the various reports prepared by a Nodal Firm on behalf of the government.[20] A Nodal Firm viz., M/s Infrastructure Professionals Enterprise Pvt. Ltd., New Delhi was appointed in 2004 to enable the implementation. An MOU was signed with e-Government Foundation (an NGO) for development of the municipal accounting software. A chartered accountant has been appointed as FBAS Coordinator. An Expert Management Group is also being developed as in-house facility to enable the DMA to implement the accounting reforms in the remaining municipalities after the withdrawal of the consultants (Chart 6.1).

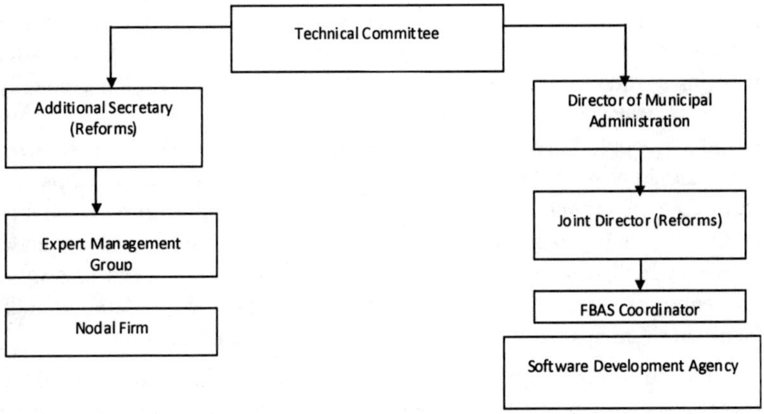

Source: Nilaya Mitash, Accounting Reforms in Karnataka, City Manager, Vol. 8, July-Sept., 2005.

Chart 6.1: Institutional Arrangements for Accounting Reforms in Karnataka.

The capacity building measures taken-up at the municipal level are commendable in Karnataka. Fresh recruitment of accountants was made through the Karnataka Public Service Commission by amending the Cadre and Recruitment Rules under the Karnataka Municipalities Act. Minimum qualification fixed is B.Com. Initially, all these accountants were trained on fund-based accounting concepts and its application in the municipalities. Five-day theatrical training and a two-day practical training was provided. This apart, each of 57 towns was provided with one IT specialist. A team of accounting specialists consisting of a chartered accountant, two accountants and one data entry operator were provided to support and help each of the town. These expert teams are appointed on consultancy basis to facilitate transfer from single entry cash basis to fund-based double entry accrual model being adopted. The ADB funded the entire computerisation project under the Nirmala Nagar Project. The initiative is extended to all the municipalities in the state under the World Bank assistance.

Municipal Accounting Reforms in Bihar[21]

The municipalities in Bihar have adopted Double-Entry Accounting on accrual basis. The state government thus has formulated the Bihar Municipal Accounts Rules, 2014. This is mainly drawn on the basis of National Accounting Manual.

Accounnting Reforms—Towards Integrated Solutions

The limitations of the cash-based single entry accounting system coupled with the increasing flow of funds and institutional funding in the urban infrastructure development sector created compulsions to shift to the double-entry accrual-based accounting system. Though the accounting reforms started with international funding through the World Bank, ADB, etc., gradually, they became part of the national reform agenda. In tune with the Government of India's recognition of accounting reforms as the critical element for streamlining and improving the municipal financial management, the municipalities across the country are slowly migrating to accrual double entry accounting system. The national urban reform agenda of the Government of India under erstwhile JnNURM and latest initiative of AMRUT accord top priority to accounting reforms. Several national initiatives, especially preparation of the National Municipal Accounts Manual facilitated the process.

The municipal corporations like Mumbai, Chennai, Bangalore, Hyderabad, etc., have taken the lead role and initiated double-entry accrual based reforms with varying degree of success. These municipalities faced several operational problems viz., lack of well qualified staff, internal resistances for change, legal and administrative constraints, etc. The systems being practiced are not totally accrual based but customised versions through experimentation over a period of time. The modifications are essential to make the system municipalised and feasible from operation point of view. Now, the systems are more stabilised.

It is a welcome feature that most of the state governments are also giving priority to the municipal accounting reforms. The Government of Tamil Nadu's initiative is worth mentioning here. The World Bank played key role while the Government of Tamil Nadu initiated a state-wide accounting reform and successfully implemented. The Karnataka Government's initiative to introduce Fund Based Accounting with double entry and accrual features is one step forward in the accounting reform movement. Formulating and notifying the Karnataka (Accounting and Budgeting) Rules, 2006 to meet the requirements of the double entry and accrual accounting system is the major legal step taken in Karnataka. Several other states are also taking similar initiatives. The JnNURM scheme provided further boost to accounting and financial reforms in municipalities in the country. And now the AMRUT is focusing on augmenting the process and determined to boost the municipal accounting reforms across the country.

The advantages of the double-entry accrual accounting systems are indisputable. The real challenge, however, is to make the municipal systems adaptable to these changes. This requires enormous capacity building efforts. The basic problem of the municipalities has been that they lack qualified and competent accountants. The officers who look after municipal accounts are semi-qualified and not trained in most of the municipalities. In this situation, building a professional cadre of municipal accounts officers at the state and municipality levels becomes essential. The Ministry of Urban Development suggested that all the state governments should create municipal cadres in different functional areas including accounting. Here also, the Karanataka Government's initiative of creating accounts cadre for the municipalities and recruiting young accountants is laudable. The approach of providing legal and administrative support, infusing young talent and refining and molding the raw talent to the municipal needs through well planned capacity building are the logical steps needed to make the initiative workable and sustainable. This is what the Government of Karnataka has attempted. The slower pace of progress should not be a cause of worry.

The IT solutions also played key role in the municipal accounting reforms. The computerisation, software development, and introducing web-based operations- all have facilitated more efficient, user friendly, transparent and faster municipal accounting and financial management. The Tamil Nadu Government's initiative and the Karnataka Government's comprehensive and long-term vision could be considered as inspiring examples in the accounting reforms. This demonstrates that the state level initiative and commitment play significant role for implementing any reforms successfully. The municipalities should also be committed for these reforms and strive hard for their realisation. The success and impact of any urban reform ultimately depends on the receptivity, commitment and absorption levels of the municipalities. Though the success of accounting reforms in larger cities is laudable, the reforms must be widened covering all categories of municipalities in a cost effective manner. The municipalities especially, larger cities, have initiated and played leading role so far. Now, it is the turn of the state governments to lead and provide necessary institutional support for further strengthening and scaling-up. The reform experiences must be consolidated, integrated and taken forward to realise the total potential of the accrual-based double entry system towards building more robust municipal financial management systems.

References

1. JnNURM Primer, Municipal Accounting-ULB level reform, Ministry of Urban Development, Government of India.
2. Steiss, Alan Walter and Nwagwu, Emeka O. Cyprian (2001): Financial Planning and Management in Public Organisations, Marcel Dekker, New York, p. 299.
3. *Ibid.*
4. *Ibid,* p. 331.
5. Joshi, Ravikant (2000): Municipal Accounting Reforms in India: A Comparative Study, submitted at the workshop on 'Accrual Accounting in the Public Sector' organised by the World Bank at Colombo, December 9-12.
6. National Institute of Urban Affair (2010):, Status of Municipal Accounting Systems in selected Urban Local Bodies in India (INDO-USAID FIRE (D) PROJECT), p. 24.

7. *Ibid.*
8. JnNURM Primer, Municipal Accounting-ULB Level Reform, Ministry of Urban Development, Government of India.
9. Ministry of Urban Development: Guidelines of AMRUT.
10. JnNURM Primer, Municipal Accounting-ULB Level Reform, Ministry of Urban Development, Government of India.
11. Vidya Sagar, Bloom Solutions, Consultant, GHMC.
12. *Ibid.*
13. Vijayawada Municipal Corporation, Andhra Pradesh
14. Narayan, S.V., Chandrasekhar, R., Srihari, P.K. and Kulkarni, Prashant (2006): Introducing Fund-based Accounting in Bangalore Mahanagara Palika: Catching Up with the Global Standards, The ICFAI Journal of Urban Policy, The ICFAI Business School, ICFAI University, Vol. I, No.1, pp. 52-69.
15. *Ibid.*
16. JnNURM Primer, Municipal Accounting, op.cit.
17. Accounting Manual for Urban Local Bodies in Tamil Nadu, Municipal Administration and Water Supply Department, Government of Tamil Nadu, pp. 4-5.
18. These illustrations are drawn from the Training Manual, Municipal Administration Department, Tamil Nadu.
19. Fund Based Accounting System (Training Material for Senior Municipal Administrative Officers), Directorate of Municipal Administration, Govt. of Karnataka.
20. G.O.No.UDD 60 PRJ 2004, Government of Karnataka, dated 18-08-2004.
21. Bihar Municipal Act, 2007; and Bihar Municipal Accounts Rules, 2014.

Chapter - 7

Recognising Urban Land as a Resource

Introduction

The urbanisation has profound impact on land and its built environment. The landscape transforms from a predominant farm based uses to a complex web of land uses to meet the requirements of non-agricultural activities of the people. Coupled with rapid population increase, the living habitat grows vertically and horizontally exerting severe pressure on land space. If the governments fail to tackle land issues rightly and efficiently, it poses critical problems for sustainable development. It is pertinent here to reiterate the statement made by the National Commission on Urbanisation that behind much of the urban chaos lies the total mismanagement of land as the principal resource for urbanisation.[1] Individuals apart, a plethora of agencies, government and private, are involved in the development and utilisation of land in urban areas. Traditionally, the government role was confined to acquisition, land use regulation and development control. Several legislations viz., Land Acquisition Act, 1894 (since revised in 2013), Urban Land (Ceiling and Regulation) Act, 1976, Municipalities and Municipal Corporations Acts, Panchayat Raj Acts, Slum Clearance/Improvement Acts, Urban Areas Development Acts, Assigned Lands (Prohibition of Transfer) Acts, Town Planning Acts, Buildings (Lease, Rent and Eviction) Control Act, etc. and a host of accompanying rules and regulations provide the administrative control mechanisms. The predominant focus on acquisition, land use regulation and development control instruments coupled with legislative and administrative loopholes, has made urban land administration uncoordinated and chaotic.

Instead of leading to orderly development, urban growth assumed myopic form and led to unintended consequences like shortage of land to poor people, slums, illegal transfers, increased land prices, land speculation and underutilisation/misuse of precious urban land. Mismanagement of urban land by the governments has resulted in the gross underutilisation and misuse of land as a resource. Unless land is recognised as a resource, urban planning cannot achieve the goals of urban development set for the nation. In recent years, there is awareness among the policy-makers and the administrators in India on this aspect. It is a welcome trend.

Two incongruent trends are discerning in the sphere of urban land in India. If realisation of the financial potential of urban land by the governments and the private sector is a positive trend, the chaotic land management practices, at the other end, are frustrating and hampering development in urban areas. In this paradoxical situation, realisation of development goals and policy objectives in the vital sectors like housing and urban infrastructure development is turning very challenging. On the developmental scene, privatisation, market orientation and price consciousness, the global themes are finding their way into the urban land policies and reforms too, in India. This chapter while analysing the Indian urban land policies and the chaotic administrative land management practices, attempts to demonstrate the potential of land as a resource for urban local governments and other urban development institutions.

Urbanisation and the Critical Issue of Land

The issue of urban land basically reflects the spatial dimension of urbanisation. Land can neither be created nor destroyed but use of the land only undergoes changes based on human requirements. Governments have control to regulate the land uses and densities. Traditionally, the focus of governments in India was on cost considerations with a view to control the costs of providing civic services and infrastructure. Underlying assumption is that increased densities increases costs, hence, the policy was to restrict the densities through FAR regulations. In controlling FAR, the goal of the Indian planners has been to indirectly limit both, job and population densities.[2] The reason being that, excessive density results in the loss of environmental quality and increased traffic congestion. In addition, higher densities would place greater demands on urban infrastructure which Indian cities with weak technical capacities and inadequate tax revenue feel ill-equipped to provide facilities in core areas. Further, the assumption is also that inadequacies in civic infrastructure could be well managed in less dense areas than in the high density areas.

With density and cost considerations in focus, the cities tend to expand horizontally leading to urban sprawl along the existing transportation corridors. The growth in urban areas thus, has an in-built tendency to expand beyond the municipal limits. The city assumes multilayer form starting from the central area expanding into farther areas. This requires proper transportation linkages across the widely spread settlements. Further, it also necessitates macro level infrastructure linking various developing areas through mega projects in sectors like roads, metro rail system, water and sewerage, solid waste management, parks and urban forestry, etc. As the population increases, the demand for housing also increases proportionately for which availability of serviced land is the critical factor.

Focus of the governments shifted towards land in the backdrop of fast urbanisation and haphazard urban growth taking place in and around the cities in the country. It is indisputable that land plays a pivotal role in the expansion of urban areas. In the spatial perspective, land is central to all the urban activities often leading to stiff competition among various stakeholders. The conversion of agricultural uses of land to the non-agricultural or urban uses is the key in this process. Land is required for various habitat needs of the people: housing; basic infrastructure facilities like roads, parks, community centres, water supply and sewerage, drainage, solid waste disposal sites; and also for commercial and industrial activities. Managing peripheral urbanising land efficiently becomes crucial in this respect as the competition among different uses leads to unhealthy trends like speculation, increased land prices, lack of access of land to deprived sections of the society and sometimes, uneconomic utilisation of scarce land.

The dated legal and administrative practices relating to urban land management have become inadequate to deal with the current problems. Though the governments have realised financial potential of urban lands, workable solutions and strategies are yet to be concretized. Short-term and partial solutions like regularisations, land sales through auction, regular increase of land values for registration purposes, etc., are being pursued focusing on one side of the issue that is, resource mobilisation. The other side of meeting housing needs and urban infrastructure issues are relegated to back. Effective national and state level urban land policies and strategies are the critical miss in India.

Urban Land Policies in India

Though India is basically considered agriculture-based country, the issue of urban land has been receiving attention of the think tank in recent years. Various committees and commissions directly or indirectly have been expressing the concerns while providing inputs for policy makers. In spite of these inputs, it is unfortunate that the national policy on urban land is yet to take concrete shape. The beginning of urban land policy can be traced back to 1960s. A Committee on Urban Land Policy was appointed by the Government of India in 1965 to evolve a comprehensive policy framework on urban lands. The Committee framed the following policy objectives:[3]

(i) to achieve an optimum social use of urban land;

(ii) to make land available in adequate quantity at the right time and for reasonable prices to both, public authorities and individuals;

(iii) to encourage cooperative effort and bonafide individual builders in the field of land development, housing and construction; and

(iv) to prevent concentration of land in a few private hands and specially to safeguard the interest of the poor and underprivileged section of the society.

The Task Force on Planning of Urban Development (1983) reiterated these objectives and added two more objectives for optimum use of scarce land in the face of rapid growth, especially in the large cities. They are:[4] (i) to encourage the socially and economically efficient allocation of urban land such that land development is done in a resource effective manner and that the magnitude of land used is optimal; and (ii) to promote flexibility in land use in response to changes resulting from a growing city.

The governments have to tackle urban land management issues effectively for realising the urban development objectives. Major land reforms are required in this respect. Repealing the Urban Land (Ceiling & Regulation) Act (ULCRA) is one such step. Foremost need is developing proper live data base on urban land. This data base must be shared across the agencies concerned and properly coordinated. For managing urban land efficiently, the National Commission on Urbanisation (NCU) recommended a Directorate of Urban Lands which will prepare land records for all the urban areas.[5] As a step toward this, the Ministry of Urban Development constituted a working group in 2008[6] to workout guidelines for implementation of guaranteed title to land and property which was a mandatory reform under JnNURM scheme launched by the Government of India. This reform was expected to reduce land related litigation, encroachments and land grabbing; facilitate urban planning; improve tax administration; and promote urban development and renewal projects in cities and towns.

There is urgent need to evolve proper land policies to manage the lands effectively and purposefully. The policies should address housing, industrial and infrastructure needs in right proportion as per the requirements and priorities. Ensuring adequate lands for slum development and shelter programmes along with tenurial rights to the slum dwellers has become essential. The National Urban Housing and Habitat Policy, 2007 provided the required action areas in this respect.

Land must be treated as resource and land use planning should also include resource planning rather than just a zoning exercise as recommended by the National Commission on Urbanisation (NCU).[7] The NCU also recommended high incidence of taxation on vacant land which will de-motivate people to hold land for speculative purposes for long years.

This however, demands effective land records with accurate ownership, land use, and other details of land and better land management methods. Unless the land record and ownerships issues are sorted out, urban land remains grossly mismanaged and the problems of housing shortage, slums, squatters and unauthorised colonies continue.

Land use control should be made development oriented rather than just a regulatory exercise which only breeds corruption and illegal development. The critical element is arriving at right balance between environmental requirements and development needs. Management approach is needed here replacing the term land use and development control with land development management as suggested in the Karnataka Urban Development Policy.[8]

The prevalent policy and administrative deficiencies, coupled with increasing influence of the market forces on urban land, have forced the policy-makers and administrative authorities to devise pragmatic land development strategies for optimisation of urban and urbanising land utilisation. This perspective is clearly visible in the strategy suggested by the National Commission on Urbanisation (NCU) way back in 1988. The NCU suggested a more progressive urban land policy which could: increase supply to the public and private individuals as against restrictive policy which will increase land prices and affect the poor; facilitate cost recovery from the public investments on infrastructure for replicating the development on a wider scale; and widening the base of land ownership not necessarily by increasing the government ownership of land.[9]

The National Urban Housing and Habitat Policy, 2007 provides useful guidelines for the public authorities to manage lands effectively. Following measures are worth mentioning here:

- Removing legal, financial and administrative barriers for facilitating access to tenure, land, finance and technology.
- Suitable restructuring for enabling both, institutions at the Central and State levels, and the private sector for increasing supply of land.
- Forging strong partnerships between public, private and cooperative sectors for accelerated growth in the housing sector and sustainable development of habitat.
- Using land as a resource for housing with special focus on the urban poor.
- Encouraging land assembly, development and disposal both, in the public and private sectors.
- 10 to 15 per cent of land in every new public/private housing project or 20 to 25 per cent of FAR / Floor Space Index (FSI) which is

greater will be reserved for EWS/LIG housing through appropriate legal stipulations and incentives.

Legal and Administrative Aspects of Urban Land

In spite of policy pronouncements and legal support aiming at optimum utilisation of urban land as per the social and economic development objectives, the status of urban land management is in a very bad shape in India. The market forces have successfully thwarted most of the government intentions. Major causes for the present crisis in urban land are lack of proper land records and administrative coordination between the public agencies involved, ineffective implementation of Urban Land Ceiling and Regulation and Land Acquisition Acts leading to artificial shortages impacting civic and social sector needs, and increasing land prices due to free market forces.

Lack of proper land records has been the major problem for the urban land crisis. The existing land records are either dated or totally inadequate to meet the fast changing urban growth. Basic data like land conversions to urban uses—legally or illegally, quality of the existing land in terms of terrain, water bodies, etc., land transfers and ownership details, etc. are not readily available for planners and municipalities. Lack of coordination between several related organisations like municipalities, urban development authorities, industrial development agencies, slum development agencies, revenue department, registration department, etc. adds further confusion to the decision makers. The governments focused more on land acquisition through Land Acquisition Act or Urban Land (Ceiling & Regulation) Act (ULCRA). The legal litigations, corruption and administrative lapses have made these laws stumbling blocks rather than promoters of development. Vast areas of land were allocated to various public sector undertakings after Independence. Bulk of this land remained underutilised while certain land encroached due to defective land record system.

In the neo-economic liberalisation era land is viewed more as an investment instrument by both, public authorities and individuals. Vast lands are being allocated to industrial needs at cheaper rates to the private sector. There has been wide criticism that the governments are acquiring land at cheaper rates from the farmers and disposing these lands to the private industrialists and developers with several concessions. Land allotment scams have become common in recent years in India. In the absence of effective monitoring, the allotted lands are not rightly used as per the conditions. Of late, the urban development authorities are also auctioning the government land at high rates for private people to mobilise resources thereby leading to further land price rise. Regularisation of illegal land occupations, transfers and construction is another area which is being used to generate revenues.

These are short sighted strategies as long-term sustainable solutions are yet to take shape.

Land use regulation or development control has been the most ineffective area in urban planning. Land uses change in spite of master plans—legally or illegally. The purpose of mixed land use is never rightly interpreted. Administrative laxity and control orientation often leads to chaotic land uses which pose complex problems in terms of pollution, congestion, health hazards, loss of lung spaces, etc. In the process, the very purpose of urban planning is lost leading to a threat to the sustainable built environment in urban areas.

The chaotic land management in urban areas has given rise to several issues and problems. Important of these are: wide encroachment of public land due to inadequate/defective land records; growth of slums due to increased land prices and governments' inability to provide land to the urban poor; increased land speculation due to fast increasing land prices; illegal land transfers and housing colonies; and lack of adequate land for government shelter projects and providing basic urban infrastructure facilities.

In the metropolitan cities, large tracts of land lie underutilised or unused. The governments at different periods allocated land to different organisations in bulk and most of such lands are underutilised. These lands belong to charities, central and state governments, public and private industries, utilities like railways, ports, etc. Though these lands were in the peripheral areas originally, with expansion of the cities, they were merged into the core cities over a couple of years. They are located in strategic places, hence, considered very valuable. Blocking of these lands for development is being considered uneconomical in the context of shortage of lands in core areas. Further, due to such blockage, cities further expand into peripheral areas compelling governments to spend more on extension of infrastructure facilities in these areas. These lands should be put to maximum utilisation so that the pressure on the other lands many be reduced. Monetisation of such lands will certainly yield considerable income which could be invested on the infrastructure development. Public authorities face a problem here from the environmentalists and heritage conservation groups. Care should be taken to safeguard the heritage and environmental issues also in allowing densification in the core areas. Crucial balancing has to be done by the planners in this respect.

Potential of Urban Land as Resource

The land values rise due to multiple reasons like population growth, industrialisation, etc. One of the main reasons, often overlooked, is the value generated by infrastructure investments made by the municipalities and other public agencies. Value capture[10] builds on the principle that the benefits of urban infrastructure investment are capitalised into land values.

Peterson (2009)[11] estimates that, on an average, investing in wastewater removal leads to gains in land value 3.03 times the cost of investment, paving roads to gains 2.58 times the cost, and providing piped water supply to gains 1.02 times the cost. Because public investments contribute for increase in land values, many land economists have argued that the government should share in the capital gain to help pay for its investment. In fact, a study (2010) of the McKinsey Global Institute estimated that India has the potential to generate upto $27 billion per annum through land monetisation. The potential would be naturally very high in the metropolitan cities.[12] Public authorities have used a variety of instruments to capture the gains in land value created by infrastructure investment. Betterment levies, which impose a one-time tax or charge on gains in land value, are one such instrument. Most countries in the world have experimented with betterment levies at some point, typically taxing away 30–60 per cent of the gain in land value attributable to infrastructure projects. Betterment charges are the common levies in the municipalities but they are one time measures. In countries like Colombia, the concept of *valorisation* is being practiced in the form of various betterment charges or taxes.

The review of land based financing instruments being practiced in India reveal that there are three major streams viz., (i) monetary exactions like development charges, betterment charges, impact fee, etc.; (ii) land exactions method include land conversion charges and town planning schemes; and (iii) monetisation of land assets by land development and disposal methods. The municipalities mostly focus on the first category of monetary exaction methods while the urban development authorities and housing boards focus on the land exaction method and the town planning departments on the land management methods. The vacant land tax is not exploited by the municipalities (in many states) though legal provisions exist due to administrative reasons. Surcharge on stamp duty is a buoyant and regular source of income to the municipalities. A look at the revenue sources of the municipalities indicates the predominance of land/property based sources.

Urban land, either directly or indirectly, has become major financial resource for the municipalities and development of the urban areas. Traditionally, urban land is subject to different types of taxation and charges. The main land based taxes of municipalities are:

(i) Municipal property tax/tax on lands and buildings.
(ii) Vacant land tax.
(iii) Land-use conversion or change charge.
(iv) Tax on the land sales and other transactions in the form of a surcharge on stamp duty.
(v) Mutation fee by the municipalities for title transfers.
(vi) Value exaction/value capturing taxes/charges

These taxes/charges, except the surcharge on stamp duty and vacant land tax in many states are levied and collected by the municipalities. The surcharge is collected by the Stamps and Registration Department of the states and the proceeds are transferred to municipalities after deducting the collection charges. Of these sources, property tax or tax on lands and buildings and the surcharge on stamp duty are the major income generating instruments. However, certain initiatives are made in recent years to exploit the sources like vacant land tax and other value capturing charges. In view of the political and social limitations on property tax, municipalities are now turning towards other land based sources.

Apart from these revenue sources, the regularisation schemes have become another good source of income to the municipalities. Regularisation of illegal buildings and lay-out schemes are being devised in many states. The government is charging heavily for regularisation process. In fact, this is used as a resource mobilisation strategy by the governments. The potential is immense in the metro cities. The income from these sources is considered capital income as they are one time charges. This additional income has boosted the financial status of many municipalities motivating them to take-up more infrastructural development works and services.

Regularisation of Unauthorised Developments

The municipalities and urban development authorities in some states are imposing penalties on illegal land transactions and regularising the unauthorised constructions. These are one time penalties but generate substantial funds to the government coffers. In Andhra Pradesh, these instruments are being used extensively to generate revenues. They extend to several illegal and unauthorised transactions in respect of Urban Land Ceiling & Regulation Act (ULCRA) excess lands, lay outs, building permissions, etc. Urban growth has been largely haphazard without following the lay-out rules and building rules. As the municipalities and development authorities are unable to regulate the growth as per the norms, the illegal lay-outs and buildings should be demolished as per the existing laws. But, this task is also becoming very difficult due to different kinds of pressures. As a way out, the Government of Andhra Pradesh devised Lay-out Regularisation Scheme and Building Regularisation Scheme to regularise illegal lay-outs and buildings and at the same time, mobilise resources by charging heavily. The Building Regularisation Scheme was later renamed as Building Penalisation Scheme due to court intervention. The process of regularisation started with buildings constructed unauthorised on 100-200 meters of plot area as back as 1992 in Andhra Pradesh first in the Municipal Corporation of Hyderabad.[13] This was later extended to all the urban development authorities and their constituent municipalities in the state in August, 1992.[14] Later, the rates of compensation were increased at different periods till 1998. In 1998, a comprehensive building regularisation scheme was introduced in the state.[15] But, the legality of this order was questioned in the High Court which quashed the order in 2002.[16] In order to overcome the legal problems, the Government of Andhra Pradesh enacted a legislation called the Andhra Pradesh Regularisation of the Un-authorised Constructions in Municipal Corporations, Municipalities and Urban Development Authorities in the state in 2003. Penalties are levied on the unauthorised buildings by which the Greater Hyderabad Municipal Corporation (GHMC) and other municipalities in the state are generating huge income. This additional money is being utilised for various development works in the city.[17] These methods though improper and not desirable from the point of orderly development of the urban areas, are indirectly generating the income to the government.

While taking advantage of the increasing land prices in the five Urban Agglomerations where urban land ceiling is in operation in the state and where the rampant illegal sale of land/plots for residential and other purposes especially, in the peripheral areas has been on high pitch, the Government of Andhra Pradesh introduced the policy of granting exemptions under Urban Land (Ceiling & Regulation) Act (ULCRA) and regularising the third party land occupations of urban excess lands. Even after repeal of the central legislation of the Urban Land (Ceiling and Regulation) Act, 1976 in 1999, the Government of Andhra Pradesh continued to apply it in the State till 2008. In the meantime, the Government of Andhra Pradesh constituted the Cabinet Sub-Committee on infrastructure to advise the government on Urban Land Ceiling Policy.[18] The Cabinet Sub-Committee recommended that the ULCRA in the state may be continued and at the same time, formulated policy guidelines to streamline exemptions and allow regularisation to end litigation, provide housing, end land mafia activities, save land and settle all issues according to a clear cut policy.[19] Based on the Cabinet Sub-Committee recommendation, the government decided to regularise the excess lands under the third party occupations and this was done with a provision for payment of regularisation charges as prescribed. This has become one good source for resource mobilisation in Andhra Pradesh. But, for the people below poverty line as defined, the regularisations are being done free of any charges.[20]

The policy of regularisation of the surplus urban lands is being continued in the state in respect of the lands saved by the Repeal Act even after its repeal in the State in 2008. The Section 3 (1) (a) of the Urban Land (Ceiling & Regulation) Repeal Act, 1999 provides that the repeal of the principal Act shall not affect the vesting of any vacant land under sub-section (3) of Section 10, possession of which has been taken over by the government or any authorised person or competent authority. Though the government has officially taken possession of the surplus lands, it has been observed that many of the lands are actually occupied by the people called third party occupants who purchased land from the land holders. As the lands are already in occupation of the third parties and constructions were also made in certain cases, the government decided to continue the regularisation scheme in the state even after repeal of the ULCRA.[21]

The validity of the benefits and the rates of compensation for granting exemptions were extended from time to time under various Government Orders. The strategy of the state government was to mobilise resources from the extension of regularisation scheme from time to time and huge chunk of land was regularised and hectares of land is still available for raising resources under the regularisation scheme[22]. The Government, in 2008 itself, estimated to mobilise about Rs 5068 crore by regularisation as well as by auction. The figures, however, may come down based on the present downtrend in the urban land prices. The entire exercise of regularisation in Andhra Pradesh is devised mainly with the objective of mobilising resources to sub serve common good. It is, in a way, regularisation of the illegal process.

As the governments, to day, are not in a position to regulate the land markets efficiently with unduly biased acquisition approach, legal loopholes and administrative laxity, they are resorting to regularisation. Though it is a short sighted approach and termed as legalising illegal activity, it appears to be one method till the municipalities and the governments become strong enough to implement the rules and regulations effectively.

Land Monetisation Method

It is pertinent to mention here the Mckinsey report (2010) observation that India could generate $27 billion per annum by land monetisation methods as against the current revenue generation of $3 billion as per 2008 prices. In order to achieve this, four strategies are recommended[23]: (i) charging development fee on FAR increases especially, in the central business district areas and major transport corridors; (ii) auction of developed sites; (iii) levy of impact fee; and (iv) maximising the potential of property tax and user charges on water supply, sewerage, transit and affordable housing. Certain mechanisms are devised in India on these lines to benefit from the increasing land values. Prominent of these are use of FAR, impact fee and land sale through auctions. We shall discuss these methods in detail.

Use of FAR as a Resource

The municipalities can use land use planning method to regulate the urban land uses and at the same it can also be used to generate additional revenues. For instance, in some areas land use regulations like FAR can be relaxed within a broader framework of development and such relaxations are charged based on the property values existing in the area. The Floor Area

Ratio (FAR) of buildings is computed by dividing a building's total floor area by the area of land parcel on which it sits, and an upper limit on this ratio effectively puts a limit on the building's height. The planning theory has been that excessive density results in a loss of environmental quality, increased traffic congestion and puts higher pressure on the civic services like water supply, waste disposal, etc., hence, they should be limited to capacities of the civic authorities which provide these services. In view of the weak technical and financial capacities of the municipalities in the country, the normal trend has been to keep the FAR limits low to manageable limits. In the last few years, however, the capacities of civic administration especially, in the large cities have improved to some extent which allows scope for relaxations. The objective of generating additional income can be built into the planning exercise itself. The relaxations could be permitted at a cost. For instance, it is estimated that in Mumbai an FAR increase from 1.3 to 4 in key commercial areas could fetch about Rs 4,000 to Rs 5,000 per square foot.[24]

Impact Fees

Levy of Impact Fees on the complexes and high rise buildings is one strategy which could generate additional income. Levy of Impact Fees is basically to mitigate the problems arising out of construction of commercial buildings by change of land use that lead to increased traffic and necessitate decongestion measures. The Impact Fees are meant to address citywide problems emanating from high-density commercial development. Different versions of the impact fee could be devised by the municipalities based on the local conditions and level of development taking place. The concept of impact fee was first introduced in Greater Hyderabad Municipal Corporation (GHMC) in Andhra Pradesh. The Government of Andhra Pradesh permitted the GHMC to levy Impact Fees to mitigate the impacts of construction of commercial buildings by change of land use that lead to increased traffic and necessitate decongestion measures. The fee is being levied since 2002. Distinction is made between on-site and off-site (local area) development cost and citywide impact. The income so collected has to be deposited in a separate account of the Municipal Corporation and utilised for implementation of the Capital Improvement and Decongestion Plan i.e., for works such as road widening, link roads, slip roads, parallel roads, junction improvements including traffic signals, flyovers, rail over-bridges, rail under-bridges, etc.

The Impact Fee as a mechanism to raise revenues from the high rise and dense areas is now made applicable in respect of all high rise buildings in the state. In Andhra Pradesh, city wide impact fee is permitted in the municipal corporation areas and the urban development authority areas. Necessary provisions are made in the Building Rules governing the municipal corporations and urban development areas in Hyderabad, Visakhapatnam and

Vijayawada.[25] As per the provisions in the Building Rules, city level infrastructure impact fees will be levied on the buildings which are above 15 m height or above five floors whichever is less. The rates range depending on the height of the buildings, use of buildings and type of municipal area (Table 7.1). A provision is also made to utilise the TDR given to the owners toward payment of impact fee. Flexibility is also provided to use a mix of both depending on the owner interests.

Table 7.1: Rates of City Wide Impact Fee in Andhra Pradesh

Use	*Height of Building and rate in Rs per sq m of built up area*			
	16 m to 21 m	*22 m to 30 m*	*31 m to 50 m*	*Above 50 m*
	Municipal Corporation Area			
Residential	350	500	1000	2000
	Other areas of UDA Area			
	175	350	750	1500
	Municipal Corporation Area			
Commercial, offices, ITES, etc.	500	1000	2000	3000
	Other areas of UDA Area			
	350	500	1000	2000
Institutional, educational and others (except industrialsheds/factories)	*Municipal Corporation Area*			
	175	350	750	1500
	Other areas of UDA Area			
	100	200	400	800

Auctioning of Urban Land

The method of auctioning of land, instead of the earlier practice of allotment on cross-subsidy, is adopted by urban development authorities in India. This has fetched very good revenues to the urban development authorities as the valuable lands were auctioned at competitive prices. The land could be sold with or without development depending on the local conditions. For instance, the Mumbai Metropolitan Region Development Authority(MMRDA) had auctioned its land assets in Bandra-Kurla in Mumbai and funded more than $4.5 billion development projects like road, mass transit, affordable housing, etc.[26] In fact, the urban development authorities are funding huge projects like roads, flyovers, bridges, etc. by this method. The urban development authorities are increasingly depending on this method in the major cities where the land values have been appreciating considerably in recent years.

The case of Hyderabad Metropolitan Development Authority (HMDA) reveals certain important aspects of land auctions by urban development agencies. Though HMDA is basically a planning agency, it has been embarking on infrastructure development projects to meet the emerging needs of expanding city. As part of resource mobilisation strategy, the HMDA preferred auctioning of prime public lands on competitive bidding basis while taking advantage of the demand for lands and increasing land prices. The HMDA auctioned 166.821 acres of plots and lands and generated about Rs 1005 crores during 2002-10 (Table-2). Further, the Government of Andhra Pradesh alienated lands to the extent of 337.29 acres to HMDA. Majority of the lands were government lands and few surplus lands under ULCR Act. The HMDA sold these lands to various private industrialists and others. Money to the extent of Rs 1702 crore was mobilised through such land sales/allocations and the proceeds were remitted to government. Further, the HMDA generated about Rs 2762 crore on sale of the lands during the last decade or so and an amount of Rs 2684 crores was remitted to government.[27]

The above data reveals that the government has alienated lands to HMDA and the development authority sold these lands either after development or they were allotted to industries by the government where the HMDA collected the proceeds. Finally, the HMDA remained as an intermediary to government for sale of the lands as most proceeds are remitted to government back. Spending on infrastructure appears to be negligible from the proceeds of the land sales. If the HMDA formed a special fund out of the land proceeds and spent such amounts exclusively on infrastructure, there would have been a better accounting. The recent recession affected the real estate business very badly and consequently the demand for high prices lands came down and the general land price levels also declined sharply. This has affected the HMDA auction programmes as the authorities had to cancel auctions due to very less demand. It is told that some resources generated through auctions are being utilised for infrastructure projects like fly-overs, rings roads, etc. Interestingly, HMDA which did huge housing projects in the 1980s has abandoned that programmes and it is more focusing on the infrastructure projects.

Table 7.2: Details of Lands Sold by HMDA by Open Auction

Year	*Extent of Land in Acres*	*Sale Amount in Rs lakh*
2002	16.67	2440.80
2004	40.32	12084.00
2006	56.62	44511.00
2007	34.20	29645.51
2008	2.43	1790.07
2009	1.29	832.50
2010	15.29	9212.33
Total	**166.821**	**1005,16.20**

The urban development authorities (UDAs) cover relatively wider areas compared to the municipalities extending into the peripheral areas. The primary function of the UDAs is preparation of master plans and development plans for proper development of the urban agglomerations. But, these agencies have entered into housing, land and infrastructure development fields too for which they need enormous funds. The UDAs are preparing and implementing mega infrastructure development projects. In order to perform these roles, they need funds. Though the governments provide basic grants, the UDAs have to generate funds on their own to fund the development projects. In recent years, they are focusing on mega infrastructure and land development projects. In this context, the role of UDAs in urban land development becomes very significant. The UDAs are authorised to levy different charges and maintain remunerative enterprises like commercial complexes, parks, markets, etc.

The example of the Hyderabad Metropolitan Development Authority (HMDA) reveals the methods of income that could be generated from lands. Apart from the development charges which are considered more stable, the HMDA is collecting the Layout Regularisation and Building Regularisation charges in the areas falling under its jurisdiction. It has been found that as the income from land auctions and borrowings increased over the years, the share of the development charges declined. The land auctions have been playing significant role in recent years.

The land-based financing offers powerful tools that can help pay for urban infrastructure investment. For an urban region considering this strategy, a logical place to start is with an inventory of land assets owned by government agencies. Such an inventory would identify current land use and the market value of land. The government can then decide which land parcels would be more beneficial to urban development if sold to private developers, with the proceeds dedicated to infrastructure investment. The municipalities and urban development authorities should have clear land development policies and dedicated funds to mobilise and invest on urban infrastructure projects.

Land Management Approaches to Meet Land Requirements for Urban Housing

Meeting the housing needs of the poor and low income groups is another crucial area of concern for the governments. The traditional method of building houses for the poor and weaker sections by the governments has met only partial success. In view of the financial constraints of the governments, complementary solutions must be worked out. The better land management practices like town planning schemes, land title registration, providing access to subsidised credit and involving the private developers are considered sustainable.

Land Pooling has been promoted as a powerful Public-Private-Partnership (PPP) method in this respect. Land Pooling and Readjustment is a process of assembling small pieces of land/plots without paying any compensation to the land owners by the development authorities. The authorities then develop the land and reallocate to the original owners and retain some part of the land. These acquired lands are either used for public purposes or sold to recover the costs of development. This method is useful when the land is at the development stage. Some states like Gujarat and Maharashtra have laws (e.g. Maharashtra Regional and Town Planning Act, 1966 and the Gujarat Town Planning and Urban Development Act, 1976) to implement the land pooling and assembly projects in the urban areas. These land pooling methods provide an effective mechanism for the development authorities to convert the fringe areas of small bits as per plans with infrastructure facilities and reduces the problems like scattered land development, poor design and development, excessive land speculation, inadequate infrastructure, etc. Most importantly, this method provides scope for generating resources for critical urban infrastructure development and housing without any funding by the development authorities.

In Andhra Pradesh, there are legal provisions in the Hyderabad Metropolitan Development Authority Act, 2008.[28] The Act provides for Land Pooling Schemes by the Hyderabad Metropolitan Development Authority (HMDA) on its own lands or it can authorise any other body or private developer to take-up land pooling scheme in the Hyderabad Metropolitan Region. The land scheme should be in conformity with the development plan and investment plan of the HMDA. The scheme should provide for necessary land development, infrastructure, transport network, drainage and other civic facilities, preservation of historical objects, etc. One important provision is the reservation of land to the extent of five per cent for the purpose of providing housing to the low income groups and weaker sections of the society. Further, this land should be handed over to the HMDA free of cost which should be used for residential and commercial purposes. Again, from this surrendered land meant for development, at least 20 per cent of the lands should be reserved for LIG and MIG equally. The minimum land required for the land pooling scheme taken-up by the HMDA or its authorised agency is 75 hectares. In respect of the land pooling scheme taken-up by a private developer, the minimum land requirement is 50 hectares.

Guided Urban Development is another method by which the development authorities can ensure planned development without incurring public expenditure. In this method, the authorities develop private lands with their consent and receive donations from the land developers in the form of land and betterment levies for servicing the land with infrastructure facilities. Such Guided Urban Development Projects were implemented in Chennai under the World Bank Urban Development Project by the Chennai Metropolitan Development Authority.

Another PPP model has been Haryana Joint Development Model. The Haryana Development Authority Act, 1977 and Haryana Development and Regulation of Urban Areas Act, 1975 allow private developers or owners of parcels of land that exceed the Urban Land Ceiling and Regulation Act to assemble and develop the land as per the approved layout by the competent authorities. The private developers will get a license to develop the land but they should pay external development charges on net sq. meter basis, reserve 20 per cent of the land to the EWS and LIG categories, sell 25 per cent of land on no-profit no-loss basis, put 30 per cent of the sale proceeds in development account to be used for development, maintain the colony for five years and return excess profit as determined to the government. Gurugram (earlier Gurgaon) Township was developed on this model.[29]

The concept of integrated townships is emerging in India to meet the requirements of the middle and higher income groups of the people. For this, the developer mobilised the land and prepares and implements the plan as per the planning standards. The government role may remain as regulator and facilitator. Integrated townships are well conceived colonies with full-fledged infrastructure and facilities like schools, health centres, offices, malls or shopping complexes, better quality civic services like water supply, sanitation and solid waste management, roads, lighting, parks, etc. The purpose of these townships is to provide housing with better services and infrastructure, decongest the central places and relieve the government from providing housing at huge costs by land acquisition.

Sustainable Framework for Land as Resource

There is need for comprehensive urban land policy, supportive legal and administrative reforms at the state level and integrated information and management systems at the local level. Monetising land assets can provide a one time opportunity to mobilize resources which is considered to be sufficient to meet infrastructure needs for at least 10 to 20 years. Charging higher FAR received much criticism from the environmentalists. Much of the problem arises due to lack of follow-up measures on account of increased FAR. This method, therefore, should be carefully handled in a way that it meets the extra land space required at the same time ensure funding for the infrastructure improvements that are needed to support the high densities in Indian cities and improve accessibility to well-located land.

Reservation of lands in the private lay-outs is necessary to meet the needs of poor. But, mere reservation will not solve the problems. Usable lands should be reserved and reserved lands should be properly developed in terms of facilities and housing. If they are kept idle, encroachments will take place.

Providing houses and good habitat to the poor and weaker sections requires developing comprehensive housing projects in active collaboration with the private developers, financing agencies, the urban development agencies and the beneficiaries each sharing part of the burden.

Private sector has to be involved in the housing as well as the infrastructure front. Discussions with the private developers reveal that they are willing to cooperate with the government but they expect faster clearances. The government in turn must be efficient in the clearances and taxing the private sector appropriately. The public-private partnerships generally suffer from suspicion, nexus for selfish interests, corruption, etc. There is need for developing clear and transparent methods to overcome these problems. The dialogue process can certainly open up opportunities for collaborative working.

This objective could be realised only when sustainable policies and strategies are devised by the state governments. The policy should invariably include following components:

(i) Rational approach in allocation of lands for industrial and public purposes with strict monitoring of the proposed land developments.

(ii) Effective land recording and information management.

(iii) Better asset management practices by the public agencies including the municipalities.

(iv) Devising land monetising methods without adversely affecting the social and environmental needs.

(v) Functional coordination between the multiple stake holders.

(vi) Well conceived PPP strategies for development of urban and urbanising lands.

Efforts must be made to evolve strategies judiciously and implementing in a transparent manner. A Report (2011)[30] of the Government of India suggested following steps to move in this direction:

(i) A systematic city-wide inventory of land assets must be made to be able to identify core and non-core land assets, and proposing the best use of public land assets must be part of comprehensive planning for the city.

(ii) A transparent and accountable mechanism for sale of public land must be put in place.

(iii) Proceeds from land sales must be used only for capital investment projects/ housing for the poor via the creation of a 'Land Capital Fund', whose governance and operational mechanisms are designed in such a manner as to ensure total transparency.

(iv) A mechanism for sharing revenues between the public agency owning the land and the infrastructure development agency must be established.

To conclude, land is a fixed asset. Unfortunately, this asset is either underutilised or used for speculative purposes in the fast unfolding urban world. Its potential is not effectively channelised for productive purposes. Land could become a good resource not only for meeting housing needs but also for infrastructure development provided it is optimally utilised and properly managed.

References

1. Government of India (1988): Report of the National Commission on Urbanisation (Vol. II), p. 117.
2. Bertaud, Alain, Consultant, World Bank and Jan K. Brueckner Analyzing Building-Height Restrictions: Predicted Impacts, Welfare Costs, and a Case Study of Bangalore, India April 2003, sourced from web.
3. Government of India(1983): Task Force on Planning of Urban Development, p. 128.
4. *Ibid.*
5. Report of the National Commission on Urbanisation (Vol. II), op.cit., p. 130.
6. Government of India (2008): Project Platinum (Partnership for Land Title Implementation in Urban Management).
7. *Ibid.*, p. 131.
8. Government of Karnataka: Urban Development Policy of Karnataka, 2009 (Draft), sourced from web.
9. Report of the National Commission on Urbanisation, Vol. II, op.cit., p. 214.
10. Peterson, George E. (2009): Unlocking land values to finance urban infrastructure-Land-based financing options for cities; in India Infrastructure Report, 3iNetwork, Oxford University Press, p. 214.
11. *Ibid.*
12. McKinsey Global Institute (2010): India's urban awakening: Building inclusive cities, sustaining economic growth.
13. G.O.Ms. No. 87, dated 12-02-1992, Municipal Administration and Urban Development Department, Government of Andhra Pradesh.
14. G.O. Rt. No. 1235, dated 14-08-1992, Municipal Administration and Urban Development Department, Government of Andhra Pradesh.
15. G.O. Ms. No. 419, dated 30-07-1998, Municipal Administration and Urban Development Department, Government of Andhra Pradesh.
16. W.P. No. 25011 of 1998, dated 25-01-2002.
17. Official records of GHMC, Hyderabad.
18. G.O. Ms. No. 460 , G.A. (Cabinet) Department, Government of Andhra Pradesh, dt: 9-11-1999.

19. Chief Commissioner of Land Administration, Government of Andhra Pradesh website and apland.ap.nic.in.
20. G. O. Ms. Nos. 455 and 456, Revenue (UC.I) Department, Dated 29-7-2002, Government of Andhra Pradesh.
21. G.O. Ms. 747, Revenue (UC.I) Department, Government of Andhra Pradesh, dt. 18-06-2008.
22. Chief Commissioner of Land Administration, Government of Andhra Pradesh.
23. McKinsey Global Institute, op. cit.
24. McKinsey Global Institute, op. cit.
25. G.O. Ms No. 86, Municipal Administration and Urban Development (M) Department, Government of Andhra Pradesh, dated 3rd March, 2006 and G.O. Ms No. 678, Municipal Administration and Urban Development (M) Department, Government of Andhra Pradesh, dated 07-09-2007.
26. McKinsey Global Institute, op. cit.
27. Hyderabad Metropolitan Development Authority, Hyderabad.
28. Hyderabad Metropolitan Development Authority Act, 2008, Sections, 24 to 27.
29. Nallatiga, Ramakrishna (2009): Potential of Land and Land-based Instruments for Infrastructure Development in Urban Areas; in INDIA INFRASTRUCTURE REPORT, 3iNetwork, Oxford University Press.
30. Government of India (2011): Report on Indian Urban Infrastructure and Services, The High Powered Expert Committee (HPEC) for Estimating the Investment Requirements for Urban Infrastructure Services.

Chapter - 8

Reforming Urban Planning

Introduction

India has entered into urban era with population projections signaling a mammoth 600 million by 2031. If centripetal growth gravitated towards major cities due to polarisation of employment opportunities in the post-Independence period, metropolitanisation or urban sprawl is the prominent feature of urban spatial transformation in recent decades. The high land costs coupled with rigid density and land use regulations in core areas created centrifugal forces pushing population growth deep into peripheral areas where the planning regulation is either absent or minimal. The high densities of about 2,450 per sq. km., in the 50 km. vicinity of seven largest metropolitan cities and the birth of a third of new towns in the 50 km. radius of urban neighbourhood of the existing cities of one million and above reflect such a growth taking shape.[1] The disturbing factor here is that the metropolitan growth is driven by density constraints in the core areas which are considered to be very low compared to international standards. Another irksome dimension of densities in core areas has been that there are blanket Floor Space Indices (FSIs) covering very large areas thus missing the opportunities to strategically increase densities around infrastructure networks.[2] This aspect is receiving attention of the reformers in recent years and the Twelfth Five Year Plan has given specific direction to go for strategic densification in urban areas.

As urban growth centering on million-plus cities is expected to intensify further, the challenge is how to transform land use and manage the high densities in peripheral areas which lack commensurate land policies and investments to enable residential and commercial development, infrastructure services and connectivity.[3] Increasing population, slums, gross infrastructure and service deficiencies and, growing inequities and social exclusions are compelling urban think tank to go ahead strongly on urban reforms aiming at orderly, inclusive and environment friendly development of urban, peri-urban and widely spread hinterland areas. Land use policy, development control, building regulation, inclusive and sustainable planning are the planning tools to achieve the goals. The issues, concerns and reforms initiated in these areas are discussed in this chapter.

Land Use and Development Control

Despite several policy discourses on components of urban land policies, comprehensive national urban land policy has not yet been formulated in India. At the other end, the competing demands from various sections of the people and sectors of development for the available land compelled the governments to take ad hoc and situation specific decisions on land allocations. In this scenario, urban planners also could not devise appropriate land use policies for spatial development. In the absence of proper urban land and land use policies, land use planning proceeded on conceptual ideas based on elaborate surveys.

Land use planning is the process by which land is allocated between various competing uses ideally to secure rational utilisation of land to achieve orderly and sustainable human settlements. But, land use planning cannot occur in vacuum or in isolation. It has to be done in the development context. Developmental needs must be identified, articulated and transformed into land use plans. The land use plans are prepared with 20-25 years perspective through a lengthy process including legal sanctions. The revision of these plans also involves same process. In a fast changing urban environment, these plans become obsolete and inadequate to meet the emerging needs.

Limitations of Master Plan Approach

The master plan approach being practiced in India is of British origin. It aims to transform physical planning into spatial planning focused on spatial distribution of land uses in cities. The Five Year Plans in the early Independence period emphasised preparation of master plans. The master plan is a statutory planning instrument to guide the orderly development of the cities and towns. It works through the system of zoning where physical spaces are compartmentalised into specific zones with some inter-connections. The master plans, instead of guiding the developments, have become rigid and static legal processes thereby missing the goals of development. Though the master plans provide for transportation network, conservation zones and economic development, the primary focus has been on land requirements and allocation of land among various uses. The master plan approach is considered failure in India due to strategic and administrative limitations viz., inability to accommodate needs of the urban poor, top down approach and lack of stakeholder participation, inability to see diversities in city environment, and missing link between master plans and the municipal, state and central development plans. As the master plans are prepared with a vision of 20-25 years and there are built-in rigidities in land uses, they become obsolete in the face of rapid urban growth. The comprehensive spatial planning updates are undertaken at almost the end of the master plan period and that too through a long process but the land use and other changes are made as and when required through bureaucratic process totally ignoring the participatory process and requirements of the poor migrants leading to growth of slums.[4] The other important limitation of the master plans is that they are prepared for municipal areas leaving the fringe areas outside the planning purview. The urban development authorities are constituted for larger urban areas to overcome this problem. But, the integration of the master plans with the other development plans has been missing. As a result, fringe area growth is taking place in haphazard manner without commensurate infrastructure, facilities and proper connectivity.

In the metropolitanisation era, land in the core areas becomes scarce and costly while the suburban areas lack adequate lands for urban uses. The conservatively low FSI or FAR stipulations in core areas and the land ceiling limitations resulted in uneconomic use of scarce land in core areas. At the other end, constraining legal environment and unresponsive land administration led to illegal and haphazard spatial form in peri-urban and hinterland areas. The progressive land use policy should therefore, focus on optimising land use in the core areas and supply of serviced lands in the suburban areas. The planning tools available are strategic increase of FSI in core areas through appropriate development controls and building regulations and, conversion of agricultural lands to the non-agricultural uses in the suburban areas in adequate quantities. Both these instruments must be judiciously used taking into consideration social and developmental needs.

Conversion of Land Use from Agriculture to Non-agricultural Uses

In the context of burgeoning urban growth backed by emerging new towns, special economic zones, industrial parks, etc., the utmost need for converting agricultural lands to the urban uses arises and, necessary administrative mechanisms must be put in place to release agricultural lands for non-agricultural uses based on the requirements. But, the land management mechanisms are cumbersome, dated and time consuming. Even in cases where the time limits are prescribed in the Acts for disposing of the applications for land use changes, actual time taken will be much longer spanning about four to six months. In cases where master plans are not prepared, it is done on plot by plot basis leading to unplanned development. Even in cases where master plans are prepared, enough care is not taken to assess the future land requirements on short-term basis. The land use conversion process involves several steps like proving the ownership of the land and large documentation to verify multiple records like tenancy, agriculture records, etc. The main impediment is the dual responsibility where enforcement of development control is the urban planning function, while procedures require it to be routed through the revenue department in several states.[5]

As the land use conversions are related to spatial development, the planning authorities must be the key agencies to deal with the land conversion issue to decide about the need for conversion, suitability of land, land use for which it could be allocated, etc. In some states like Maharashtra and Karnataka, the laws specify for consultation with the town planning authorities but in other states, the laws are unclear. It is alleged that more time is consumed in the revenue departmental procedures than by the town planning department but it should have been the other way.[6] In view of these problems, the JnNURM stipulated simplification of the process of conversion of land use from agriculture to non-agricultural uses as an optional reform. The JnNURM primer suggested that the land owner should approach the planning authority for land use changes and the planning authority should serve as single window for processing the application from the planning perspective in the background of the existing development control regulations and urban development needs.[7]

The focus of the reform should not only be on planning and development aspect but also on reducing the overall time taken for approvals. It is suggested that all the three major tasks viz., identification of land parcel, application of planning principles/scrutiny against Master Plan and issuance of the order, should be bound by outer time limits in the statute.[8] The conversion of agricultural lands to non-agricultural uses should be guided by the main objective of supply of adequate lands to meet the requirements of urban growth. In a broader perspective, the land use conversions should be able to balance the urban development and resource generation needs at one end, and the environmental requirements at the other end.

Development Controls

The development controls are formulated to achieve orderly and safe urban life. But, the urban spatial policy distortions led to disorderly and disjointed growth of cities and towns. Urban planning missed socio-economic dimensions of development. With over emphasis on physical planning, development control became constraining pushing people outside the core cities and decline of the core areas. On other side, the governments lacked adequate funds to take-up required infrastructural and other urban services to the growing urban populations especially, in the peripheral areas. This led to twin critical problems—one, growth of slums without any civic services and unhealthy environment and secondly, haphazard urban growth as people try to overuse scarce land while floating development controls and building regulations and construct buildings—residential and non-residential without minimum facilities like roads, drains, lighting, water, sanitation, etc. The developmental forces and pressures override planning regulations leaving the authorities helpless losing control over the urban growth and, resulting in contradictions in the pronounced policy and implementation thereby missing the direction.

Building Regulations

The building bye-laws and regulations aim at safe, healthy and environment friendly construction of the buildings. In order to achieve this, building regulations should regulate various aspects of the building construction and occupation. Important aspects of controls relate to zoning, land use, building plan, structural design, building safety, and other special needs like protecting the heritage buildings/areas. The attributes of building regulation are open space, set-backs and FSI/FAR, loading capacity, fire safety, use of buildings, etc.

Building Code and Model Building Bye Laws

In order to guide the building construction activity, the National Building Code, 2005 was formulated. This code provides required guidelines for construction of different categories of buildings mainly from the point of safety, public health and welfare of the people who dwell in the buildings. The central Town and Country Planning Organisation[9] formulated Model Building Bye Laws in 2004 to guide the state governments and the municipalities in devising appropriate building controls and regulations suitable to local situation and requirements. These Model Bye Laws are tools to regulate various aspects of buildings viz., residential and non-residential buildings, structural safety and services, fire protection, conservation of heritage sites and facilities for handicapped persons. The model bye laws provide certain guidelines for buildings in hill areas, norms for parking places, needs of non-residential buildings such as public places, local shops, markets, district markets, hotels, swimming pools, industrial plots, bus terminals, educational institutions, etc. They prescribe standard space requirements of various parts of a building and those of light and ventilation called general building requirements, and norms for working out occupant load of buildings constructed for various uses like residential, educational, institutional, business, etc. Internal norms are also provided for various parts of the buildings like main building, staircase, courtyard, projections, habitable rooms, etc. In order to maintain standard structural strength of the buildings, norms are prescribed for use of materials, alternate materials, plumbing and sanitary services, water supply arrangement, etc. Specific norms are prescribed for fire protection and heritage buildings.

The municipalities are primarily responsible for implementing the building regulations and development control in municipal areas. These functions are shared between the municipalities and urban development authorities (UDAs) wherever they are constituted. Historically, they are governed by a host of laws, bye-laws and building regulations. These laws and regulations have been undergoing modifications in view of the fast changes taking place in the cities and towns. Comprehensive building regulation reforms are attempted in Andhra Pradesh in recent years. These initiatives are discussed to understand the emerging scenario.

Building Regulation Reforms in Andhra Pradesh

The Andhra Pradesh (Andhra Area) Town Planning Act, 1920 was the older legislation governing town planning function in municipalities of Andhra Pradesh. It provides for the preparation of General Town-Planning Schemes (GTPS) and Detailed Town-Planning Schemes (DTPS) in respect of all lands within the municipal area to ensure regulated development of towns to secure their present and future inhabitants, sanitary conditions, amenity and convenience. The Act operates through the Director of Town and Country Planning and the respective municipalities. The Andhra Pradesh Municipalities Act, 1965 and the Hyderabad Municipal Corporation Act, 1955 which is operational in the Hyderabad municipal limits and other municipal corporations in the state such as Visakhapatnam, Vijayawada, Guntur, Rajahmundry, Warangal and Kurnool give wider powers to the elected bodies. Most town planning and building regulation/development control functions under AP Urban Areas (Development) Act, 1975 are delegated to the Municipal Corporation of Hyderabad and other municipal corporations in the state.[10] Attempts were made to streamline the issuance of building permissions by fixation of timeframes for quick and prompt disposal of building regulations.

The government amended rules made under the Town Planning Act to make provision for the levy and collection of development charges for the implementation of essential schemes like traffic improvement, construction of bridges and development of green belts, parks, etc.[11] The government declared certain major roads in Hyderabad city as commercial roads and authorised the Municipal Commissioner of Hyderabad Municipal Corporation to grant building permissions for proposed construction of residential/commercial/institutional use except industrial use buildings at the option of the owners on the sites abutting these roads subject to payment of impact fee to Greater Hyderabad Municipal Corporation (GHMC) apart from other fees and charges.

In order to streamline the entire building regulation and approval process in the state, a comprehensive effort was made in 1998.[12] Formulation of common building rules applicable to all categories of municipalities in the state is considered an important reform in the country aimed at standardisation of building rules and FAR. The FAR and other standards of building plan requirements for different types of buildings, uses, and occupancies in municipalities, municipal corporations and urban development authorities were streamlined. The rules lay down the procedure for obtaining building permit, specify the time limit for grant or refusal of building application and, prescribed requirements for sites and buildings under different land uses in terms of plot size, maximum coverage, FAR and setbacks. Specific norms for lighting and ventilation, mandatory inspections and issue of fit-for occupation certificate, conservation and harvesting of rain water are provided. The government also empowered the Municipal Corporation of Hyderabad to relax zoning and building regulations in cases where people surrender land for widening of roads free of cost. It provided for additional FAR of 1.00 in the extent of land affected in road widening and surrendered free of cost and, relaxations to set-back and coverage to certain extent when the permissible FAR cannot be achieved. The Municipal Commissioner is also competent to permit change of land use to enable commercial construction except in the case of recreation zone by allowing additional FAR up to 0.50 on the land surrendered free of cost.

It was realised soon that the building regulations issued in 1998 in the state were inadequate to deal with the fast growing urban areas, increasing land prices and emerging building technologies. The building rules were again revised in 2006.[13] Initially, these rules were made applicable to the Hyderabad Urban Development Area but later, extended to all other municipal corporations and urban development authorities in the state. Provisions were added to meet specific requirements in the subsequent years too. For instance, minimum setback area of seven meters was stipulated in respect of all high rise buildings to provide necessary fire services in case of fire accidents.[14] Further, for encouraging affordable housing for the poorer sections of the people, specific concessions were provided for developing housing colonies in partnership. Concessions like automatic exemption from the Urban Land Ceiling Act, concession on registration fee, automatic land use conversion from agricultural land to urban uses and 50 per cent waiver on development charges were provided.[15] Another change brought-out relates to providing barrier free environment in the public buildings for persons with disabilities as per the National Building Code of India, 2005.[16]

The building regulation reforms brought revolutionary impact on the building activity in the cities as the FSI regulation was liberalised considerably. The developers started new projects with optimism. Another big impact has been reduction in the unauthorised construction activity as the most builders voluntarily followed the rules due to liberalisation of FSI. Further, compliance with the 10 per cent mortgage clause and obtaining of occupancy certificate trend increased in recent years.[17]

The building regulations issued in 2006 were again revised in 2012. The JnNURM reform requirements and other changes made subsequently were integrated into the 2012 Building Regulations.[18] The main features of new building regulations are outlined below:

(i) The developer shall provide at least 20 per cent of developed land for Economically Weaker Sections (EWS) and Low Income people.

(ii) Some incentives in the form of TDRs are provided to encourage the owners to develop parking complexes / parking lots.

(iii) Compliance of National Building Code provisions for amenities and facilities in all buildings is made mandatory. In case of high rise buildings, in addition to NBC, fire safeguards are also prescribed.

(iv) Benefits in the form of TDRs, additional floor construction and concessions in setbacks including front setback are provided for road widening purposes.

(v) Levy of City Level Infrastructure Impact Fees to ensure development of infrastructure facilities.

(vi) Incentives are provided in the form of rebate in property tax (10 per cent) for owners leaving more setbacks and installing solar heating, lighting, rainwater harvesting and recycling of waste water systems.

(vii) Occupancy Certificate (OC) is made mandatory for all buildings constructed on plots of above 100 sq. mtrs. The requirements for issue of OC are also simplified by prescribing simple parameters viz., no. of floors, external setbacks, usage of the building, parking space provision and abutting road width.

(viii) Provision is made to constitute a Town Planning and Building Tribunal for dealing with all town planning issues.

(ix) Provision for institution of a Building Ombudsman for dealing with all complaints of building violations, shortfall in building standards, services, specifications and safety aspects.

(x) Financial institutions are to extend loans only as per sanctioned building plans.

(xi) Registration Department has to register only as per sanctioned building plans.

(xii) Provision for compounding minor deviation upto 10 per cent setback violations is made.

(xiii) Flexibility is given in layout pattern/site planning and design parameters to encourage group housing/group developments and gated communities viz., row type, semi-detached, cluster type, group of apartment blocks and residential enclaves.

(xiv) Simplified building form by prescribing setbacks and heights. The FSI or FAR norms are liberalised considerably.[19]

Impact of FAR or FSI

The FAR or FSI determines the total built up space that a plot is allowed to hold subject to availability of land, household densities/density of dwelling sizes and availability of parking.[20] One of the important urban planning reforms brought-out in Andhra Pradesh relates to liberalisation of the FSI regulation. The FSI regulation is double edged reform. At one end, it can boost building activity contributing for the growth of urban areas and more compliance of the building regulations by developers. The experience of Andhra Pradesh has undoubtedly demonstrated this point. But, the other end of the spectrum is that increased densities lead to environmental problems. The environmentalists always argue that indiscriminate increase of densities only results in congestion, strain on limited infrastructure, pollution and environmental disasters in the long run. The debate assumes much significance in Indian cities and towns which already suffer from over population and lack of basic infrastructure facilities.

The suggestion made by the Twelfth Five Year Plan[21] is pertinent to mention here. While suggesting on urban planning reforms to meet urban development needs, the Plan recommended strategic densification in the fast growing areas. The critical issue is the term 'strategic densification' which is vague susceptible for different interpretations. It has to be standardised and applied appropriately while taking into consideration local environmental and infrastructure needs. Unless this is done, the issue remains controversial. The Government of Andhra Pradesh constituted in 2003 a committee to examine the issues relating to liberalisation of the FSI rules.[22] The FAR or FSI issue continues to be critical as different views are expressed and studies are also inconclusive. One aspect is proved by studies that increasing FSI or FAR levels has impact on increased densities. But, the inter-connections between employment, infrastructure and environment are unclear at present in Indian situation. This aspect requires further studies and insight to arrive at concluding planning standards.[23]

Sustainable Urban Planning

Urbanisation and environment are intrinsically linked aspects of development phenomenon. Traditionally, the process of urbanisation has impacted the environment by disturbing the natural state. As urbanisation process itself is irreversible, the alternative is controlling negative factors which affect the urbanisation-environment balance through a set of town planning devices. The master plans, development controls and building bye laws/regulations are the instruments to achieve this. The exercise of control over city development is essential for ensuring planned development, conserving architecture, human safety and protection of environment. Further, regulating urban built environment is necessary to promote growth of cities and towns in spatially balanced and environmentally sustainable manner.

The Problem of Unregulated Urban Growth

Unplanned and unregulated urban growth driven by population pressure and migration coupled with the greed of urban reality sector to maximise the returns has resulted in unsustainable urban form. Haphazard building activity, lack of control over land use, high rise buildings and high densities without commensurate infrastructure and, environmental pollution are the resultant symptoms of uncontrolled growth. At the other end, disasters like earthquakes, floods and fire accidents are increasingly threatening the safety of the urban dwellers. The existing town planning and building regulations have become inadequate and ineffective in dealing with these problems.

Need for Sustainable Urban Growth

In the context of growing environmental pollution, haphazard and chaotic growth of urban areas, threatening slums, exclusion and social conflicts, transportation blockades and a host of other problems associated with unregulated urban growth, there is utmost need for looking into sustainability aspect of urban growth. The concepts like integrated development, safe and healthy built environment, urban forestry, etc. must be converted into achievable planning objectives and targets. The development controls and building regulations should adequately meet the environmental, human life safety and public health requirements at the one end, and the developmental needs of the people at the other end. The challenge is how to devise appropriate urban plans and building regulations which could effectively meet the needs of development, deal with natural disasters and address environmental concerns.

Urban Planning and Sustainable Development

The National Mission on Sustainable Habitat (NMSH) was launched by the Government of India in order to draw necessary action plans to achieve sustainability of human habitats.[24] The NMSH is one of the eight missions formed under the Prime Minister's National Action Plan for Climate Change (NAPCC). As one of the four Sub-Committees under the NMSH, the Sub-Committee on Urban Planning has addressed the issue of sustainable habitats from a planning perspective. This document attempts to present a broad approach to achieving and enhancing the sustainability of human habitats in India.

The Brundtland Commission (1987) defined sustainable development as development that meets the needs of the present without compromising the ability of future generations to meet their own needs.[25] The Expert Committee of the Ministry of Urban Development has identified two main characteristics that define sustainable development viz., efficiency and equity: efficiency, in functioning and equity, when distributed within the generation and between generations. Applied to the urban planning and development, the fundamental elements of sustainability are the utilisation of natural resources in a city region most efficiently, most equitably across sections of society and in such a manner that the resources are conserved and renewed for future generations to meet their needs and aspirations.[26] The sustainable urban planning framework primarily consists of five interrelated components: (i) resource optimisation; (ii) integrated spatial planning linking land use with transportation needs; (iii) environmental conservation and green building; (iv) inclusive social sector development; and (v) participatory planning process.

There is increasing awareness in recent years to make the urban planning effective and environment friendly. The NMHS initiative clearly brings-out the intention to put in place steps that would lead to a desired outcome or result on the urban planning front. While the end-situation will be affected by several variables, known and unknown, a scientific approach of measurement leading to review may be undertaken which will enable the plan and policy-makers to gauge performance and revise the steps as required for the next planning cycle. In order to make this process implementable, parameters and indicators are identified.[27]

Building sustainable habitat is one of the important dimensions of sustainable development. Sustainable habitats strive to optimise energy consumption. In view of the energy constraints across the globe, conservation of energy has become a demanding concern and countries are devising various means to conserve energy. Urban planning should become energy sensitive to achieve this objective. On the lines of global awareness and steps to conserve energy, certain initiatives are made in recent years in India. Formulation of the Energy Conservation Building Code (ECBC), 2007 and GRIHA Rating System are the two important steps taken. If ECBC aims to incorporate energy consciousness into building design and construction technology, GRIHA promotes green building initiatives to optimize energy consumption.

Energy Conservation Building Code

The Energy Conservation Building Code (ECBC) provides minimum requirements for the energy-efficient design and construction of buildings. The ECBC formulated in 2007 is made applicable to buildings or building complexes that have a connected load of 500 KW or greater or a contract demand of 600 KVA or greater.[28] It includes building envelopes, mechanical systems and equipment heating, interior and exterior lighting, electrical power and motors. The buildings which have a conditioned area of 1000 sq.m. or more fall under this category.[29] The Government of India launched the ECBC for voluntary adoption in the country. The Bureau of Energy Efficiency, Ministry of Power revised the ECBC in 2008 based on initial feedback. The ECBC building envelop requirements are based on the climatic zones in which a building is located, viz., composite, hot-dry, warm-humid, temperate or moderate and cold.[30]

The expectation is that by adopting the ECBC for high energy consuming buildings considerable energy could be saved. It is estimated that building sector represents about 33 per cent of electricity consumption in India, residential sector contributing 25 per cent. The studies show that ECBC compliant buildings could save 40 to 60 per cent energy compared to conventional buildings. On the national scale, it is estimated that mandatory enforcement of the ECBC could save approximately 1.7 million KW annually.[31]

Griha Rating for Integrated Habitat Assessment (GRIHA)

The green building technology assumes much significance from the sustainable development point of view. The building activity impacts the environment during construction as well as maintenance. Buildings consume lot of natural resources like land, soil, trees, metals, plastic, water and electricity and generate wastes like site waste, construction waste, sewage, organic and inorganic waste, chemical waste, etc. From the pollution angle, it is estimated that building activities contribute 50 per cent of the world air pollution, 42 per cent of its green house gases, and 50 per cent of all chlorofluorocarbons (CFCs).[32] In India too, the waste generated by the construction industry constitutes about 25 per cent of 48 million ton solid waste being generated.[33]

As buildings drain out considerable share of scarce resources and produce waste impacting the environment adversely, the concept of green building has emerged across the globe to build sustainable habitats. The purpose of green building in simple terms is to build energy efficient sustainable habitat. The green building could be considered as a way of life. It aims at: optimising demand for electricity, water and other natural resources in construction and operation; generate all electricity on-site through renewal means; cater to water demands by sustainable processes like rain water harvesting; grow its own food on-site; and recycle and reuse all its waste and reduce the burden on environment.[34] A building is considered green if it provides optimal comfort, uses minimum resources, consumes minimum energy and generates optimum waste over its entire life cycle.[35] Though the initial cost of a green building is a bit higher, its life cycle cost is considered to be much lower. Green building rating system is an evaluation tool to measure environmental performance of a building through its life cycle. The system is operated voluntarily across the globe to develop environment awareness in building technology. Some of the international building rating agencies are: HK-BEAM—The Hong Kong Building Environmental Assessment Method, LEED—Leadership in Energy and Environmental Design (USA), CSBEE—Comprehensive Assessment System for Building Environmental Efficiency (Japan) and BREEAM—Building Research Establishment's Environmental Assessment Method (UK).

GRIHA is the national rating system for green buildings devised in India. The system is developed by The Energy and Resources Institute and endorsed by the Ministry of New and Renewal Energy. The purpose is to minimize building resource consumption, waste generation and, reduce the environmental impact. The system is based on five 'R' philosophy of sustainable development, viz., refuse, reduce, reuse, recycle and reinvest. It is a method of quantifying energy consumption, water consumption, waste generation and renewal energy integration to facilitate measurement and rating.[36] The rating is done on a five star scale beginning from 1 star for 50-60 points scaling up to 5 star for 91-100 points with ten points variation for each star added. The GRIHA certification will get faster environmental clearances. Some of the buildings rated by GRIHA include the Common Wealth Games Village, New Delhi, and Centre for Environmental Science and Engineering Building, IIT, Kanpur.

Institutional Aspects

Achieving the planning goals ultimately depends on how effectively the planning standards, regulations and rules are implemented. This again rests on the institutions dealing with urban planning and the administrative processes and methods adopted by them to implement the plans. A complex institutional framework has emerged in India to deal with urban planning. The core institutions include municipalities, urban development authorities, local planning authorities and town and country planning departments. Major administrative issues relate to building approval process, inter-organisational coordination, and people's involvement in the planning process. Key reforms initiated in these areas are discussed hereunder.

Reforms to Strengthen Municipal Role in Urban Planning

One of the major functions of the municipalities has been town planning in India. But, after Independence, role of the municipalities in town planning is relegated to back. Urban planning was confined mainly to physical planning and town planning is considered purely technical exercise requiring specialised skills. As the municipalities lacked adequate expertise, state governments created special purpose bodies namely, town and country planning departments and urban development authorities for carrying-out this vital function of planning in urban and metropolitan regions. The role of municipalities in town planning is limited to implementation of building regulations at the local level. This divided responsibility ultimately led to ineffective regulation and haphazard growth of cities and towns.

Beginning from the Constitution 74th Amendment, there has been focus on strengthening of the role of municipality in urban planning. Transfer of urban planning function to the municipalities and constitution of the District/ Metropolitan Planning Committees are the important reforms envisaged in the Constitution and later on pursued under JnNURM. The ward committees provided in the Constitution 74th Amendment Act also need to take active role in preparing development plans and monitoring of building activities in the local municipal areas. Setting-up of Wards Committees in every municipality with more than three lakh population involving local councillors for decentralised local governance is one of the mandatory provisions under the Constitution 74th Amendment. The Model Nagara Raj Bill prepared by the Ministry of Urban Development aims to further strengthen the ward committee system. The proposed model reinforces emphasis on community participation in local governance. Every ward in each municipality should have a ward committee and area sabhas/ward sabhas for involving the people. New legislation has been brought out in many states to implement this model. One of the functions entrusted to the ward committees under the new system is preparation of the ward development plan and monitoring of the building activities to assist the town planning officials.

Peri-urban Growth and Role of Urban Development Authorities

With the expansion of urban growth into peripheral areas in the fast developing cities engulfing several surrounding villages and outgrowths, the need has arisen to regulate land use and building activity in these areas. The institution of urban development authority (UDA) is created in India to deal with regulation of urban growth in the peripheral areas including the municipal areas. The UDAs have become main planning agencies in the urban agglomerations. They prepare master plans for the declared urban development areas and regulate developments. The municipalities and panchayati raj institutions (PRIs) confine themselves to the respective local areas and perform planning functions to the extent delegated by the state government.

The urban development authorities (UDAs) are playing significant role by preparing master plans and devising appropriate development controls and building regulations in the UDA areas. In fact, the role of the UDAs is more in respect of the land use regulation and less in the building regulation. The building regulations to a large extent are implemented by the municipalities. Urban development authorities undertake large scale land development and housing, development of satellite towns and growth corridors, and plan for large scale infrastructure projects in the areas like water supply, industrial estates, parks, transportation, conservation of lakes, urban forestry, etc. The state governments have made separate legislations to govern the activities of UDAs. However, Bihar is the only state which has abolished the regional development authorities and repealed the respective law and transferred their functions to the municipalities.

Major issues confronting the UDA areas are illegal land divisions, unauthorised land conversions from agricultural and forest uses to urban uses, and unplanned development of industrial and residential areas without minimum standards and basic civic amenities. Lack of effective control, resource constraints to provide urban services and greedy reality are the major problems in the metro regions. Operation of several central, state and local level agencies with no proper coordination is another important concern.

Inter-organisational Coordination

In the urban agglomerations, the metropolitan/urban development authorities (UDAs) prepare master plans and the local bodies implement the master plan. In respect of other urban areas, the master plans are prepared by the state level organisations like the directorates of town and country planning while they are implemented by the municipalities. In many cities, the building permission powers are also shared between the urban development authorities and the municipalities. Clearances from the pollution control boards, departments of environment, fire services department, etc., are also required to approve special category building plans. This is considered as one of the reasons for delays in the issue of building permissions.

At the municipal level, the corporation/council is the main policy-making agency. The corporation/council enjoys certain delegated functions in regard to urban planning and building regulations. The master plans and town planning schemes though prepared by the technical agencies like UDAs and town planning departments, the municipalities are sanctioning authorities. In the urban areas, the importance of the municipalities has been relegated to back with the emergence of the UDAs and the local planning authorities. As the expertise at the municipal level is very weak to deal with the emerging problems, the UDAs are being strengthened with more planning role. The urban planning functions are shared between the town planning departments/

UDAs and the municipalities with the latter focusing more on building permission role and the former on development/land use control and planning. The National Commission on Urbanisation, as back as 1988, recommended for transferring the town planning function from the special purpose agencies to the municipalities as they are ultimately responsible for overall development of the cities and towns. The Constitution 74th Amendment Act (1992) later reiterated this view and included town planning and urban planning including economic development planning as one of the main functions of the municipalities as per the Twelfth Schedule of the Constitution. Transfer of the urban planning function to the municipalities was one of the mandatory reforms under the JnNURM scheme also. Yet, most of the states are reluctant to transfer urban planning function totally to the municipalities under one pretext or the other. At the other end, several organisations perform regulatory and developmental functions in urban areas. The municipalities, cantonment boards, panchayati raj institutions in rural areas, revenue department, urban development authorities or local planning authorities and pollution control boards belong to the regulatory category, while the housing boards, industrial and infrastructure development corporations and a host of other sector specific agencies are engaged in the developmental activities. The coordination mechanism being weak, the urban planning and development functions remained totally disjointed and disarticulated even after the Constitution 74th Amendment Act and the reform oriented JnNURM.

Reforms in Building Permission Process

The process of building permissions is considered cumbersome and time consuming due to various technical details involved and procedural requirements to be fulfilled during the construction. Several stages are there in building permission process viz., preliminary approval, approvals from line departments like Fire Services, Pollution Control Board, Environment Department, etc., final approval, occupancy certificate, etc. People in general complain that the officers are delaying the permission for unnecessary reasons. There exists wide communication gap between the town planning officers and the builders. General tendency of the people to flout rules during construction normally delays the process of issuing occupancy certificate. In fact, people construct buildings but fail to take occupancy certificates for years together. As taxation and town planning organs and other service departments like electricity work in isolation, people generally give less priority to getting occupancy certificate. This leads to unauthorised construction. There is hence, need for simplifying the procedures and educating the people on the need for constructing buildings as per the regulations. This is however, a herculean task. The building approval process is lengthy and time consuming. For instance, it is brought-out by studies that in cities like Bengaluru, Indore, Thiruvananthapuram, it requires 30 days and in cities like Delhi, Kolkata, Mumbai, Ranchi and Surat, it requires 60 days as per building bye-laws. On an average, 51 approvals from various agencies are needed.[37]

The important aspect is the cost of delays in the building approval process which is as much as 1528 per cent of income per capita as against the average cost of doing business in India which is 49.8 per cent of income per capita across sectors and this cost is naturally passed on to the consumers.[38] In order to avoid delays in building approvals, certain reforms are needed to simplify the process. Revision of building bye laws to streamline the approval process for construction of buildings and development of sites was an optional reform under JnNURM and is now focused under AMRUT. Some reforms like green channel, single window method, citizen's charters and automation through IT applications are already being implemented by the municipalities and urban development authorities in many states. The city level reforms have been devised and implemented to streamline the process of building approvals. The thrust of the reforms has been on reducing the time taken for granting necessary building approvals and transparency in the procedures. The cities like Bhubaneswar and Hyderabad have implemented green channel reform, a fast track building approval system.

The Greater Hyderabad Municipal Corporation (GHMC) and the Hyderabad Metropolitan Development Authority (HMDA) introduced green channel method in 2010 for fast track approval of the G+4 houses while eliminating malpractices. Excellent groundwork was done to enable the applicants to get permissions within seven clear working days for non-multistoried buildings and layouts and, 30 days for multistoried buildings. A panel of certified architects and document auditors who have necessary knowledge of land laws has been prepared to scrutinise the applications. However, the scheme has not picked up due to lack luster response from the builders and also due to various administrative problems. The GHMC is again reformulating the scheme with online application and approval features. The objective is to clear the building permissions within 15 days upto ground and four floors at the circle and zonal office levels and clearing of high rise buildings at the headquarters level within 30 days. The proposal is to use customized version of the AutoDCR software which is being used successfully by the municipal corporations like Pune, Chennai, Coimbatore, Nagpur, etc., to scrutinise and approve building permissions. The proposed system has several automatic features like up-loading and reading of the building drawings, verifying their properness as per the building rules, generating fees and area statements, etc. This online system will be an integrated one linked to the issue of occupancy certificates and property tax department.[39]

The Bhubaneswar Development Authority (BDA) adopted the Green Channel to dispose building permits within 72 hours to cater to the needs of a growing city. Prior to this initiative, the BDA used to receive about 150 applications for plan approval in a month and it took over 60 days to approve. The individual owners used to initiate construction without getting a formal approval due to abnormal delays, and in the process violated development norms, and building rules. The Bhubaneswar Development Authority (BDA) introduced a fast track scheme known as building permits within 72 hours to facilitate the plot owners to build houses as per the approved plan in time. This scheme was applicable to the cases where all development parameters were adopted by the plot owner and the building application was submitted along with necessary documents and an additional fee of Rs 1,000 towards processing of application on a priority basis. Such applications were allowed to be processed immediately and the building permissions were issued within 72 hours, which helped to obtain loans from banks to start construction. The scheme has received good response from the building owners as they receive sanction in prescribed time and are willing to pay the extra fee.[40]

Single Window Approval System (SWAS)

One of the reforms much discussed in recent years is the Single Window Approval System (SWAS) where the builder gets all related sanctions and approvals at one counter. This system has potential to reduce the delay to a large extent. Some initiatives made by a few city governments to implement this system are explained here.

Bengaluru—Automatic Plan Sanction

Automatic Plan Sanction (APS) is a single window initiative of Brihat Bangalore Mahanagara Palike (BBMP) to sanction residential building plan within three working days, provided the plan prepared and documents enclosed are in order and in compliant with the building bye laws of BBMP. This scheme simplified the application process by making the applicants to file applications and all subsequent letters/permissions at the single window. This scheme is open to all properties falling under BBMP limits and is restricted to residential plans on site area not exceeding 4000 sq ft and not exceeding G+3 floors (height not exceeding 15 m) including additions or modifications.[41]

Under the APS, the owner or builder will have to take the assistance of an engineer, architect or a supervisor, duly registered with the BBMP, to prepare the plan as per the Building Bye Laws, 2003. The registered engineer/architect/supervisor will guide the owner in filing the APS application form together with all required documents and affidavits. The application fee in the form of demand draft is payable at the single window counter itself and the fee structure details are provided in the application form. The filled in application form addressed to Assistant Director of Town Planning has to be handed over at the Single Window Countre of respective jurisdictional (Zonal) office, which will be duly acknowledged and this has to be shown at the time of receiving the permission. If the application and documents are complete and the plan is as per the building bye laws, it can be automatically sanctioned and the sanctioned plan can be collected on third working day from the date of submission. As the owner/proposer needs to obtain other certificates like commencement certificate and occupancy certificate, the procedures for obtaining these certificates is also simplified.[42] The BBMP not only simplified the building permission process to a simple exercise, but also reduced the time taken for the issue of certificates, i.e., commencement and occupancy certificates. Provisions were also incorporated for fresh submission of application/plan for sanction at the expiry of permit validity and for considering minor deviations, if they were justified and found valid by the BBMP through Single Window Counter.

Indore—Automatic Building Plan Approval System

Automatic Building Plan Approval System (ABPAS) is an integrated online system developed by the Urban Administration and Development Department of Madhya Pradesh Government and successfully implemented in Indore with an aim to simplify the building approval process and get faster and accurate approvals at one stop. It is web-based and integrated system. The software products are designed to get faster and accurate approval of building proposals, bring uniformity in reading, drawings and eliminate lengthy manual process and human errors; attain standardisation and transparency in the total process of building permission and reduce paper work, valuable time and effort of architects as well as the authority. Initially, the architects submit building plans along with basic data over the internet for approval. After initial scrutiny of the documents, a date for site visit is to be fixed and intimated via SMS to the architects.[43]

The basic feature of the ABPAS is that it is totally automated. It reads data from CAD drawing, establishes relationship between various entities, imposes specific conditions, and geometrically maps each and every entity by corresponding with complex and interlinked rules. Analysis is done as per development control regulations automatically by the system. Tables and reports are saved along with the drawing and reports are sent to consultant/ officials online. System automatically generates various scrutiny reports dynamically based on the DC Rules defined by the respective Authority. Generated report shows the failed/passed items with their rules in a very user friendly viewable and printable format. Final sanctioned drawing is generated in non-editable PDF format. It has gained popularity and acceptance among the citizens in Indore.[44]

Pune—Automated Building Plan Approval

The Pune Municipal Corporation has re-engineered their building permit processes integrating computerised procedures for submission and verification of documents and drawings through 'Single Window' system. It ensures smoother approval of building permission within a fixed time by involving the stakeholders through a transparent mechanism. Instead of running from pillar to post for days together, the residents can get their plans approved by filing an application online (by using Bio-Metric login verification) and submitting the hard copies through a single window system. The online system is applicable to both, ordinary citizens and the builders. Not only applications, the builders and citizens can also make payment online. It could be done through credit cards or electronic clearing system. The approval of the building is also done online.[45]

The reformed model led to some concrete and systemic advantages viz., (i) by migration from paper-based system to Auto-DCR system, the workforce efficiency is increased from 60 per cent to 95 per cent; (ii) time required for sanctioning of building proposal is reduced to 21 days from 45-50 days; (iii) revenue generation increased to Rs 218 Cr. (FY 2006-07) from Rs 26 Cr (FY 2002-03); (iv) the architect's/authority's effort for drawing and calculations is reduced; (v) the automated system produces accurate reports by eliminating the human errors and manipulation; and vi) standardised the submission and scrutiny process.[46]

It is pertinent to mention here that the Ministry of Urban Housing and Poverty Alleviation constituted a committee in 2003 to suggest measures for streamlining and fast tracking the building approval processes in the country and exploring the introduction of the 'single window approval system' or 'one stop shop'. The committee observes that India has reduced the average time from 227 to 196 days to process building permit applications by implementing strict time limits in municipalities but there are still 34 procedures to be followed.[47] Main causes for delay are rigid planning process, coordination problems, lack of transparency, ambiguity in laws, corruption, etc. After studying different models, the committee has recommended that the state governments should initiate the process of Single Window Approval System (SWAS) enabled by IT.[48]

The Single Window Approval System has potential to reduce the delay to a large extent. But, the system should be properly administered. The system should take care of the following important aspects.[49]

(i) Provision of the list of approvals required, checklist of documents needed and the timeline prescribed for every approval online for the knowledge of the builders.

(ii) Providing for fast track systems for affordable housing projects and green building projects.

(iii) Prescribing fee structure for each clearances and procedure for online payment with single payment to be made electronically at the receiving counter.

(iv) Interconnecting all regulatory authorities at the central, state and local government levels and sharing of information/status electronically.

(v) Providing for forward linkages with organisations requiring NOCs and backward linkages for various inputs from concerned organisations.

(vi) Providing facility for online tracking application to know status of approval by logging in.

(vii) Sending alerts by SMS or E-Mail to the applicant after each clearance.

(viii) Automatic generation of sanction order after approval.

(ix) Facility for attachment of AutoCAD drawings and other documents in PDF format in off line mode and sending with online applications.

(x) System of correlating soft plan and drawings with Master Plan/ CDP/Zonal plan.

(xi) Automation of entire scrutiny process.

(xii) Facility to bring plinth/completion/occupancy certificates in automation.

(xiii) The system should facilitate green building scrutiny checks that enable municipality to provide green ratings to the buildings.

(xiv) After making online payments and digitally signed certificates, final certificates should be made available online.

(xv) Facility to get the sanctioned building data mapped to GIS system.

Tracking of Unauthorised Constructions

Tracking of building activity is the major town planning function of the municipalities. But they are not able to perform this function effectively in the large cities where growth is fast. Shortage or lack of town planning staff at the municipal level is one of the main constraints. Even the existing staff is burdened with many other activities. As a result, unauthorised building constructions are growing in metropolitan cities creating problems for the authorities. Effective building regulation demands regular and constant tracking of the construction activity. For, once the building comes-up, its demolition virtually becomes impossible in the current politic-administrative and judicial scenario in the country. Unable to track unauthorised buildings and take timely corrective action, the governments are resorting to regularisation process. This is a wrong solution to the problem. The need, therefore, is developing spatial and building data base, and effective tracking methods.

Development control and building regulation activities are most significant in terms of shaping the future of cities and towns. Several IT and GIS solutions are being developed to help the local planning authorities and municipalities to turn out more productive and quality work with limited human resources. IT and GIS solutions could be utilised for a variety of planning activities like issue of building permits, building construction monitoring, developing geospatial data and information dissemination on building activity to the urban communities. This apart, GIS-based planning is also useful in respect of infrastructure planning, slum planning, land use mapping, mapping of properties, civic facilities, and heritage places, identifying potential flood prone areas, identification of dangerous buildings, etc. In fact, GIS has vast potential in the large metropolitan cities where city form changes very fast affecting the wellbeing and safety of the people.

In order to cope with this problem, the municipalities should build good data base on buildings. The building activity must be monitored online. The methods like OSRT in GHMC will help in tracking the building activity. In this method, the town planning officials regularly check the unauthorised constructions in various areas and send the images of such unauthorised constructions to the head office by a mobile. The authorities would be in better position to take timely action on unauthorised constructions.

Urban Planning and Urban Development—the Missing Link

The town or urban planning in India is governed by the master plans. The limitations of master plans to regulate growth and promote development have amply demonstrated in cities and towns. As the master plans remained mere regulatory instruments, their implementation suffered due to administrative lapses. At the other spectrum, most of the master plans focused on the core cities while neglecting the growth in the peripheral areas. Most distressing aspect of urban planning in India has been isolation of the master plans from the development initiatives in urban areas. The Twelfth Plan has thus commented: "The 'Master Plan' approach generally focuses on only the core area of the city, has little linkages to any financial and operating strategy and, in many cases has been used as a regulatory tool instead of being a blue-print for development of dynamic and smart cities. A master plan typically freezes the land use pattern and building bye-laws and determines the permissible limits of Floor Space Index (FSI) and minimum setback areas. Often these provisions do not take into account the potential of the city to grow, especially where trunk infrastructure has been laid. This results in sub-optimal use of land as well as the infrastructure."[50] It is further commented that even the City Development Plans prepared under the JnNURM scheme lacked overall perspective including the issues of peri-urban areas and suggested to adopt a holistic approach and participatory planning process.

The need for having a Development Plan with a time perspective of 10 years with effective master plan linkage is emphasised. It is therefore that the provision of preparation of State Annual Action Plan (SAAP) has been enshrined under AMRUT.

District/ Metropolitan Planning Committees : Yet to Take Off

The Constitution 74th Amendment Act (CAA) provides for setting-up of District Planning Committee (DPC) at the district level and Metropolitan Planning Committee (MPC) in the million plus cities for preparing integrated development plans. The DPC is provided at the district level to integrate all local plans at the district level while the MPC is entrusted with the same function at the metropolitan region level. They comprise of two-third members representing the constituent local governments-rural and urban, in proportion to their population and the remaining to be nominated. These institutions are expected to ensure orderly, integrated and accountable development in rural as well as urban areas. The 74th CAA provides that while preparing the draft development plan, the respective DPC or MPC shall consider the following:[51]

(i) The plans prepared by the municipalities and panchayats in the specified area;

(ii) Matters of common interest between municipalities and panchayats including coordinated spatial planning of the area, sharing of water, physical and natural resources, integrated development of the infrastructure and environmental conservation;

(iii) Overall objectives and priorities set by Government of India and the State Government; and

(iv) The extent of investments likely to be made by agencies of the central and state governments and other available resources whether financial or otherwise.

These planning committees should hold consultations with institutions and organisations related before finalising the plans. The chairperson of every DPC or MPC shall forward the development plan, as recommended by the committee, to the Governor of the State. Unfortunately, the state governments are not serious in implementing this reform. The DPCs are constituted but they are not functioning effectively to realise the objective of integrated planning. For instance, a study conducted in southern states brought-out that lack of clarity of their role, absence of mechanisms for preparing and consolidating the plans of local institutions and absence of technical support are some of the reasons for the failure of this constitutional institution except in Kerala where decentralised planning is institutionalised and working effectively.[52] Though the legislations were made, the MPCs are not constituted except in a few metropolises.

In this context, the Twelfth Plan suggested that the DPCs/MPCs must be strengthened to provide better linkages between rural and urban areas. It has recommended that "the MPCs/DPCs should create the spatial development plan for the region including any rural areas that may lie within the district boundary but outside the municipal limits of an urban area. Such plans should take a longer 20+ year perspective with a formal review every five years. The broad Spatial Development Plan should then be used by the municipalities as a guiding framework to create the second level detailed plans for the city. The timeline of the Metropolitan Development Plans and the District Development Plans should be synchronised with Municipal Spatial Plan."[53]

There has been wide discussion on the role of the Metropolitan/Urban Development Authorities. Lack of clarity on this aspect is also one of the reasons for not implementing this reform. While clarifying this, the Twelfth Plan suggested that the Metropolitan Development Authority should work under the aegis of the Metropolitan Planning Committee and should be vested with the responsibility of enforcing and regulating the metropolitan development plan. It should be the appellate authority for conflict resolutions on spatial plans for all Local Planning Authorities in the metro region, in keeping with the letter and spirit of the Constitutional Amendment Act.[54]

Towards Strengthening of Urban Planning

A key weakness of India's urbanisation and urban development efforts as remarked by the Twelfth Plan is that the urban development agenda is being implemented through disjointed projects/activities with inadequate or no planning for the urban area as a whole. In spite of the recommendations of the National Commission on Urbanisation way back in 1988 and the Constitution 74th Amendment Act which provided for transferring of the urban planning function including the economic planning in urban areas to the municipalities in totality, the situation appears to be discouraging. The over dependence on the master plan concept has failed in the country due to mere bureaucratic or technocratic approach. The master plans generally focus on the core area of the city, without proper urban growth vision and strategy to connect/integrate the peri-urban and rural areas within a regional framework. The approach lacks integration of spatial planning (including transportation and land use planning) with sectoral planning. The zonal plans have not been completed due to over emphasis on details even after years of plan preparation exercise.[55] It is suggested to change the nomenclature of the master plan to spatial and development plan to capture spatial and socio-economic dimensions of planning.[56]

There is need for preparing state level strategies of spatial development plans to realise state and central governments' economic, infrastructure and social development priorities and facilitating resource allocations. Similarly, the spatial development plans must be prepared by the DPCs and MPCs including sector specific plans. The Steering Committee of the Twelfth Plan[57] recommended for Unified Metropolitan Transit Authority for Metropolitan Areas for preparing integrated transport plans and transit oriented land use zoning and development. The municipalities should also prepare similar spatial development plans at the local levels. The spatial plans in general should include land use and development controls at different levels. The concept of spatial development plan is a paradigm shift from mere land use based master plan approach to a comprehensive urban planning integrating land use, development control, transport and sector specific development plans.

Urban planning reforms under the erstwhile JnNURM suggested for enacting Urban and Regional Development Planning Act (URDP Act) in every state to formulate an integrated vision, strategies and spatial structures covering State, District and Taluka level development plans. All other instruments e.g. Central SEZ Act, 2006, State Industrial Development Corporation Acts, etc., dealing with access to land and its use from the central, concurrent and state lists should be readjusted accordingly. The URDP Act should become an umbrella Act in the state for all land uses. The State Urban and Regional Development Planning Board (SURDP Board) has to be constituted to take the land use decisions. The existing State Planning Boards should get upfront agreement from the SURDP Board on all land use related issues before clearing and allotting funds for sectoral and other land and built space related projects in the State. The District Planning Committees, Metropolitan Planning Committees and the municipalities and the panchyati raj institutions should prepare spatial development plans under the above overall arrangement. These institutions should prepare 20-25 years perspective plans and 5 years development programmes. The overall planning framework envisaged is regional level land use structure plans, zonal level matrix of land use and simplified development controls for each land use level.[58] The ward committees should prepare ward level development plans through active people's participation.

The Twelfth Plan observed that Development Plan should fit into city's natural endowment, economic potential and promote clean and green strategies in the growth of cities. It suggested several sector specific plans viz., city mobility plan, city sewerage and sanitation plan, city water plan, economic and commercial activity plan, infrastructure plan, affordable housing plan, environment conservation plan, urban poverty reduction strategy and plan for development of peri-urban areas. The Development Plan and Financial Plan was a necessary precondition for receiving assistance under the JnNURM.[59] However, all these instruments were used by the state governments just for seeking the central assistance and nothing satisfactory happened at the grass-root level, in many cases.

The Twelfth Plan envisaged that India's urbanisation should be guided towards inclusive, equitable and sustainable growth of towns and cities. The urban planning should therefore address the concerns and issues in these areas. Inclusive city development entails a paradigm shift in the urban policies, planning and regulation to incorporate the working poor and other vulnerable in urban development agenda. These sections should find their voice in the urban planning process. The planning should become participatory and people centered. The participatory planning basically concerns with the involvement of beneficiaries, civil society organisations, stakeholders, etc. The participatory model being practiced in Kerala is a good example in this respect. In Kerala, local development plans are prepared by the area sabhas and ward committees and are being integrated at the town and state levels through a well connected institutional mechanism. People's participation is emphasised much in the slum rehabilitation, shelter and urban poverty alleviation programmes in India. These initiatives are unfortunately not integrated with spatial and infrastructure plans.

Urban planning can contribute to sustainable development only when it is effectively integrated at different levels—local, regional, state and national. The bigger question is how to integrate land use, economic, infrastructure and social dimensions into a spatial framework. The foremost need is to make land use and development control environment and people friendly for sustainable development. If green building approach makes urban planning environment friendly, the inclusive principles make urban planning people centered. Urban planning is less and less about defining land use and zoning but more about addressing ways to create cities that work for all citizens and, offer opportunities to rich and poor alike, where communities and their concerns are at the heart of decision-making on development and where risk and vulnerability is effectively addressed for all citizens, particularly the urban poor. The key challenges are learning to work with informality, addressing climatic change by bridging the green and brown and, dealing with safety, security and disasters.[60] The reforms initiated in recent years are on right path but they should be put on fast track through necessary institutional mechanisms, administrative reengineering and commitment.

Constant monitoring of the building activities is essential to ensure orderly growth of the cities and towns. In this respect, computerisation of the data and developing better building information system are useful to monitor the activities. Manual inspections are time consuming and difficult in the larger cities where building activities are fast. The geographic information system (GIS) is one of the tools to develop up-dated building information. It is a computer-based tool for mapping and analysing things that exist and events that happen on earth. GIS technology integrates common database operations such as query and statistical analysis with the unique visualisation and geographic analysis benefits offered by maps. The GIS maps for all the wards should be prepared including details of the layouts, open plots, buildings and amenities. The building information could be developed on certain building parameters like site area, location, details of permit, status of construction, deviations observed, etc. Geographical Informational System has been a proved tool to implement, study, monitor and analyse any geo-database framework. Urban governance systems have recently made progress in exercising GIS oriented databases. GIS tools can be of greater significance to monitor building activities and to study impact of building regulations on urban growth.

To recap the discussion, following urban planning agenda should be carried-out strictly to achieve the goals of development.

(i) Transfer of the total planning functions and delegation of powers to the municipalities.

(ii) Adopting an integrated approach wherein the spatial plans and development plans are intertwined effectively.

(iii) The city level development plans and the financial plans must be made mandatory for all municipalities.

(iv) Revitalising/creating the planning cells of municipalities by injecting necessary professional expertise including staffing.

(v) Strengthening of the DPCs/MPCs on the lines provided in the Constitution 74th Amendment Act and the directions given in the Twelfth Plan.

(vi) Involving the ward committees and the area sabhas actively in the local planning process.

(vii) Simplifying and rationalising the building byelaws and norms for conversion of agricultural lands to urban uses.

(viii) Promoting green building technology and energy efficient buildings for sustainable urban development by mandating ECBC and GRIHA rating for high energy consuming projects.

(ix) Establishing effective inter-agency coordination mechanism at the local levels.

(x) Promoting land-readjustment strategies for orderly development of the peripheral areas and slum areas.

To conclude, urban planning reforms in India reflect the concerns of policy-makers to the emerging urbanisation, environmental, economic and social developmental issues. The crucial aspect is how to transform these concerns into appropriate governing processes through regulatory, promotional and institutional mechanisms for achieving the pronounced goals of environmental sustainability, social inclusion and economic growth.

References

1. The World Bank, Urbanisation beyond Municipal Boundaries, op. cit., p. 1.
2. *Ibid.,* p. 5.
3. *Ibid,* p. 1.
4. Ministry of Urban Development, Government of India, Integration of City Planning and Delivery Functions, State Level Reform under JnNURM, p. 4.
5. Ministry of Urban Development, Government of India, Simplification of Legal and Procedural Framework for Conversion of Agricultural Land to Non-Agricultural Purposes, JnNURM Primer, pp. 2-3.
6. *Ibid,* p. 4.
7. *Ibid,* p. 5.
8. *Ibid,* p. 7.
9. Town and Country Planning Organisation, Ministry of Urban Development, Government of India (2004): Model Building Bye-laws, July.
10. Government of Andhra Pradesh, GO Ms No. 503, Municipal Administration and Urban Development (M) Department, dated 20.4.1987.
11. Government of Andhra Pradesh, GO Ms. No. 158 MA, Municipal Administration and Urban Development (M) Department, dated 22.3.1996. (Further, under GO Ms. No. 508 MA dated 15.11.2002).
12. Government of Andhra Pradesh, GO Ms. No. 422 dated 31.7.1998 and GO Ms. No. 423 dated 31.7.1998, Municipal Administration and Urban Development (M) Department.
13. Government of Andhra Pradesh, G.O. Ms No. 86, Municipal Administration and Urban Development (M) Department, dated 3rd March, 2006.
14. Government of Andhra Pradesh, G.O. Ms. 17, Municipal Administration and Urban Development (M) Department, dated 10-01-2007.
15. Government of Andhra Pradesh, G.O. Ms. No. 589, Municipal Administration and Urban Development (M) Department, dated 25-09-2009.
16. Government of Andhra Pradesh, Memo No. 789/M1/2009, Municipal Administration and Urban Development (M) Department, dated 21-06-2010.
17. Enadu (Telugu daily news paper) Hyderabad, 1st June, 2013, p. 15.
18. Government of Andhra Pradesh, G.O.Ms.No.168, Municipal Administration and Urban Development (M) Department, dated 07.04.2012.
19. However, the Government is considering reintroducing the concept of FSI to facilitate infrastructure development in the congested areas. Discussions are going on to re-think about FAR or FSI liberalisation.
20. Sridhar, Kala Seetharam, (2007); Impact of Land Use Regulations on Suburbanisation, Evidence from Indian Cities, Institute for Social and Economic Change, p. 6.
21. Planning Commission, Government of India, 12th Five Year Plan, p. 328.
22. Eenadu (Telugu daily newspaper), Hyderabad, 1st June, 2013.
23. Sridhar, Kala Seetharam, op. cit.

24. Government of India (July, 2011): Report of the Sub Committee on Development of Sustainable Habitat Parameters in the Field of Urban Planning, Town and Country Planning Organisation, Ministry of Urban Development, p. 5.
25. United Nations (1987): Report of the World Commission on Environment and Development-Our Common Future.
26. Report of the Sub Committee on Development of Sustainable Habitat Parameters, op. cit., p. 7.
27. *Ibid,* p. 14.
28. Ministry of Power, Government of India (2007): Energy Conservation Building Code, p. 2.
29. *Ibid.*
30. USAID ECO-III Project and Bureau of Energy Efficiency (2011): Energy Conservation Building Code User Guide, p. 26.
31. *Ibid,* p. 1.
32. Ministry of New and Renewable Energy, Government of India, and Association for Development and Research of Sustainable Habitats, the Little Book of GRIHA rating, p. 2.
33. *Ibid.*
34. Ministry of New and Renewable Energy, Government of India, and Association for Development and Research of Sustainable Habitats, the Little Book of GRIHA rating, p. 8.
35. *Ibid.*
36. *Ibid.,* p. 11.
37. Town and Country Planning Organisation, Ministry of Urban Development (2010): Streamlining Procedures for Clearance of Building Projects, pp. 4-5.
38. Ministry of Urban Housing and Poverty Alleviation, Government of India (2013): Report of the Committee on Streamlining Approval Procedures for Real Estate Projects, p. 16.
39. Times of India, 1st September, 2013, Hyderabad.
40. Centre for Good Governance, Administrative Staff College of India, JnNURM Rapid Training Programme Governance and Reforms, pp. 18.
41. Centre for Good Governance, Administrative Staff College of India, JnNURM Rapid Training Programme Governance and Reforms, pp. 15-16.
42. The owner/proposer has to file intimation of commencement of work at the Single Window Counter of the concerned Zonal Office. After completing the building up to the plinth level, one has to file an application for Commencement Certificate at the Single Window Counter. Site inspection will be made by an authorised BBMP officer within 3 working days from the date of submission of application. If there are any objections, the site engineer will inform the same to owner/proposer; if there is no objection, Commencement Certificate can be collected by owner/proposer on the next day after inspection. After the building is completed, according to the sanctioned plan and the provisions of building bye laws, the owner/proposer has to file application in prescribed format at the Single Window Counter for Occupation Certificate. The Assistant Director of

Town Planning or the designated engineer will inspect the building for its adherence to the sanctioned plan and, if found to be in order will issue the Occupancy Certificate within 15 working days.

43. Report of the Committee on Streamlining Approval Procedures for Real Estate Projects, op. cit., p. 102.
44. *Ibid.*
45. *Ibid,* p. 103.
46. *Ibid.*
47. *Ibid.*
48. *Ibid,* p. 19.
49. *Ibid,* pp. 70-72.
50. Planning Commission, Government of India, 12th Five Year Plan, p. 327.
51. The Constitution 74th Amendment Act, Articles 243 ZD and Article 243 ZE.
52. Prasad, D. Ravindra and Gnaneshwar, V. (2004): URBAN GOVERNANCE REFORMS IN INDIA-A Study of the Impact of Constitution (74th Amendment) Act, 1992 on the Working of Urban Local Bodies in Andhra Pradesh, Karnataka, Kerala and Tamil Nadu (Vol. II), RCUES, Osmania University, Hyderabad, p. 124.
53. Planning Commission, Government of India, 12th Five Year Plan, p. 328.
54. *Ibid,* pp. 328-329.
55. Ministry of Housing and Urban Poverty Alleviation, Government of India (2011): Report of the Working Group on Urban Strategic Planning 12th Five-year plan Steering committee on Urban Development and Management, pp. 7-8.
56. *Ibid,* p. 14.
57. *Ibid,* pp.19-20.
58. Integration of City Planning and Delivery Functions, State Level Reform under JnNURM, op. cit., p. 6.
59. Twelfth Five Year Plan, op. cit, p. 128.
60. UN-HABITAT (2010): Planning Sustainable Cities, Nairobi, pp. 7-8.

Chapter - 9

Housing the Poor—Shelter Reforms in India

Introduction

Proper shelter is not only a basic physical need but also an index of dignified and healthy human living. But, the most visible disturbing manifestation of India's urbanisation has been the proliferation of squatters and slum areas depicting deplorable dehumanised living conditions. It has become a herculean task for the poor urban households and a great challenge to the governments. The economists view increasing housing demand as a positive factor contributing much to the nation's economic growth. It is considered an opportunity to the housing industry. The housing sector has strong backward and forward linkages with several ancillary industries viz., construction workers, builders, developers, suppliers, engineers, property consultants, plumbers, electricians, etc.[1] It has multiplier effect occupying third place in the linkage effect. In India, it is the second sector next only to agriculture providing employment to about 33 million people and the number is expected to increase to 83 million by 2022.[2] The housing amenities where poor people live in general are considered far below the international minimum standards in India. As per the NSSO 2009 estimates, 11 per cent of the urban households do not have adequate sanitation while four per cent do not have electricity.[3] The Census 2011 brought-out that 18.6 per cent of urban households do not have latrine, seven per cent do not have electricity and 30 per cent do not have tap water.[4]

The shelter problem basically concerns with three main issues: (i) how to increase shelter supply; (ii) how to improve and upgrade slums; and (iii) how to conserve the existing housing stock.[5] Shelter related problems received much attention of the governments since Independence. Various slum improvement and economically weaker section housing programmes are being implemented aiming to provide proper shelter and basic services to the urban poor. Yet, the situation appears to be dismal. Several causes like urbanisation, migration, land and economic growth strategies cumulatively contributed for this situation. New perspectives on shelter related problems emerged in India beginning with the National Commission on Urbanisation (NCU) in 1988. Several reform initiatives and strategies are being pursued by the Government of India and the state governments in the post-1990 periods to mitigate the shelter problems of urban slum dwellers. This chapter critically discusses the efficacy of the policies, strategies and reform initiatives so far pursued, the lessons learned and the future perspectives.

Housing Shortage in India

Housing received top priority in the successive Five Year Plans in the country since Independence. Yet, housing shortages continue in some form or the other to bother the policy makers and administrators. It is not just the number of units but the quality of houses, congestion and amenities of the habitat which have become crucial problems. The National Buildings Organisation (NBO) estimated in 1981 that about two lakh units were needed for the houseless and 31 lakh more units required to people who live in kutcha units.[6] Further, the NBO commented that even with best possible efforts, just 10 per cent of the backlog was being met. According to 2001 Census, there was a net addition of 19.52 million dwelling units in the urban housing stock, amounting to average annual construction of 1.95 million houses during the period 1991 to 2001. The share of ownership housing in urban areas has increased from 63 per cent in 1991 to 67per cent in 2001. Further, the households having one-room accommodation declined significantly in urban areas from 39.55 per cent to 35.1per cent during the period 1991–2001. This was a result of upward shift of accommodation and accelerated supply of housing stock.[7]

Table 9.1: Housing Shortage according to Income Groups

Sl. No.	*Category*	*2007*		*2012*	
		No in millions	*Per cent*	*No in millions*	*Per cent*
1.	EWS	21.78	88.13	10.55	56.18
2.	LIG	2.89	11.69	7.41	39.44
3.	MIG/HIG	0.04	0.18	0.82	4.38
4.	Total shortage	24.71		18.78	100

Source: Rebekah Kurien: Innovations in Affordable Housing in India, Shelter, Vol. 14, No.1, HUDCO, New Delhi, April, 2013, p. 56.

Yet, the housing shortage especially, for the poor continues to accentuate in its multidimensional form in urban India. The Technical Group on

Estimation of Housing Shortage[8] estimated the total shortage of dwelling units in urban areas in 2007 to be 24.7 million growing currently at a rate of 36 lakhs per annum. Bulk of the shortage i.e., 99 per cent of this pertained to the EWS and LIG segments of the urban population. This is an indication of upward movement of poor people from EWS to LIG considered to be positive from the poverty reduction point of view (Table 9.1).

The housing requirement for the Eleventh Five Year Plan period (2007 to 2012) was estimated to be 26.53 million. The Technical Group on Urban Housing Shortage appointed for the Twelftth Five Year Plan estimated the housing shortage of 18.7 million — 0.99 million for households living in non-serviceable kutcha houses, 2.27 million for households living in obsolescent houses, 14.99 million for households living in congested houses requiring new houses and 0.53 million for homeless households.[9] Though the situation improved in respect of the kutcha and obsolescent houses since 2007, the congested way of living increased further in 2012 (Table 9.2).

Table 9.2: Housing Shortage (in millions)

Sl. No.	*Households that require new houses*	*2007*	*2012*
1.	Living in non-serviceable kutcha houses	2.18	0.99
2.	Living in obsolescent houses	2.39	2.27
3.	Living in congested houses	12.67	14.99
4.	Homeless	Not included	0.53
5.	Sub-total (1+2+3+4)	17.24	18.78
6.	Housing deficit (total no. of households minus housing stock)	7.47	Not included
7.	Total housing shortage	24.71	18.78

Source: Rebekah Kurien: Innovations in Affordable Housing in India, Shelter, Vol. 14, No. 1, HUDCO, New Delhi, April, 2013, p. 56.

The homeless people are considered most vulnerable requiring housing on priority basis. Further, it has been estimated that 73 per cent of the shortage in self-occupied housing is in bottom 40 per cent of the urban households.[10] The urban housing shortage mainly pertains to the people in the low income category. In the context of very high land prices and growth of luxury housing in recent years, there is urgent need for focusing on shelter programmes for the urban poor.

Urban Housing Policies in India—An Overview

As housing falls in the Concurrent List in the Constitution, both, central and state governments have the responsibilities on the subject. Housing and slum rehabilitation received attention in the successive Five Year Plans. Several policies, strategies and reforms are introduced to meet the housing and habitat needs of the people. For a more focused discussion, these initiatives are broadly grouped into five periods, viz., Post-Independence Period, Urbanisation Period, Economic Reform Period, JnNURM Period and Post-JnNURM Period.

Housing Policies in the Post-independence Period (1950-1970)

In the wake of World War II, partition and the consequent refugee influx, housing became a crucial issue in the urban areas in post-Independent India. Concern for housing shortage was hence, expressed in the very First Five Year Plan. Initially, the focus was on house construction for government employees and weaker sections. The State Housing Boards and Improvement Trusts devised schemes to construct houses, especially for the employees. In respect of slums, the policy was slum clearance and rehabilitation. Housing for the weaker sections was confined to more about slum rehabilitation. The Central Government provided loans for government employees and provided assistance to state governments for taking up housing on subsidy basis. The National Buildings Organisation was set-up in 1954 as a technical wing to the Ministry of Works and Housing. The Government of India sponsored a number of schemes, viz., Integrated Subsidised Housing Scheme for the Industrial Workers and Economically Weaker (or poorer) Sections of the Community (1952) and, Low Income Housing (1954). The focus slowly shifted to providing shelter for urban poor. The Slum Clearance (1956) and Middle Income Housing (1959) schemes were launched to provide shelter for the slum dwellers.

The attention of the governments on weaker sections began from the Second Five Year Plan onwards. Initially, it was institution building, funding and coordinating the slum development programmes. The rising land prices especially in larger cities and industrial townships, forced the policy-makers to attend to the land related issues primarily to solve the problems of slum rehabilitation. In respect of house construction, the emphasis was on reducing costs to make it affordable to weaker sections and low income people.

Housing Policies in the Urbanisation Period (1970–1990)

The faster urban bound migration in the post-1970 period in India caused proliferation of slums. In the context of rising land prices and cost of construction, there was search for alternate solutions to solve shelter problems. The institutional support for housing received prime attention in the post-1970 period. Institutions like the Housing and Urban Development Corporation (HUDCO-1970), Housing Development Finance Corporation (HDFC-1978) and National Housing Bank (NHB-1988) were created at the national level to cater to the financial needs of housing. If HUDCO focused on low-income and economically weaker sections of the society, HDFC and

Life Insurance Corporation of India catered to needs of middle and higher income groups of people. At state level, Housing Boards, Slum Clearance Boards and Cooperative Societies implemented the housing programmes. Public sector housing was initially confined to the provision of housing to specific categories of people, viz., industrial workers, plantation workers and other poorer sections. But, soon it was discovered that the contribution of public sector housing was fractional compared to the demand.

The policy of direct provision of housing in urban areas had become unfeasible in the face of rapidly increasing demand for housing, spiraling land prices and increasing cost of construction. The housing strategies, therefore, shifted to the provision of basic services and rehabilitation in the slum areas. The government role in housing had per force to be promotional as suggested in the Seventh Five Year Plan.[11] As major effort is expected to come from the private sector, government attention needs to be focused on the priority sector, i.e., urban poor. In fact, the Seventh Plan clearly stated that "the government role will have to be restricted to the improvement of slums, direct provision of housing to the weaker sections of the society and encouragement and support of housing finance institutions that promote channeling of private resources into housing in a constructive way." The strategy thus changed from formal housing to the 'sites and services' approach involving land acquisition, development and disposal and, basic service provision.

In slum areas, the earlier policy was slum clearance that envisaged re-housing of slum families with service improvements. This approach was first adopted in 1956. Slum Clearance Boards were created at the state level to take-up the task of slum clearance. In view of the difficulties and hardships encountered in relocating the slums, the emphasis has shifted to improvement of the existing slums by providing security of tenure, construction assistance in terms of both, technology/expertise and finance and provision of civic services like lighting, drainage, water supply, sanitation, etc. Schemes like Environmental Improvement of Urban Slums (1972), Low-Cost Sanitation (1981), Urban Basic Services (1986) and Urban Basic Services for the Poor (1990) were implemented to carry-out these tasks.

The Environmental Improvement of Urban Slums (EIUS) scheme focused on improving civic services in slum areas with a view to upgrade the environmental status of the slum areas. This scheme was later transferred to the state sector. The UNICEF launched social development programmes in small and medium towns in certain states in India during the early 1980s focusing on water supply, sanitation, women and child welfare, health, etc. Inspired by these efforts, the Government of India launched the Urban Basic Services (UBS) scheme for providing similar services in small and medium towns in India. Later, this scheme was renamed as Urban Basic Services for the Poor (UBSP) to give more attention to the needs of the poor in urban areas.

Basically, all these schemes aim at providing civic services and, improving the health and education aspects of poor communities in slum areas. The strategy of linking shelter with services thus began slowly in India. From the Seventh Five Year Plan (1985-90) onwards, the core responsibility of housing construction shifted to the private sector, the government role was confined to mobilisation of resources, subsidies, land and provision of basic services in slum areas.

In view of the migration of people from rural areas to large cities, the Government of India launched the scheme for Integrated Development of Small and Medium Towns (IDSMT) from the Sixth Five Year Plan onward. The scheme aimed at arresting the migration to large urban centres by developing small and medium towns in between. The focus of housing shifted from direct housing to sites and services development especially, in small and medium towns. The institutions like HUDCO played critical role in financing housing schemes for LIG and weaker sections. Self-help strategy was promoted as far as housing is concerned. But, the sites and services schemes failed due to land acquisition problems. The housing especially, for the poor and low income groups continued to receive attention of the governments but the strategic failures were noticeable and glaring. The major issue of housing revolved around land related issues. The land acquisition, socialising urban land through Urban Land (Ceiling & Regulation) Act (ULCRA) and rent control legislations—created a constraining regulatory environment which impacted the housing initiatives in urban areas adversely. If the direct provision of housing failed due to financial limitations, the land development initiatives failed due to the obstructive legal system and administrative deficiencies.

It is true that the problem of housing in the third world countries is basically a problem of the low income groups. The governments and the international agencies like the World Bank, Asian Development Bank and USAID are, therefore, focusing more on the housing needs of the poor in the third world countries. Considerable funds are now available from these international agencies. But the real problem appears to be conditionalities imposed by these agencies on the loan receiving governments. For instance, the World Bank was instrumental in bringing-out institutional changes in the project areas in Indian cities like Kota, Chennai and Mumbai. In Mumbai, a comprehensive policy reform was initiated at the instance of the World Bank to carry-out housing projects including areas like finance and management of the local government, staff training, partial rent decontrol and revision of building standards. In fact, cost recovery, affordability and project approach are the guiding principles of the World Bank aided projects. If cost recovery invites public protests, the concept of affordability is riddled with administrative problems. The role of international agencies though a problematic area is a blessing in disguise. The governments in the third world countries are now exposed to international perspectives and are able to experiment with various methods of financing urban services.

The need for a changed perspective and new values in the thinking among the slum dwellers as well as the public authorities was suggested as back as 1988 by the National Commission on Urbanisation (NCU). The major thrust of this new approach was that providing housing to the poor should not be considered as a great challenge and insurmountable problem but a manageable situation and opportunity for making it a productive resource. A paradigm shift is required[12] in approaches from: demolition of deficient housing stock to conservation and improvement; public housing agencies as providers to promoters; and finished houses to massive serviced land and amenity development. The housing policies and strategies devised in the later periods incorporated these core ideas.

The National Housing Policy in the country was first formulated in 1988 that incorporates these emerging ideas. Main objectives of the Housing Policy this is reference no. as laid on the table of both the Houses of Parliament on 12th May, 1988, were:

(i) To motivate and help all people and in particular, the houseless and inadequately housed, to secure for themselves affordable shelter through access to land, material, technology and finance;

(ii) To encourage investment in housing in order to achieve a sustained growth of the nation's housing stock, and its proper conservation, renovation and up-gradation;

(iii) To create an enabling environment by eliminating constraints and developing an efficient and accessible system for the delivery of inputs to maximise housing efforts;

(iv) To improve the environment of human settlements with a view to raising the quality of life through the provision of drinking water, sanitation and other basic services;

(v) To promote vernacular architecture and preserve nation's rich heritage in the field of human settlements.

The priority sections identified in the policy document were houseless and disadvantaged groups like scheduled castes and tribes, rural landless labour, widows, economically weaker sections, low income groups and the victims of natural calamities. In the urban sector, the focus was on slums and squatter settlements. Measures like providing basic services, access to urban land and finance, developing low cost housing and skill training were suggested to improve the urban housing.

In spite of well-conceived policy objectives, urban housing suffered from chronic administrative problems, viz., low budgetary allocations; rigid land use and building standards; legal and bureaucratic obstructions; lack of coordination in the provision and maintenance of basic services amongst city and state level agencies; and above all, insufficient involvement of the people. Housing should not be treated in isolation as a mere social responsibility. The housing policies must be effectively linked to the building technologies, urban land and economic development. Most of the above problems could be solved through an integrated perspective and coordinated efforts.

Housing Reforms in the Economic Reform Period (1990-2005)

Economic reforms pursued in the post-1990 period had profound effect on the housing policies in India. There was a shift in the housing strategies re-aligning urban sector reforms with the emerging macro-economic context. Urban land was viewed as a resource for urban development and shelter. The Government of India framed national housing policies to guide the governments to provide housing to all people. The focus of these policies however varied based on experiences and emerging needs. The reframed National Housing Policy, 1992 focused on controlling land prices. The Housing and Habitat Policy, 1995 emphasised the need for improving housing stock through housing revolution.[13] The National Sample Survey Organisation (NSSO) in its 61st Round reported that the number of urban poor has risen by 4.4 million persons during 1993-94 to 2004-05. In this context, the National Urban Housing and Habitat Policy carefully analysed ways and means of providing the 'Affordable Housing to All' with special emphasis on the EWS and LIG housing.

In the post-1990 period, apart from formulation of the national housing policies, several schemes were launched to develop slums and provide housing for poor. They were:

(i) EWS housing scheme for beedi workers and hamals-1991.

(ii) National Slum Development Programme -1997.

(iii) A Two Million Housing Programme for EWS/LIG-1998.

(iv) Valmiki Ambedkar Awas Yojana (VAMBAY)-2001.

The Two-Million Housing Programme launched in 1998-99 was a loan scheme. It aimed to facilitate construction of seven million dwelling units in urban areas every year apart from 13 million units in rural areas. The Valmiki Ambedkar Awas Yojana (VAMBAY) was a centrally sponsored scheme with in-built subsidy for undertaking dwelling units and sanitation units in the slum areas. It had a 50 per cent subsidy component. The VAMBAY scheme was designed to provide shelter either by new construction or by upgradation to the slum dwellers that were below poverty line. In view of the importance of sanitation in slum areas, 20 per cent of the total allocation was specified for sanitation and community latrines. The toilets were to be maintained by the local community with a contribution of Rs 20 per family. The state governments implemented the scheme by providing land and arranging for the credit component for house construction. The upper limit was Rs 30,000 per unit in mega cities, Rs 25,000 in million plus cities and Rs 20,000 in other cities and towns. The National Slum Development Programme (NSDP) was launched in 1997 to improve urban slums. The Central Government provided special assistance to state governments under this scheme.

The post-1990 period witnessed economic liberalisation and privatisation drives in India aiming at faster economic growth in the globalising world. The housing sector too required massive expansion of housing finance and administrative reforms at the state and local levels. Though the national housing policies provided broader framework to the governments, commensurate administrative action was missing. The changed macro development context needed fundamental urban governance reforms. Urban governance reforms picked-up with the Constitution 74th Amendment and were broad based and put on fast track through JnNURM. Concrete and committed reforms, including housing and shelter sector, were initiated under JnNURM in urban India since 2005.

Housing Reforms in the JnNURM Era (2005-2015)

The urban housing shortage for the Eleventh Five Year Plan period (2007-12) was estimated at 26.53 millions including backlog. It was observed that due to low affordability among poor people, still a sizeable number were forced to live in slums.[14] However, most of the institutional financing had been in the MIG/HIG sector.[15]

The National Urban Housing and Habitat Policy, 2007 is the latest policy initiative of the Government of India which incorporates the emerging ideas on urban housing with the central theme of affordable housing to all. The National Urban Housing and Habitat Policy, 2007 talks much about supply of land, affordable shelter and services. Equity, affordability and integrated development of slums are the prime drivers of the recent policy. It aims to achieve the following:[16]

(i) Removing legal, financial and administrative barriers for facilitating access to tenure, land, finance and technology.

(ii) Careful review of authorised Floor Area Ratio (FAR) in line with international practices for allowing more efficient use of scarce urban land by construction of high rise buildings.

(iii) Facilitating accessibility to serviced land and housing with focus on economically weaker sections and low income group categories.

(iv) Suitable restructuring for enabling both institutions at the Central and State levels, and the private sector for increasing supply of land.

(v) Special efforts for catering to the housing and basic service needs of Scheduled Castes, Scheduled Tribes, Backward Classes, minorities, disabled persons, slum dwellers, street vendors other informal sector workers and other vulnerable sections of the society.

(vi) Forging strong partnerships between public, private and cooperative sectors for accelerated growth in the housing sector and sustainable development of habitat.

The JnNURM was an ambitious mission mode reform oriented scheme launched in the urban sector in 2005. It aimed to rebuild urban India to meet the globalisation demands through a committed reform agenda. The JnNURM, apart from Urban Infrastructure and Governance (UIG) and Urban Infrastructure Development Scheme for Small and Medium Towns (UIDSSMT), launched two urban poverty oriented schemes, viz., Basic Service for the Urban Poor (BSUP) for mission cities and Integrated Housing and Slum Development Programme (IHSDP) for the non-mission cities. The earlier slum development and shelter programmes, namely, NSDP and VAMBAY were subsumed in the BSUP and IHSDP schemes. These schemes were formulated to realise the objectives of the National Urban Housing and Habitat Policy, 2007. The BSUP and IHSDP schemes emphasised provision of basic services including shelter in integrated manner in slum areas. They cover security of tenure at affordable prices, improved housing, water supply, sanitation, education, health, and social security. The aim was to construct one million affordable dwelling units with 0.25 million units for EWS category during the seven year scheme period. The performance however was far from satisfaction as only 41.3 per cent of sanctioned units were completed till December, 2012 under the BSUP and IHSDP schemes.[17]

The key reforms contemplated through these two schemes were:

(i) Internal earmarking within local body budgets for basic services to the urban poor (in proportion to the share in total city/town population).

(ii) Implementation of 7-Point Charter, i.e. provision of basic services to urban poor including security of tenure, affordable housing, water supply, sanitation and ensuring delivery of already existing universal services of the government for education, health and social security within the Mission period as per agreed timelines.

(iii) Earmarking at least 20-25 per cent of developed land in all housing projects (both, public and private agencies) for EWS/LIG category with a system of cross-subsidisation.

(iv) Ensuring 20-25 per cent FAR reservation for EWS/LIG housing units.

(v) Making additional FAR available in the transportation corridors of outer zones through TDRs.

(vi) Using land as a resource for meeting housing needs of the urban poor.

(vii) Promotion of land assembly, development and disposal through effective PPPs.

(viii) Reservation of 10 to 15 per cent of land in every new public or private housing projects or 20 to 25 per cent of FAR/FSI for EWS/LIG housing.

(ix) Promotion of beneficiary led housing development.

A number of other schemes were also launched by the Government of India to supplement the JnNURM efforts. These schemes are discussed hereunder.

Urban Statistics for HR Assessment (USHA)

The Government of India launched Urban Statistics for HR Assessment (USHA) scheme for the development and maintenance of a national database, MIS, and knowledge repository on urban poverty, slums, housing, construction, and other urbanisation-related statistics. Its key objective was to support the government with an information base and knowledge inputs for planning, policy-making, project design, formulation, implementation, monitoring, assessment, and evaluation, particularly in the context of programmes relating to urban poverty, slums, and housing. The scheme was earlier known as MIS on Housing Data, Research and Survey.[18] It sought to specially support the effective implementation of two schemes of JnNURM, viz., Basic Services to the Urban Poor (BSUP) and Integrated Housing and Slum Development Programme (IHSDP).

The four pillars of USHA were: database including MIS and sample surveys; action research; impact assessment; and capacity building/training. The scheme was aimed to coordinate applied research and capacity building activities pertaining to urban poverty, slums, housing, construction and other urbanisation-related statistics in collaboration with reputed research and training institutions at national, regional and state levels and experts. The components of USHA were: data centre and MIS on urban poverty, slums, housing, building construction and related urbanisation statistics; Knowledge

Centre/National Resource Centre for urban poverty and slums ; sample surveys in areas of urban poverty, slums, housing and building construction; socio-economic research studies in areas of urban poverty, slums, housing and building construction; capacity building and training in areas of urban poverty, slums, housing and building construction statistics; and publicity and awareness.

Affordable Housing in Partnership

In view of the budgetary constraints to provide affordable housing, the Government of India launched a new scheme called Affordable Housing in Partnership under JnNURM covering 65 Mission cities initially. This scheme was launched in 2009 aimed at building partnerships across various sectors — private sector, cooperative sector, financial sector, parastatals, municipalities etc., for providing affordable housing to all. Its focus was on stimulating economic activities among the urban poor through partnerships. It covered primarily mission cities. The norms recommended by Deepak Parekh Committee on affordability were adopted in the scheme. In order to attract the private sector, incentives were contemplated in the land use conversion, FAR relaxation, etc. The construction cost was to be on 'no-loss, no-profit' or 'at reasonable cost'. The financial viability could be achieved by cross-subsidy with HIG and commercial spaces. The institutional funding from banks must be linked. The scheme[19] expected that the states to reduce the stamp duty to a maximum of two per cent for LIG and nil for the EWS categories. The objective was to promote affordable housing projects for a total of 10 lakh units with a central assistance of 25 per cent of cost of provision of civic services with a central outlay of Rs 5,000 crore. The central assistance per unit was restricted to Rs 50, 000. The other features were: EWS/LIG beneficiaries were to be enabled to access loans under Interest Subsidy Scheme for Housing the Urban Poor (ISHUP) which provided five per cent interest subsidy up to rupees one lakh; title would be in the name of women to the possible extent; and 25 per cent of built-up area of the projects could be allocated for EWS/LIG units.

Rajiv Awas Yojana to Achieve Slum Free India

While recognising the need for increased focus on shelter needs of slums, the Government of India devised a new scheme called Rajiv Awas Yojana (RAY) as a follow-up to the JnNURM. The goal of the scheme was to achieve Slum Free India. In this respect, land became critical. In view of the past failures to provide required land for slum development, the RAY scheme focused more on land requirement. It emphasised two aspects: one, providing tenurial rights to the slum dwellers in the housing projects and second, taking appropriate measures to ensure that adequate lands were made available for slum shelter programmes. The land requirement estimations for meeting housing needs of the slum dwellers reveal an interesting fact. It is estimated that in the mega cities of India viz., Mumbai, Delhi, Kolkata, Chennai, Bengaluru, Hyderabad and Ahmedabad, the land requirement to meet the housing need of slum dwellers is less than five per cent of the total land of the city. In smaller towns, this figure would be still lesser.[20] The need, therefore, was not very high.

Affordable housing was the underlying theme of the RAY scheme. The purpose was to make housing affordable to all sections of the people including the poor, EWS, LIG and MIG categories. The Government of India issued guidelines to implement the project. The reforms contemplated in the National Urban Housing and Habitat Policy, 2007 were targeted for implementation through the RAY projects. The specific legislative and administrative reforms envisaged under RAY scheme were:[21]

(i) Central Assistance under RAY was to be predicated on the condition that States/UTs assign legal title to slum-dwellers over their dwelling space to provide security of tenure;

(ii) Legislation for property rights to all slum dwellers;

(iii) Reform to the rent control laws regarding urban housing;

(iv) Review and amendment to the legislations, rules and regulations governing urban planning and development structures and systems;

(v) Internal earmarking within local body budgets for basic services to the urban poor;

(vi) Provision of basic services to urban poor at affordable prices through convergence;

(vii) Earmarking at least 20-25 per cent of developed land in all housing projects (both public and private agencies) for EWS/LIG category with a system of cross-subsidisation;

The Government of India initiated Rajiv Rin Yojna (RRY) as a sequel to RAY for providing housing finance on low interest to the urban poor.

Housing Reforms in Post-JnNURM Period (2015 Onwards)

With the advent of a new regime at the centre in May 2014 the whole process of urban governance in the country was overhauled. The new government reviewed the earlier development approach and a few innovative initiatives were taken to bridge the gaps. The Government of India launched a couple of schemes based on the experiences of earlier ones in 2015. The main schemes were Atal Mission for Rejuvenation and Urban Transformation (AMRUT), Smart City Mission, and Housing for All-Prime Minister's Awas Yojna (PMAY).

Atal Mission for Rejuvenation and Urban Transformation (AMRUT)

The Government of India felt that providing basic services (e.g. water supply, sewerage, urban transport) to households and build amenities in cities which will improve the quality of life for all, especially the poor and the disadvantaged is a national priority. Learnings from the earlier Mission have shown that infrastructure creation should have a direct impact on the real needs of people, such as providing taps and toilet connections to all households. It is therefore the AMRUT focuses on infrastructure creation that has a direct link to provision of better services to people. The Thrust Areas under the Mission are: water supply, sewerage facilities and septage management, storm water drains to reduce flooding, pedestrian, non-motorised and public transport facilities, parking spaces, and enhancing amenity value of cities by creating and upgrading green spaces, parks and recreation centres, especially for children. This reform-based Mission adopts whole city approach with special emphasis on marginalised groups in the society. In contrast to earlier approach of project-based financing, the Mission provides for preparing State Annual Action Plan for improving the services by consolidating the same prepared by individual municipalities.[22]

Smart Cities Mission[23]

The purpose of the Smart Cities Mission is to drive economic growth and improve the quality of life of people by enabling local area development and harnessing technology, especially technology that leads to Smart outcomes. Area-based development will transform existing areas (retrofit and redevelop), including slums, into better planned ones, thereby improving liveability of the whole city. New areas (green field) will be developed around cities in order to accommodate the expanding population in urban areas. Application of Smart Solutions will enable cities to use technology, information and data to improve infrastructure and services. Comprehensive development in this way will improve quality of life, create employment and enhance incomes for all, especially the poor and the disadvantaged, leading to inclusive Cities.

Some typical features of comprehensive development in Smart Cities are : Promoting mixed land use in area-based developments — planning for 'unplanned areas' containing a range of compatible activities and land uses close to one another in order to make land use more efficient; Housing and inclusiveness — expand housing opportunities for all; Creating walkable localities — reduce congestion, air pollution and resource depletion, boost local economy, promote interactions and ensure security; Preserving and developing open spaces — parks, playgrounds, and recreational spaces in order to enhance the quality of life of citizens, reduce the urban heat effects and generally promote eco-balance; Promoting a variety of transport options — Transit Oriented Development (TOD), public transport and last mile para-transport connectivity; Making governance citizen-friendly and cost effective — increasingly rely on online services to bring about accountability and transparency, especially using mobiles to reduce cost of services and providing services without having to go to municipal offices; form e-groups to listen to people and obtain feedback and use online monitoring of programmes and activities with the aid of cyber tour of worksites; Giving an identity to the city — based on its main economic activity, such as local cuisine, health, education, arts and craft, culture, sports goods, furniture, hosiery, textile, dairy, etc; Applying Smart Solutions to infrastructure and services in area-based development in order to make them better.

The strategic components of Area-based development in the Smart Cities Mission are city improvement (retrofitting), city renewal (redevelopment) and city extension (greenfield development) plus a Pan-city initiative in which Smart Solutions are applied covering larger parts of the city.

Housing for All (HFA) – Prime Minister's Awas Yojna (PMAY)[24]

"Housing for All" Mission for urban area is being implemented during 2015-2022. The mission will support construction of houses up to 30 square meter carpet area with basic civic infrastructure. Slum redevelopment projects and Affordable Housing projects in partnership should have basic civic infrastructure like water, sanitation, sewerage, road, electricity etc. Municipalities should ensure that individual houses under credit linked interest subsidy and beneficiary led construction should have provision for these basic civic services. The mission has since been launched to provide housing to all by 2022. The mission seeks to address the housing requirement of urban poor including slum dwellers through following programme verticals:

- ***In-situ slum re-development***— Rehabilitation of slum dwellers with participation of private developers using land as a resource, extra FSI/FAR/TDR to make project financially viable.
- ***Promotion of Affordable Housing for weaker section through credit linked subsidy***— The Mission, in order to expand institutional credit flow to the housing needs of urban poor will implement credit linked subsidy component as a demand side intervention. Credit linked subsidy will be provided on home loans taken by eligible urban poor (EWS/LIG) for acquisition, construction of house. Interest subvention subsidy for EWS and LIG for new house or incremental housing - EWS: annual household income up to Rs 3 lakh and house sizes up to 30 sq.m - LIG: annual household income between Rs 3-6 lakhs and house sizes up to 60 sq.m. Beneficiaries of Economically Weaker section (EWS) and Low Income Group

(LIG) seeking housing loans from Banks, Housing Finance Companies and other such institutions would be eligible for an interest subsidy at the rate of 6.5 per cent for tenure of 15 years or during tenure of loan whichever is lower. The Net Present Value (NPV) of the interest subsidy will be calculated at a discount rate of 9 per cent. The credit linked subsidy will be available only for loan amounts up to Rs 6 lakhs and additional loans beyond Rs 6 lakhs, if any, will be at nonsubsidised rate. Interest subsidy will be credited upfront to the loan account of beneficiaries through lending institutions resulting in reduced effective housing loan and Equated Monthly Installment (EMI).

- ***Affordable Housing in Partnership with Public and Private sectors***—with private sector or public sector including parastatal agencies-Central Assistance per EWS house in affordable housing projects where 35 per cent of constructed houses are for EWS category. To increase availability of houses for EWS category at an affordable rate, States/UTs, either through its agencies or in partnership with private sector including industries, can plan affordable housing projects. Central Assistance at the rate of Rs 1.5 Lakh per EWS house would be available for all EWS houses in such projects. The States/UTs would decide on an upper ceiling on the sale price of EWS houses in rupees per square meter of carpet area in such projects with an objective to make them affordable and accessible to the intended beneficiaries. For that purpose, State and cities may extend other concessions such as their State subsidy, land at affordable cost, stamp duty exemptions etc.
- ***Subsidy for beneficiary-led individual house construction*** — The fourth component of the mission is assistance to individual eligible families belonging to EWS categories to either construct new houses or enhance existing houses on their own to cover the beneficiaries who are not able to take advantage of other components of the mission. Such families may avail of central assistance of Rs 1.5 lakh for construction of new houses under the mission. A beneficiary desirous of availing this assistance shall approach the municipality with adequate documentation regarding availability of land owned by them. Such beneficiaries may be residing either in slums or outside the slums. Beneficiaries in slums which are not being redeveloped can be covered under this component if beneficiaries have a Kutcha house.

The foregone discussion reveals that the housing policies in India basically revolve around slum development and rehabilitation (Table 9.3). The issues relating to accessing urban land and finance for housing the urban poor received prime attention of the urban policies. The approach has been shifted from slum clearance and improvement to an ambitious goal of slum free cities. The reforms could play a very significant role if implemented effectively, in achieving the goal of slum free cities. The reforms basically focus on three aspects, viz., strengthening the role of municipalities, strategic interventions for slum development and shelter provision and, the reforms targeting elimination of land and housing market distortions. The municipality is expected to play key role for participatory planning and development at the local levels. The reforms like internal earmarking of budgets for slum development, reservation of 20-25 per cent of developed land to weaker sections and tenurial security are aimed to strengthen the slum rehabilitation and development process. The reforms like amending rental laws, reduction of stamp duty, property titling, etc., are meant to correct the housing market distortions. As the entire subject of housing the poor related to slums and their development, a detailed discussion is attempted in the following section.

Table 9.3: Housing Strategies and Initiatives in India

Post-Independence Period (1950-70)	
Strategies	***Initiatives***
(i) Support through subsidies and loans	(i) Ministry of Works and Housing
(ii) Assisting state governments to provide housing for low income housing	(ii) National Buildings Organisation (1954) to provide technical assistance
(iii) Land acquisition for housing schemes	(iii) Setting-up of State Housing Boards to take-up housing projects
	(iv) Urban Community Development (UCD) programmes (1958) to involve communities
Urbanisation Period (1970-90)	
(i) Providing of housing finance to low income households	(i) Housing and Urban Development Corporation (HUDCO-1970) to provide finance to low income Housing
(ii) Socialisation of urban land	(ii) Urban Land (Ceiling and Regulation) Act (ULCRA-1976) to socialise urban land
(iii) Promotion of housing in the small and medium towns	(iii) Setting-up of Housing Development Finance Corporation (HDFC) in 1978 to provide housing finance in retail sector
	(iv) Environmental Improvement of Urban Slums (EIUS 1972) scheme to provide basic services in slums
(iv) Improving living conditions in slums	(v) Integrated Development of Small and Medium Towns (IDSMT) to provide housing sites
(v) Government role in housing to be promotional leaving house construction to private sector and households	(vi) National Housing Bank (NHB) to mobilise finance and to promote housing finance companies
(vi) Promoting low cost housing technologies	(vii) Building Materials and Technology Promotion Council to develop low building Materials

(Contd...)

	(viii) Low Cost Sanitation (LCS) in 1981 for liberation of Scavengers
	(ix) Urban Basic Services (UBS) in 1986 to provide basic services
	(x) Urban Basic Services for the Poor (UBSP) scheme in 1990 targeting the urban poor
	(xi) Housing and Shelter Up-gradation (SHASHU) scheme to provide employment for people involved in housing
Economic Reform Period (1990-2005)	
(i) Housing policy development	(i) National Housing Policy (1994)
(ii) Providing incentives to housing finance companies to promote housing	(ii) National Housing Policy (1998)
(iii) Development mortgage market controlling and prices	(iii) Allowing of commercial banks in housing sector
(iv) Land as a resource for housing programmes	(iv) National Slum Development Programme (1996)
	(v) Two million housing programme for EWS & LIG (1998)
	(vi) Valmiki-Ambedkar Awas Yojana (VAMBAY) scheme in 2001with subsidy element for housing and services to slum dwellers
	(vii) Allowing FDI in housing sector
JnNURM Era (2005-2015)	
(i) Urban planning as a tool for promoting shelter and optimum use of urban land	(i) National Housing and Housing Policy, 2007
(ii) Increased role for the private sector for housing	(ii) BSUP & IHSMP schemes under JNNURM (2005)
(iii) Forging partnerships among public, private and cooperative sectors for sustainable habitat development	(iii) Rajiv Awas Yojana for slum free cities in 2009
(iv) Reservation of land for EWS/LIG sectors by legislation	(iv) Affordable Housing in Partnership (2009) under JnNURM
(v) Beneficiary led promotion for the poor	(v) Repeal of Urban Land Ceiling & Regulation) Act
(vi) Providing land tenure in slum rehabilitation projects	(vi) Interest Subsidy Scheme for Housing
(vii) Promoting affordable housing for the poor	(vii) USHA Scheme for developing statistics and capacity building on slum development under JnNURM
(viii) Financial incentives in the shape of interest subsidy for promoting weaker section housing	(viii) Rajiv Rin Yojna (RRY) for providing loan to poor people for housing activities on subsidized rate
Post-JnNURM Period (2015 onwards)	
(i) Consolidating the fruits of earlier	(i) Atal Mission for Rejuvenation & Urban Transformation (June 2015)
(ii) Bridging the gaps in implementation of Initiatives earlier schemes	(ii) Smart Cities Mission (June 2015)
(iii) Ensuring the benefits reaching to real beneficiaries	(iii) Housing for All (Urban) by 2022-Prime Minister's Awas Yojana (2015)
(iv) Efficient delivery of services	

(Contd...)

(v) Attempting bottom-up approach in Planning	
(vi) Ensuring beneficiary participation in planning and delivery of services	
(vii) Adopting 'action-plan-based' approach of financing instead of earlier approach of 'project-based'	
(viii) Active association of all stakeholders viz. state government, municipality, parastatals, CSOs, CBOs, beneficiaries	
(ix) Incentive instead of penalty for implementing reforms	

Slum Development--the Priority Sector

Globally, slums exist but under different forms. Slums have some common features, viz., they lack basic services like water and sanitation, the areas contain dilapidated or poor quality structures; overcrowded and densely populated; the residents have insecure land tenure; and the residents are poor and socially excluded. They are viewed as the products of failed policies, bad governance, inappropriate regulation, dysfunctional land markets, unresponsive financial systems and fundamental lack of political will.[25] In simple terms, they are the result of lack of appropriate land and shelter policies or failure of these policies. Yet, their inner layers of life are highly resourceful. In the fast urbanising world, slums become the main challenge for urban development sector.

A Report of the United Nations on Millennium Development Goals (2010)[26] comments that over the past 10 years, the share of the urban population living in slums in the developing world has declined significantly from 39 per cent, in 2000, to 33 per cent, in 2010. On a percentage scale, this may be cause for optimism. But, in absolute terms, the slum dwellers in the world are 828 millions now compared to 657 millions in 1990 and 767 millions in 2000. An important area of concern expressed was non-availability of exact figures on slums in different countries. The Report further remarked that when the international community adopted the Millennium Declaration and endorsed the 'Cities without Slums' target in 2000, experts had underestimated the number of people living in substandard conditions. They had also determined that improving the lives of 100 million slum dwellers was a significant number but a realistic target to be achieved within the next 20 years. But, three years later in 2003, new and improved data sources showed for the first time that 100 million was only a small fraction, about 10 per cent of the actual global slum population. The Report concluded that the target would require redefinition based on correct data and baseline figure for getting serious commitment from national governments. In the South Asian countries, the average slum population, however, is observed to be decreasing. According the UN Report[27], the share of urban population living in slums in South Asian countries decreased from 57 per cent, in 1990, to 35 per cent by 2010. Similar to the global trend, the information on slum population in India also suffer from accurate data. National level information on slums was not available till the 2001 Census.

The Census of India defines slums on three separate criteria: (i) compact areas of at least 300 persons or about 60-70 households of poorly built congested tenements, in unhygienic environment usually with inadequate infrastructure and lacking in proper sanitary and drinking water facilities; (ii) the areas notified as slums by the state governments and municipalities under an Act; and (iii) all areas recognised as slums by the state governments and Housing Boards either notified or not.[28] The NSSO defines slum in general as a compact settlement with a collection of poorly built tenements, mostly of temporary nature, crowded together usually with inadequate sanitary and drinking water facilities in unhygienic conditions and such an area is considered non-notified slum if at least 20 households lived in that area. The areas notified as slums by the respective municipalities, corporations and urban development authorities are treated as notified slums.[29] The UN Habitat defines slum as a contiguous settlement where the inhabitants have inadequate housing and basic services and, they are not treated as integral part of the city.[30] Dr. Pronab Sen Committee appointed by the Government of India to estimate slum population in the country defined slum on the lines of the NSSO. According to this Committee, 'slum is a compact settlement of at least 20 households with a collection of poorly built tenements, mostly of temporary nature, crowded together usually with inadequate sanitary and drinking water facilities in unhygienic conditions'.[31] The Sen Committee has not differentiated between the recognised and unrecognised or notified and non-notified slums. These apart, various state governments defined slums under respective slum Acts describing slums in a similar manner but without prescribing any strength of the households to notify or declare slums.

The comprehensive data on slums is not available in India. Earlier estimations were based on survey by different organisations like NSSO and covered selected towns and cities. Even the Census of India 2001 covered the towns having population of 20,000 and above in its slum survey. The available statistics show that slum population was 28 million in 1981 constituting 17.5 per cent of the total urban population which increased to 45.7 million by 1991 forming 21.5 per cent of the total urban population.[32] The 2001 Census of India in its slum survey estimated that 52.4 million people (10.2 million households) lived in slums in the 1743 towns of 20,000 and above population as per 1991 Census. The slum population was 23.5 per cent of the population of the towns where slums were reported.[33] These estimations are understated as the slum definition adopted by the Census 2001 is considered unrealistic leaving several smaller settlements with less than 60 households. This appears to be true as the UN Population Report 2001 estimated a slum population of 158.42 millions in India.[34] The Town and Country Planning Organisation estimated a slum population of 61.8 million in 2001 on full count based on 2001 census. The Committee on Slum Statistics/Census (2010)[35] estimated 75.2 million slum population covering all 5161 census towns in 2001 constituting 26.3 per cent of the total urban population and projected a slum population of 93.05 million in 2011 and 104.66 million by the year 2017. The slum population in India continues to increase despite several government measures to develop them.

The problem of slums received attention in the very First Five Year Plan in India with the comment that 'slums are disgrace to nation and it is better to pay for the cost of clearing slums rather than to continue to pay the mounting cost of slums and suffer destructive effects upon human lives and property indefinitely'.[36] The slum development focus shifted from clearance and relocation to improvement in the country due to land and financial constraints. The emphasis has been on provision of basic civic services like sanitation, water supply, roads, street lights and health. The schemes like Environmental Improvement of Urban Slums (EIUS), 1972, Low Cost Sanitation (LCS) 1981; and National Slum Development Programme (NSDP), 1996 were implemented to provide better services in slum areas. The living conditions in slums are still discouraging. The slum development schemes suffered from several lapses viz., (i) non-listing of all habitations which should be classified as slums; (ii) lack of information on the number of households in slums; (iii) absence of master plans; (iv) divided institutional responsibilities with no proper coordination arrangements; (v) failure to recognise slum settlement needs in urban plans; (vi) land related disputes; and (vii) lack of convergence among stakeholders.[37] These deficiencies led to rethinking on slum development and radical shifts in approach and strategies in recent years in India.

The National Urban Housing and Habitat Policy, 2007 similar to earlier thinking preferred development of the existing areas or in situ type development rather than relocating. The specific strategies recommended are:[38]

(i) Specially designed slum improvement programmes which focus on upgrading of basic services and environment improvement of urban slums with a participative, *in-situ* slum rehabilitation approach.

(ii) Inner-city slum redevelopment programmes for creating a better environment with cross subsidisation and special incentives.

(iii) Land pooling and sharing arrangements to facilitate land development and improvement of basic amenities in slums.

(iv) Release of Transferable Development Rights and additional FAR for accelerating private investment in provision of shelter to the poor.

(v) Involving the Community Based Organisations (CBOs), Non-Governmental Organisations (NGOs) and Self-Help Groups (SHGs) in partnership with the Private Sector.

(vi) Giving primacy to provision of shelter to the urban poor at their present location or near their work place and ensuring that rights provided are non-transferable for a period of 10-15 years.

(vii) Only in cases, where relocation is necessary on account of severe water pollution, safety problems on account of proximity to rail track or other critical concerns, relocation of slum dwellers is preferred while ensuring fast and reliable transportation to work sites.

(viii) Encouraging income generating activities in slums, which are non-polluting on a mixed land use basis while integrating such activities with housing and habitat projects.

(ix) Formation of Group Cooperative Housing Societies of urban poor and slum dwellers across the country for providing better housing serviced by basic amenities through thrift and credit based CBOs.

Land is significant component of the cost of housing. But the master plans, state governments, development authorities and municipalities failed in the past for ensuring adequate supply of serviced land for EWS and LIG housing. In fact, some states and urban development authorities have resorted to auction of the limited land available with them in cities, setting exorbitant benchmarks for the market price of land. In this context, there is need for a well-defined policy for allocation of land to EWS and LIG housing in the master plans. In line with the practice followed globally in upgrading slums, the occupied land or a part thereof should be allocated to the slum-dwellers to enable them to have access to housing and basic amenities.[39]

Issues and Predicaments

The observations of the UN Report are worth noting here which explain the current problems in India too. The UN Report observed that the public authorities faced four critical problems in meeting the goal of Cities without Slums by 2020 as targeted by UN in 2000. They were: lack of land titles and other forms of secure tenure; cutbacks in funds for subsidised housing for the poor; lack of land reserves earmarked for low-income housing; and an inability to intervene in the market to control land and property speculation. Low incomes in the face of rising land prices virtually rule out the possibility that the working poor can ever own land, contributing to the problem of urban slums.[40] The housing market is interplay of demand and supply factors. The growing income levels of people especially, the middle income people and rapid growth of metropolitan and other cities are considered the factors causing demand in India. The supply side is suffering from constraints like lack of availability of land and credit facilities, legal and regulatory deficiencies and, limitations of the private sector to provide low income housing.[41] Even if the land is available, the prohibitive prices, lack of clear titles, legal regulations, poor land records, etc., are the other problems which constrict supply of land for housing.

The housing financial companies largely serve the middle and high income people. The critical element of land is indisputable from the above observations. The governments have to focus on dealing with land problems. As the poor people cannot afford high cost dwelling units, the concept of affordable housing is introduced into the housing projects in India. Interestingly, affordable housing concept is now popular not only in the public sector but also in the private sector as there has been drastic decline for luxury housing due to the global recession. The issues relating to housing finance, affordability and land continue to hamper the housing programmes in the country. The goal of the slum free cities cannot be achieved unless these issues are effectively addressed.

Housing Finance and Access of Credit for Low Income Housing

Urban housing is primarily in the private domain in India. The government role in housing finance is more in the nature of subsidy provision for EWS and LIG income groups and promotion of low income housing finance market. Several tax incentives and subsidy oriented housing schemes are devised to meet the low income housing needs.

Tax Incentives

The Government of India has provided several tax incentives for promotion of housing to the companies as well as individuals. For individuals, under section 24 of IT Act, the interest paid on housing loans is exempted up to Rs 1.5 lakhs for owner occupied houses and full exemption for the rented dwellings. The capital gains tax concession is also provided when a residential property is sold and another residential property is acquired under Section 54 of IT Act. In respect of profit derived from business of providing housing finance, up to 40 per cent is deducted from income provided if it is carried to special reserve (section 31 (1) of IT Act). A deduction for bad and doubtful debts equivalent to 10 per cent of doubtful and loss assets is available to the banks. The risk weight on housing loans has also been increased from 50 to 75 per cent for the banks provided the capital adequacy requirement is maintained at 12 per cent for the housing finance companies.

Interest Subsidies to Expand Credit

The Government of India launched Interest Subsidy for Housing the Poor Scheme in 2009 to promote housing in the EWS and LIG categories. This enabling initiative aimed to provide expanding credit flow to the housing sector and increasing home ownership in the country.[42] The scheme was innovative in the sense that both subsidy and loan features were built into it. The loan was provided to the EWS and LIG households for either acquisition or construction of a dwelling unit. The families who owned a plot in the above categories were also covered under the scheme. The subsidy loan for EWS families was Rs 1,00,000 with a repayment period of 20-25 years for a house of at least 25 sq. mtr. with a provision of additional loan if needed on subsidised interest rates. For LIG families, a loan up to Rs 1,60,000 was given for dwelling units of at least 40 sq mtr. but the subsidy part was Rs 1,00,000 only and over and above this, it was at subsidised interest rates.[43] The subsidy on loans was five per cent on the interest charged on the administrative loan amount. The subsidy was calculated on the notional discounted interest of the nine per cent. The scheme was administered through National Housing Bank (NHB) and HUDCO, the recognised nodal agencies. The subsidy was not be given to the beneficiaries directly but to the banks and other housing finance companies who agreed to become part of the scheme. No collateral was needed for the minimum Rs 1,00,000 loan and no prepayment charges was levied for repayments.

The Government of India launched another incentive scheme of one per cent interest subvention for stimulating housing credit to the lower and middle income individuals. The loan ceiling for availing the benefit which was Rs 10 lakhs maximum for a house of Rs 20 lakhs initially in 2009 was increased to Rs 15 lakhs in 2011-12 for a house of Rs 25 lakhs. The commercial banks and housing finance companies implemented the scheme and it covered rural as well as urban areas.[44]

Other Financial Instruments

In order to promote foreign investment in housing sector for development of townships and housing-cum-infrastructure projects, the Government of India has permitted foreign direct investment (FDI)[45] up to 100 per cent in order to create employment opportunities, generate economic activity and build housing and infrastructure stock in the country. The stipulation is that each project should have a minimum area of 10 hectares for developed and serviced housing plots and in case of construction and development projects, minimum built up areas should be 50,000 sq. meters. The foreign investment of minimum capitalisation of US$ 10 million for wholly owned subsidiaries and US$ 5 million for joint ventures with Indian partners is allowed. Further, External Commercial Borrowing for affordable and low cost housing was allowed in the budget. The developers who construct majority units for LIG and EWS housing could utilise such borrowing facility.

The Government of India also introduced a credit risk guarantee scheme. Under this, the lending agencies were provided credit risk for EWS/LIG category of borrowers up to Rs 5 lakhs in urban areas without any collateral security. The credit amount of default covered was up to 90 per cent for loans up to Rs 2 lakhs and up to 85 per cent for default amount between Rs 2-5 lakhs.[46]

The Deepak Parekh Committee made certain recommendations to expand the flow of credit to housing. At the outset, 100 per cent increase in the outlay for affordable housing under JnNURM was suggested. The levy of a cess of 0.5 per cent in all the central government taxes to be credited to dedicated Shelter Fund to be managed by the National Housing Bank was suggested to mobilize finances. This fund had a budgetary support of equal amount to increase the impact on affordable housing. The Government of India, in the budget 2013-14 proposed an Urban Housing Fund to be set up by NHB with an outlay of Rs 2,000 crore to ensure institutional finance for urban poor to implement this recommendation. Other recommendations to mobilize funds included : permitting housing finance institutions to access long-term External Commercial Borrowing market; reinstating section 80-IB of Income Tax Act for developers engaged in affordable housing; reduction of stamp duty rates and registration fee for affordable housing to two per cent *ad valorem* uniformity in all states as the five per cent provided in JnNURM was considered too expensive; and levy of flat charge of Rs 1000 on registration of equitable mortgages.[47] Further, the Task Force also recommended for a housing finance company to provide micro housing loans to the poor, schemes to provide household savings for housing, and finally, it suggested for including affordable housing as infrastructure to facilitate RBI, SEBI, IRDA, etc., extend facilities for affordable housing for housing financial agencies.[48]

Land—the Critical Element in Low Income Housing

Land is abundant but its utility and access are the main factors which make urban lands much valuable and scarce. Land values reflect the capitalisation of future expected income stream in urban context as against the more static farming environment. The change of land uses from agriculture and allied to urban has propensity to increase land prices mostly on speculative factor. Unless the governments effectively take steps to control land speculation, the prices will spiral leading to land cost inflation. The high land prices make urban land inaccessible to poor and low income people. Some feel that the process of globalisation has made urban land almost inaccessible to the urban poor and all national level policies are based on market solutions for the poor.[49]

Modern governments have multiple policy instruments available to deal with urban land development and related social issues. The options are: land acquisition; direct government investment in urban infrastructure and housing; public-private-partnership; monetisation; and fiscal measures like taxation and charges. Despite availability of these options, the governments have been largely biased towards acquisition and direct investment. Emphasis on powerful legislations like the Land Acquisition Act, 1894 and the Urban Land (Ceiling & Regulation) Act 1976 (ULCRA) and on resource mobilisation for direct government investment amply demonstrate this trend.

ULCRA - A Failed Experiment

The objectives of the Urban Land (Ceiling & Regulation) Act, 1976 (ULCRA)[50] were to prevent the concentration of urban property in the hands of a few persons and speculation and profiteering therein; to bring about socialisation of urban land in urban agglomerations to ensure equitable distribution; to discourage construction of luxury housing leading to conspicuous consumption of scarce building material; and to secure orderly urbanisation. The ULCRA provided for imposition of a ceiling on both, ownership and possession of vacant land in urban agglomerations and acquisition of the excess vacant lands. These objectives, though laudable could not be achieved for obvious reasons. The implementation of the ULCRA failed mainly due to:[51]

- Absence of clarity and too much discretionary powers given to the state governments for granting exemptions.
- Compensation provided for the acquired land was very little, which often led to lengthy litigation disputes. The maximum compensation was Rs 10 per sq. meter and the total compensation could not exceed Rs 2 lakhs per owner. This made landowners reluctant to declare their vacant land as surplus.
- Absence of a mechanism to encourage the entry of the vacant urban land into the land market through appropriate fiscal measures. Land prices in cities reached astronomical heights due to artificial scarcity of land created by ULCRA.

It is interesting that the ULCRA led to land blockage, increased prices due to speculation and created scarcity of urban lands denying access of urban land to the poor and low income people, the very objective of land socialisation which it aimed to achieve. The performance of the ULCRA clearly indicates that the social purpose could not be realised. For instance, the state governments could physically acquire only nine per cent of the total estimated excess vacant lands.[52] The Andhra Pradesh data shows that lands allocated for weaker section housing purposes were negligible; about only 30 per cent of the lands were allocated for housing purposes.[53] With the realisation of the failure of the ULCRA, it was repealed by the Government of India in 1999 and later in many states as part of the JnNURM reforms.

Land Acquisition—A Cumbersome and Socially Sensitive Process

The process of land acquisition for public purposes is riddled with complex legal, social and economic issues. At the outset, people oppose the land acquisition proceedings as it disturbs their livelihood/assets. The compensation and rehabilitation procedures are also cumbersome and not attractive. Several housing and infrastructure projects in towns and cities are held-up and sometimes shelved due to prolonged procedures and court litigations. The sites and services projects taken-up under schemes like IDSMT in the small and medium towns were found to be failures because of land acquisition problems. Success was seen only in respect of housing projects where government/municipal lands were readily available. In cases where alternative sites are offered, the owners express their inability to accept either due to livelihood problems or due to inadequate compensation. In spite of the government intentions to increase the compensation and providing tax incentives, the response appears to be not encouraging. In the context of increased land prices, land acquisition method is clearly proved to be ineffective and costly proposition.

Urban land is neither effectively used as an economic resource nor as a planning instrument for orderly development. On the supply side, valuable land especially in the core city areas is either locked-up in litigations or inefficiently used both, making urban land underutilised. In the peripheral areas, agricultural lands are being converted to urban uses unauthorisedly due to non-availability of land and land use regulatory restrictions. In respect of land for shelter, neither the demand nor the supply of land was effectively managed. It is necessary that land should be made accessible to all groups of people and the prices also should be under check to control unwarranted speculation. The NCU made a very pertinent recommendation that the new developments in the public or private lands should be permitted only if substantial percentage, at least 15 per cent of the total layout, is made available to the duly constituted authority.[54] These lands could be used for shelter projects of the governments. These observations indicate the prospective policy of reservation of urban lands to the shelter programmes in India.

The Strategy of Reservation of Land for the Poor

The concept of reservation of land to the poor in urban areas was introduced as back as 1976 in Gujarat. The Gujarat Town Planning and Urban Development Act (GTPUD), 1976 has a provision of reserving land for the urban poor defined as socially and economically weaker section housing. The GTPUD Act[55] provides for *"reservation of land to the extent of ten per cent; or such percentage as near there to as possible of the total area covered under the scheme, for the purpose of providing housing accommodation to the members of Social and Economically Weaker Sections of Society (SEWS)"*. The idea of reserving land for the Social and Economically Weaker Sections of Society (SEWS) housing is to minimise the distance between residential location of the poor and the distance from the work area by providing land for the poor in all the town planning schemes in a city. The successive national housing policies, JnNURM reform agenda and the RAY scheme gave top priority to the strategy of reservation of land to the weaker sections housing in the public and private sectors. This was a mandatory reform to be implemented by the state governments. The Ahmedabad Municipal Corporation in 2009 allocated 135 hectares out of 1692 hectare of land for SEWS housing. The allocation of the public lands for slum up-gradation is considered to be small around eight per cent compared to other public uses giving indication that slum housing is not a priority of the public authorities.[56]

Some research studies bring-out that though planning schemes such as Town Planning Scheme in Gujarat does provide lands for the poor, the local government did not take advantage of this provision and build SEWS housing on it due to various administrative and legal reasons. The underutilisation of SEWS housing land is considered to be a serious mismanagement of resources and inefficient land management problem. Further, it is brought-out that lands are available for poor people housing contrary to the claims of public authorities that there are no lands available for housing the poor in the city of Ahmedabad according to one study.[57] The experience of the Ahmedabad Municipal Corporation should be taken as a lesson and the municipalities should deal with the reservation of land for shelter programmes in a more effective and purposeful manner.

Resolving Land Related Problems

Social housing is the most priority area enunciated in the successive housing policies in the country after Independence. But, the efforts so far made appear to be very inadequate and feeble. Though governments have announced the policies and willingness to provide affordable housing, the commensurate administrative commitment and implementation mechanisms are found to be deficient. The Urban Land Ceiling and Regulation Act, 1976 (ULCRA) which was meant for socialising urban lands, could not realise the objective and instead, it led to unnecessary land blockage and increase of land prices. The land use conversion is riddled with legal and administrative problems. In the post economic liberalisation period in India, demand for urban land became much more competitive. The speculative feature dominated rather than housing and urban development. The major issue appears to be increasing competition for the scarce and costly urban lands where commercial interests dominate. Land mafias in metropolitan cities and industrialist lobbies outplay in this competition. Public authorities often succumb to private sector pressures.

Reservation of lands for the urban poor and other economically weaker sections is a welcome feature. But, the real problem is commitment of the public authorities to use such lands effectively to realise the objective of providing affordable housing. In respect of private lay-outs, it requires workable public-private strategies to construct affordable houses. Further, the lands reserved should be usable for housing. If reservation is done purely to satisfy the legal provisions, it may not serve the purpose.

Land title registration is yet to take proper shape in urban sector in India. This is important for assuring security of tenure for the urban poor. It will also curb misuse of land/house allotments to the poor. Several cases of misuse by the administration and vested interests are surfacing in several parts of the country mostly reported in the media. For instance, in Hyderabad, allotment of houses constructed under the JnNURM were reported[58] to be made overlooking the basic norms leading to allotment of double houses for the same beneficiaries in respect of about 100 beneficiaries and double registration of same house to two beneficiaries falling in two districts viz., Hyderabad and Ranga Reddy in case of at least 54 persons and allotment of houses to the persons not eligible as per the norms. This is not a solitary case in the country.

Such a practice only indicates that the benefits under social housing are not going to the needy persons due to gross neglect of the rules/norms and defective property information and registration system. One basic need is streamlining the land record and information system in urban areas. Once, such an exercise is done, reservation of lands to the poor becomes purposeful. The planners can identify viable lands for housing the poor and reserve such lands for housing. Until land records are in proper shape, public authorities can neither correctly identify potential lands for social housing nor streamline beneficiary selection process.

The Concept of Affordability

The concept of affordability is the most complex dimension especially, with regard to housing for the poor. The concept was first mentioned in the First Five-Year Plan itself in India. Even today the concept is not properly understood and applied to the housing sector in the country. The measurement of affordability is fraught with both, conceptual and administrative problems. However, affordability can broadly be measured by taking into consideration the factors like capital cost of dwelling unit, household income and the terms of finance, each one being highly variable and related to other factors. In administrative terms, affordability in housing is viewed as a ratio of price/rent of housing to income of the household. In simple words, it reflects the financial ability of the households to own a house. It is relative and differs based on income groups.

The US Department of Housing and Urban Development defined affordable housing as a "housing project which has at least 20 per cent of the units available to families making less than 50 per cent of the median income in the area where the project is located to be considered affordable."[59] This definition indicates that affordability is also related to local economic conditions. The Deepak Parekh Committee[60] defined the affordability ratio for the EWS-LIG categories of households as not exceeding 30 per cent of household's gross monthly income in respect of EMI/Rent Income Ratio and cost not exceeding four times household's gross annual income with respect to cost of housing to income ratio. The ratios for the MIG category of the households are not exceeding 40 per cent and cost not exceeding four times respectively. The Ministry of Housing and Urban Poverty Alleviation defines affordable houses in terms of area — 300 q. ft. (super built area) for EWS, 500 sq. ft. for LIG and 600 to 1200 sq. ft. for MIG category of households. The carpet area is also prescribed as 25 sq. meter for EWS, 48 sq. meter for LIG and 80 sq. meter for MIG (Table 9.4).[61] With respect to ability to repay home loans, monthly installments not exceeding 30-40 per cent of the monthly income of the buyer is considered.

The Task Force on Promoting Affordable Housing (2012)[62] desired that the carpet area criteria should be standardised with built-in flexibility. Freedom should be given to the state governments for adoption of these standards with about 10 per cent variation to suit local conditions. The affordable housing area criteria on the whole should range between 21 sq. mtr. for EWS and 60 sq. mtr. for LIG categories of households (Table 9.4). It is also suggested that in the event subsidies are tied to affordable housing scheme, the carpet area for EWS could vary between 21 to 25.2 sq mtrs. and for LIG-B or upper lower income group, it could range between 35.9 and 45.1 sq. mtrs.

A closer look reveals that affordability is not only relative to the different categories of households, but also to the level of taxes, charges, house maintenance costs, locations, etc. Closely related to this is adequacy concept which refers to the size of the family and size of the dwellings. In this perspective, it is argued that affordable housing should take into account not only income criteria but also other criteria like location, tenure, size of households, migrants, single male/female family, etc. Another important aspect is that it should be fixed locally and not uniformly at the national level.[63] Some critics comment that though the governments succeeded to some extent

Table 9.4: Affordable Housing Parameters

Income group	*Affordability in terms of size*		*Affordability of EMI/ Rent in per cent of HH gross monthly income*	*Affordability of cost in multiple of gross annual income*	*Affordability in terms of Rs*
	Super built up area sq. ft.	*Carpet area sq. mtrs.*			
			Deepak Parekh Committee		
EWS and LIG		300 to 600	> 30	> 4 times	
MIG		>1200	40	> 5 times	
			Ministry of Housing and Urban Poverty Alleviation		
EWS	300 minimum	25 minimum	30 to 40	-	Rs 3,300 per month
LIG	500 maximum	48 maximum	-	-	Rs 3,301 to Rs 7,300 per month
MIG	600 to 1200 maximum	80 maximum	-	-	
			Task Force on Promotion of Affordable Housing (MHUPA)		
EWS		21-27 sq. mtr. carpet area		5 times	Rs 8,000 per month or Rs 1,00,000 per annum
LIG-A		28-40 sq. mtr. carpet area		- Do -	Rs 16,000 per month or Rs 2,00,000 per annum
LIG-B		41-60 sq. mtr. carpet area		- Do -	

Source: Compiled from Deepak Parekh Committee Report, Guidelines for Affordable Housing in Partnership (2009), Guidelines for Scheme of Interest Subsidy for Housing the Urban Poor (2009) and Task Force on Promoting Affordable Housing, November, 2012, Ministry of Housing and Urban Poverty Alleviation, Government of India.

in providing adequate housing to the poor through various schemes, there has been always problem with affordability issue in terms of income, quality of housing, land tenure, etc.[64] The affordable housing therefore, remains a controversial issue and problems persist unless the policy-makers adopt cautious approach. One solution is, as affordability issue is area specific, the affordability levels should be fixed based on local economic factors. From the point of size, flexible norms could be fixed based on actual size of the household. The approach should be flexibility for adoption at local levels as against the 'one size fits all' strategy.

The Task Force on Affordable Housing[65] also made several financial and non-financial recommendations on the lines of the Deepak Parekh Committee for effective implementation of affordable housing projects. The financial reforms suggested are: service tax exemption for low cost houses upto 60 sq. mtr. carpet area; rebates under section 35 AD of Income Tax Act for affordable housing in partnership scheme; tax deduction under section 80-IB (10) of Income Tax Act; and including affordable housing as infrastructure facility. It has also recommended for a viability gap funding for affordable housing projects similar to infrastructure projects. The other suggestions include reducing the area limit under FDI for the projects from 50,000 sq. mtr. to 20,000 sq. mtr. built-up area and, reducing the maximum investment limit from five to two million dollars; increasing the income limit for interest subsidy of 5 per cent available under Subsidy Scheme for Housing the Poor to at least Rs 5 lakhs instead of one lakh; allowing a moratorium of three years for payment of EMI; and appropriate reduction of stamp duty for registration of property for the affordable houses for EWS and LIG categories. On the non-financial side, the committee suggested for reducing timelines for approval of housing projects, increasing FSI and providing TDR facilities, revising planning and building rules and adopting liberal density norms subject to a minimum of 350 DUs/hectare.

Legal Constraints

The laws governing the land and properties viz., Urban Land (Ceiling & Regulation) Act (ULCRA), Land Acquisition (LA) Act, Rent Control Act and a host of revenue Acts in the states have become out dated requiring radical changes. The housing and shelter problems get accentuated due to various legal constraints. The governments are very slow in reforming the legal environment to make it more poor friendly. The abolition of the ULCRA has released considerable urban lands into the market but these lands may be cornered by the middle and higher income groups unless appropriate land allotment policies are devised. This danger is perceived in the light of the recent commercialised trend of the governments and urban development agencies. For instance, the Hyderabad Metropolitan Development Authority has auctioned the prime government lands allocated to them for funding the mega infrastructure projects. The states are lacking proper legislations to develop residential housing on partnership basis similar to Gujarat.

Yet another area of concern is rent control laws. Rent control laws were introduced to prevent landlords from exploiting tenants after the World War II. Unfortunately, these legislations became permanent fixtures in the Indian urban sector. As these laws are tenant biased, the owners lost interest in these properties leading to more dilapidated buildings and court litigations. The rent controls act as disincentive for the development of rental housing. Further, they lead to several unlawful practices, waste of scarce resources, and exclusion of the low income people into the rental market and court litigations. Indirect effect has been decline of inner city areas and growth of slums. It is in this backdrop, the Government of India had made rent control reform mandatory under the JnNURM. The rent control reforms could release more and quality rental housing into the market thereby reducing the rental levels and indirectly arresting slum formations. It was also expected to result in more effective allocation of the existing stock and ultimately contributing for the overall economy of the country. Four states have already amended the rent control laws in India. The West Bengal brought out its new Rent Act in 1997, Maharashtra and Karnataka in 1999, and Rajasthan in 2001.

The reforms in general introduced exemptions to certain categories of buildings like high rent properties and newly constructed properties and, provided for periodic increase of the rents and limited period tenancy. These reforms positively impact the rental housing. This becomes very important in view of the emphasis being laid on promoting rental housing. The Ministry of Urban Development has formulated Model Rent legislation to standardise the reform process in the states. The main features of this legislation are: providing exemptions for certain special category properties and new properties; including maintenance cost in the rents; compulsory registration of rental agreements; limiting inheritability of tenancy; and faster judicial processes, etc.

The legal system is burdened by a significant social cost resulting in lack of clear title to property. Civil courts are clogged with longstanding land-related cases which by some estimates account for more than 60 per cent of total cases. The government is often mired in land-related litigations. A guaranteed land title system will substantially reduce the burden on the civil courts by enabling faster dispute resolution. The Government of India has initiated measures to develop model property title certification system and if it is implemented properly, it will facilitate streamlining the land and property record system to a large extent.

Perspectives and Prospects

The housing and shelter reforms have come a long way in India especially, in the post-1990 periods. The formulation of the National Urban Housing and Habitat Policy, 2007, launching of BSUP and IHSDP under JnNURM in 2005, the RAY scheme focusing on slum redevelopment in 2009, and finally, Prime Minister's Awas Yojna (PMAY) under Housing for All by 2022 (urban) in 2015 not only broadened the reform base but also fast tracked administrative action on housing and slum rehabilitation front. These reforms are crystallised formulations in the light of experiences gained in the implementation of several schemes and projects targeting the slum dwellers and the urban poor in the country since Independence. As urban poverty issues are complex, diverse and always assuming a new shape with changing times, the emphasis should be on devising innovative solutions with integrated and inclusive strategies. The Indian planning framework through Five Year Plans has been providing the national level guidance and support to solve the emerging issues and problems. The Twelfth Five Year Plan recommended a multi-pronged strategy to meet the shelter and housing needs of the urban poor while considering the experiences gained in the JnNURM. The main components of the strategy are:[66]

(i) Augmenting affordable housing with private capital;

(ii) Increasing ownership of the programme by encouraging beneficiary contributions;

(iii) Promoting societies and self-help groups;

(iv) Expanding slum rehabilitation ;

(v) Creating and maintaining rental housing including dormitories for migrant and floating populations;

(vi) Encouraging low cost housing.

Certain concrete measures are recommended to strengthen and reformulate the slum rehabilitation strategy. They are:[67]

(i) Of the total stock created, at least 30 per cent should be rental, and 5 per cent should be dormitories;

(ii) Providing affordable housing for the urban poor in peri-urban areas accompanied by basic services and functional transport linkages;

(iii) Developing customized approaches for slum rehabilitation through community participation methods and institutions;

(iv) Expanding incremental slum rehabilitation in other towns;

(v) Incentivising the private sector to augment affordable housing supply;

(vi) Revitalising Housing Boards and development authorities to mobilise institutional funding through networking with private sector.

(vii) Accrediting community based organisations for effective community participation and social audit;

(viii) Increasing corpus of the Credit Risk Guarantee Fund to cover all slum rehabilitation;

(ix) Simplification of the process of project approvals of affordable housing.

It is seen that the JnNURM reforms relating to housing and shelter were implemented slowly by the state governments and the municipalities. The ULCRA has been repealed in all but one state and 46 cities have internal earmarking of funds for services to the poor.[68] The strategies adopted so far for providing affordable housing for the poor revolve around subsidies, financial incentives and provision of civic services. The direct provision of housing through subsidies led to the benefits actually passed on to the higher income groups as the beneficiaries prefer to speculate on the benefits rather than retaining the units. This has been a common trend in urban India despite rules and restrictions on transfers. The proposed rental housing concept has problems relating to maintenance of dwelling units especially, old dwellings. The rental housing without appropriate O & M package will in all possibility convert them into slums over a period of time.[69]

The award of tenurial rights is another crucial aspect which needs attention. The existing legal and regulatory environment governing urban land in India is fragmented, negative and constraining hence, not conducive for awarding tenurial rights especially in case of the lands belonging to private people. The state governments should deal the land related problems effectively by committed institutional reforms. A balanced approach in land allocation, integrated land use policy and streamlining of the land records by an appropriate method of property title certification are some of the essential measures required. These measures could be supplemented by computerisation and developing connectivity across the stakeholders. The process of awarding tenurial rights should be transparent and efficient. The critical areas to be addressed are: identification of the rights of the holders in case slums are on private lands, delineation of the plots by proper surveys, determining compensation, assigning titles, land registration and resolution of disputes. Care should be taken to ensure that the tenurial rights are not misused by the beneficiaries. The government should facilitate creation of different mid-way tenures to suit affordability levels of different categories of population.[70] It is desirable to give limited tenurial rights to meet essential requirements like collateral security for housing loans, protection from eviction and providing better homes for the urban poor. The UN suggested for a collective tenurial rights on the lands to overcome the problem of speculative market forces and to develop unity among the slum communities.[71]

The housing finance is constrained by inability to mobilise deposits for medium and long-term micro financing needs for housing sector by the Housing Finance Companies (HFCs) and access international funding in spite of various tax incentives and allowing FDI in housing sector. A multipronged strategy is required to develop a robust housing finance sector by way of proper mix of subsidies, tax incentives, saving and deposit mobilisation schemes and, strengthening of microfinance and cooperative housing institutions. If the tax incentives will work for the higher and middle income people, the saving and cooperative approaches will benefit the low income people. The indiscriminate subsidies in different sectors have had a negative impact on the economy of the country since Independence. The economic reforms have opened-up innovative ways of financing. The housing subsidy system should be streamlined and made need based to serve the poor more effectively. Care should be taken to ensure that the subsidies reach the truly needed people and they are not misused. Otherwise, the subsidies will be hijacked by the higher income groups and the poor remain poor as has been the trend so far. In order to promote flow of credit to the poor, the subsidy could be linked to saving schemes. Appropriate saving mechanisms on the lines of the SEWA Bank could be explored in the housing sector to mobilise deposits, develop saving habits and ownership among the urban poor targeting their shelter needs. This will not only facilitate proper use of the subsidies but also willingness to pay for their houses by the poor which in turn could make the housing finance market more sustainable. This is quite essential as indiscriminate subsidies result in misuse and do not to lead to realisation of the goal of housing for all. The initiation of Aadhaar Card linked release of subsidy has paid a good dividend. It has not only ensured the transfer of subsidy directly to the bank account of the beneficiary but has also helped curbed corruption and thus saved huge public money from siphoning off by the intermediaries.

Alternate methods should be adopted in view of the land acquisition problems being faced by the public authorities. The Transferable Development Rights (TDRs) which are being promoted should be managed effectively to avoid TDR hoarding and scams. In order to manage the TDRs, TDR Banks to be managed by the municipalities or the urban development agencies could be set-up. It is expected that these TDR banks will bring more transparency in the TDR process and reduce speculation on TDRs.[72]

Active private sector involvement is essential to meet affordable housing needs. At present, the Housing Finance Companies (HFCs) are not contributing much for affordable housing. For instance, the HFCs disbursed only 20 per cent of loans to the Rs 3-10 lakh bracket people in the year 2011.[73] Tax incentives could motivate private developers to take-up housing projects for the affordable housing. But, the crucial problem is undue delays in project approvals. It is alleged that on an average, the approvals for pre-construction process takes about 2 to 2.5 years involving about 40 departments of central government, state government and municipalities. The delays in project approvals are estimated to escalate the project costs to the extent of 25-30 per cent.[74] As per a World Bank Report 'Doing Business 2013', India ranks 182 out of 185 economies, on 'ease of dealing with construction permits'.[75]

Streamlining of the building approval process was an optional reform under JnNURM. Some states and cities implemented this reform in some form or other. Further, 33 mission cities and 172 towns undertook implementation of the e-governance reform to submit the building plans online while bringing down the processing time to seven days maximum.[76] Though these reforms brought some improvement, yet the building permission process is stated to be lengthy. The single window method of clearance is introduced in cities like Hyderabad to overcome the delays in housing project approvals. The Hyderabad Metropolitan Development Authority introduced Green Channel mechanism to speed-up the process. But, this reform did not pick-up due to lackluster response from the builders. These mechanisms must be made more effective by involving the concerned departments. Inter-departmental coordination is still a neglected area in Indian administration. Reforms for quicker housing project approvals must be implemented fast to involve the private sector in the cost sensitive areas like affordable housing.

The slum development rarely finds place in the urban planning in India. Most of the housing and slum rehabilitation projects have failed due to isolated approaches, improper selection of sites, restrictive land use plans and lack of amenity development plans. The slum relocation, to the extent possible, should be avoided as the slum dwellers primarily cannot afford to be dislocated. If relocation is unavoidable, their employment needs should be properly addressed along with the basic amenities. This is possible only when slum rehabilitation is attempted in an integrated and city wide planning framework.

To conclude, the future of slum rehabilitation and shelter progammes depends on how effectively the state and local governments deal with three major critical areas viz. security of tenure; expansion of housing finance market to make it accessible and affordable to urban poor; and spatial planning and reorganisation of slum spaces in a more environmentally sustainable and integrated manner. The bigger question is arresting the formation of new slums by continuous supply of affordable housing to low income people. This requires urgently that the increasing census towns are converted as statutory towns. Such an observation was made by the High Power Committee on Infrastructure and Services (2011) as well. This move not only help in bringing these areas under different urban laws but would expand the urbanisation in the country. Urbanisation in the country warrants a new definition and approach for providing housing to all, especially economic weaker and marginalised sections of the society.

References

1. National Housing Bank (2012): Report on Trend and Progress of Housing in India, p. 119.
2. *Ibid.*
3. *Ibid.*
4. Office of the Registrar General, Census Operations, India, Government of India, Census 2011. (www.censusindia.gov.in).
5. Government of India (2008): Report of the National Commission on Urbanisation, Vol. II, p. 225.
6. *Ibid,* p. 87.
7. Government of India: Planning Commission, Eleventh Five Year Plan-2005-12, p. 407.
8. Government of India: Ministry of Housing and Urban Poverty Alleviation: Report of the Technical Group (11th Five Year Plan-2007-2012) on Estimation of Urban Housing shortage, and 11th Five Year Plan, Planning Commission, p. 411.
9. Government of India:Ministry of Housing and Urban Poverty Alleviation (2012): Report of the Technical Group on Urban Housing Shortage-2012-17, and National Buildings Organisation, p. 4.
10. Planning Commission, 12th Five Year Plan, p. 336.
11. Planning Commission, Seventh Five Year Plan, pp. 294-295.
12. National Commission on Urbanisation, 1988, op. cit, p. 212.
13. Mahadevia, Darshini (2009):Urban Land Market and Access of Poor, in INDIA: URBAN POVERTY REPORT 2009, Ministry of Housing and Urban Poverty Alleviation, Government of India, Oxford University Press, p. 199.
14. Karnad, Girish T.G.: Affordable Housing for All: An Overview of Housing Policies in India, Shelter, HUDCO, op. cit., p. 3.
15. 11th Five Year Plan, Planning Commission, Government of India, p. 412.
16. Government of India: National Urban Housing and Habitat Policy, 2007.
17. Karnad Girish, op. cit., p. 4.
18. Ministry of Housing and Urban Poverty Alleviation, Government of India: Guidelines for Urban Statistics for HR and Assessments (USHA), 12th Five Year Plan, op. cit., p. 409.
19. Ministry of Housing Urban Poverty Alleviation, Government of India: Guidelines for Affordable Housing in Partnership.
20. UN-HABITAT, and UN Economic and Social Commission for Asia and the Pacific (2008): Housing the Poor in Asian Cities, Nairobi, p. 4.
21. Government of India (2009): Guidelines of Rajiv Awas Yojana.
22. Government of India (2015): Guidelines of Atal Mission for Rejuvenation and Urbn Transformation (AMRUT).

23. Government of India (2015): Guidelines of Smart Cities Mission.
24. Government of India (2015): Guidelines of Housing For All Mission (Urban): Prime Minister's Awas Yojana (PMAY).
25. United Nations (2010): The Millennium Development Goals Report 2010, New York.
26. *Ibid.*, p. 64.
27. United Nations (2010): *op.cit.*
28. Government of India:Ministry of Housing and Urban Poverty Alleviation, and National Buildings Organisation (2010): Report of the Committee on Slum Statistics/Census, p. 2.
29. *Ibid,* p. 3.
30. *Ibid.*
31. *Ibid,* p. 7.
32. Planning Commission, Tenth Five Year Plan, pp. 627-628.
33. Report of the Committee on Slum Statistics/Census, op.cit., p. 9.
34. *Ibid,* pp. 21-22.
35. Report of the Committee on Slums, op. cit.
36. Government of India: Planning Commission: First Five Year Plan.
37. Government of India: Planning Commission: Tenth Five Year Plan (2002-07), p. 629.
38. National Urban Housing and Habitat Policy, 2007.
39. Mahadevia, Darshini, op. cit., p. 201.
40. UN-HABITAT: The Millennium Development Goals Report 2010, op cit., p. 63.
41. National Housing Bank, op. cit., p. 121.
42. Government of India: Ministry of Housing and Poverty Alleviation: Interest Subsidy for Housing the Poor (ISHUP-2009) Guidelines.
43. *Ibid,* pp. 1-2.
44. *Ibid,* p. 131.
45. Government of India (2005): Ministry of Commerce and Industry, Department of Industrial Policy and Promotion; Press Note 2, 3rd March, 2005.
46. *Ibid,* p. 132.
47. Deepak Parekh Committee, op. cit., pp. 20-21.
48. *Ibid,* p. 22.
49. Mahadevia, Darshini (2009): Urban Land Market and Access of Poor, in INDIA: URBAN POVERTY REPORT (Summary of Report) 2009, Ministry of Housing and Urban Poverty Alleviation, Government of India and UNDP. http://undp.org.in/india-urban-poverty-report-2009.
50. Statement of Objects, Urban Land (Ceiling and Regulation) Act, 1976.
51. Government of India: Ministry of Urban Development: Repeal of Urban Land Ceiling and Regulation Act (ULCRA), State Level Reform, JnNURM Primer.

52. Government of India: Press Information Bureau: 'Performance of Repealed Urban Land (Ceiling & Regulation) Act, 1976'. http://pib.nic.in/focus/fojan99/fo1201991.html.
53. Chief Commissioner of Land Administration, Government of Andhra Pradesh.
54. *Ibid.*
55. Gujarat Town Planning and Urban Development Act (GTPUD), 1976, Section 40(j).
56. Joshi, Rutul and Sanga, Prashant (2009): Land reservations for the urban poor: the case of town planning schemes in Ahmedabad, Working Paper 4, Centre for Urban Equity CEPT University Ahmedabad, December.
57. *Ibid.*
58. Eenadu (Telugu daily), Hyderabad, 15th April 2011.
59. Quoted in Ahuja, Hitesh, Mandhyan, Ritika, Joshi, Hemand and Jain, Davan Sh (2012): Housing Affordability for the Urban Poor : A Case Study of Slums in Bhopal, (Draft, January 27, 2012, p.19 (Available on : meshfresh.com/GS12), p. 21.
60. Government of India: Report of the High Level Task Force on Affordable Housing for All, December, 2008, Deepak Parekh, et.al., Ministry of Housing and Urban Poverty Alleviation, p. 8.
61. Government of India: Ministry of Housing and Urban Poverty Alleviation: Guidelines for Affordable Housing in Partnership, April, 2009.
62. Government of India (2012): Ministry of Housing and Urban Poverty Alleviation: Task Force on Promoting Affordable Housing, pp. iv-vi.
63. *Ibid.*
64. Ahuja, Hitesh, Mandhyan, Ritika, Joshi, Hemand and Jain, Davan Sh (2012): op.cit., p. 19.
65. Task Force on Promoting Affordable Housing, op.cit., pp. v to xi.
66. 12th Five Year Plan, op. cit., p. 336.
67. 12th Five Year, Plan, op. cit., pp. 49-340.
68. Planning Commission, Mid Term Appraisal for the Eleventh Five Year Plan (2007-12), Urban Development, p. 382.
69. Editorial, Shelter, op cit., p. 2.
70. NRV, SPA, Delhi, op. cit., p. 37.
71. UN-HABITAT, op.cit., p. 7.
72. Joshi, Pratima: Guidelines for Crating Affordable Housing in Urban Areas (Working Draft), Shelter, op. cit., p. 12.
73. KPMG and NAREDCO, Bridging the Urban Housing Shortage in India, pp. 9-11.
74. KPMG and NAREDCO, *op. cit.*
75. Quoted in the Ministry of Housing and Urban Poverty Alleviation: Report of the Committee on Streamlining Approval Procedures for Real Estate Projects in India-Volume I, January, 2013, p. 16.
76. *Ibid.*

Chapter - 10

Reflections and Way Forward

Introduction

Emerging globalisation is linking the national economies on the principles of free trade, comparative advantage and global competition. Cities of the developing countries are striving to compete and prosper in the changed scenario. Economic restructuring of regions and cities is taking place where few cities with attractive location, economic dynamism, political commitment, good connectivity, high quality labour and urban infrastructure are able to succeed in the global competition by innovative development strategies and robust governance. The gap between the smart cities which are progressing well and the lagging cities is widening. Even in the fast developing cities, spatial inequities and enclaves of rich and poor communities are surfacing contributing to social tensions. The current urban space is becoming complexly dynamic and fragmented. Understanding and articulation of these complex processes is leading to new insights and perspectives.

Some scholars consider this trend as metropolitanisation or metropolitan regional growth. The metropolitan regions are territorial platforms of intense economic activity. K. C. Sivaramakrishnan (2013)[1] described them as multi-municipal and non-municipal arrangements challenging us to rethink traditional concepts of agglomeration economies and positive externalities. Recognising the need to maintain strategic flexibility as well as democratic participation, nation states in advanced capitalist countries and developing countries have been experimenting with rescaling and reorganising their functions. Ashima Sood (2013)[2] conceptualised the emerging phenomenon as multiplicities. The multiplicities that plague urban governance in India reflect deep interconnections between the challenges posed to and by urbanisation—political, economic, ecological, cultural, and social. It is further observed that the nature of these interdependencies presage the welcome inauguration of new research agendas across the social science disciplines, especially on sustainable urban systems and mapping of labour and resource flows into and out of city regions.

The urban crisis was imminent in India with inexplicable prolonged neglect of the urban governance and development problems in the face of rapid urbanisation and metropolitanisation. Realisation of this crisis in public domain is taking place slowly and in a more focused manner in recent years. Dimensions of urban crisis are far reaching and huge embracing the whole national development spectrum constituting productivity and quality of life for not only the swelling urban populations but vast number of urbanising and rural populations who either directly or indirectly depend on urban areas for livelihood and better services. The increasing slums are physical manifestations of poverty and deteriorating quality of life. The myopic urban growth with least concern for environmental issues at the other end is the index of utter governance failure.

Though there are several aspects of urban governance, the crucial neglect relates to the funding of urban development sector. The size of the national urban finance sector is very small compared to the many developed and developing countries and in relation to the public services that the urban local governments are mandated to deliver. It is reported that the total municipal revenue in India account for about 0.75 per cent of the country's GDP as against 4.5 per cent in Poland, five per cent in Brazil and six per cent in South Africa. In terms of both revenue and expenditure, the municipalities account for little above two per cent of the combined revenue and expenditure of Central Government, State Governments and urban local governments. This is in contrast to the situation obtaining in advanced countries, where local bodies normally account for 20-35 per cent of the total government expenditure.[3] It is true that the municipalities around the world are facing a paradox: the need for a strong fiscal base to create and maintain infrastructure and services; and the pressure to lower taxes to attract foreign and domestic investment. Despite those constraints, there were around the world many experiences, reforms, innovations and partnership arrangements which could serve as a remedy.[4]

The unfolding urbanisation is challenging to the policy-makers and administrators in India similar to the international experiences. Key challenges of urban development could be summarised as:

- Polarisation of economic growth leading to metropolitanisation;
- Weak public institutions and bureaucracy;
- Increasing urban infrastructure gaps leading to wide inequities;
- Rising urban poverty, informal sector and social exclusion;

- Low quality of services and inefficient delivery systems;
- Poor revenue base of Urban Local Governments;
- Threatening environmental concerns.

The urban policy response and reform agenda in India must be viewed in this backdrop. The outlook may be bleak but it is essential to be optimistic and continue to search for feasible solutions. In the backdrop of the urban policy and reform initiatives, emerging scenario mapping is attempted in this concluding chapter.

Reflections on Urban Policy Development

A recap of the policies and urban development programmes reveals changing emphasis with shifts in the development paradigms. After Independence, modernistic and socialistic ideologies attracted Indian think-tank which found their way into the urban development programmes in the form of subsidies and free dole outs. But, most of the initiatives were of mitigation nature devised to deal with newly emerging urban issues arising out of population growth. The pressing housing and growing slum problems due to industrialisation received attention and programmes were devised to address the housing shortages and needs of the slum areas. Initially, the idea was to relocate slums but the experiences soon compelled the authorities to change the policy to environmental improvements and provision of basic services. On the housing front, it was employee housing including low income housing with cross-subsidies. Urban policy as such was not the priority. The need for formulating comprehensive urban policies was felt only in the 1970s and after. Emerging imbalances in urbanisation received attention leading to policy thought to rectify the imbalances, especially, spatial imbalances. Though several commissions and committees examined the urban issues and proposed remedial measures, the setting-up of the National Commission on Urbanisation (1988) paved way for concerted policy dialogue. The economic importance of urban growth and land related issues were slowly recognised. Though the dimension of urban land became critical, the policy stance was welfare oriented till economic liberalisation movement picked-up during 1990s. The draconian Land Acquisition Act and the Urban Land (Ceiling & Regulation) Act (ULCRA) are examples of ill-conceived legal instruments driven by socialistic principles but ended-up in administrative power and legal tangles. The environmental issues also found place in the urban agenda and environmental protection became one of the functions of the municipalities as per the 74th Constitution Amendment.

The neo-liberal paradigm era (post-1990's) in India is associated with the thinking that democratic governance, free markets and dominant private sector are the recipe for growth and prosperity of the nation. But thinkers like John Williamson (2006) argued that market liberalisation by itself is insufficient—it must go hand-in-hand with and perhaps even be preceded by capacity building, market regulation and more effective law enforcement.[5] The free market and dominant private sector have become controversial policy discourses due to ideological differences and they remain so even today. Effective market regulation and strong governance, however, are being pursued in some form or other. The urban policies, programmes and reforms reflect these shifting paradigms in the country. The resultant scenario depicts a picture of unholy compromises on controversial issues oscillating between socialistic and capitalistic ideologies with political undercurrents.

Several experiments are being made since Independence to evolve better policies, strategies and institutional arrangements to meet the urban demands and challenges. The Five Year Plans provided policy directions and development framework for urban development in the country. The urban development strategies and reforms are devised based on the changing urban context and emerging urban needs. If housing, slums and provision of basic civic services received attention during earlier periods of Independence, the strategy changed to environmental improvement of urban slums and development of small and medium towns in the context of rapid growth of the metropolitan cities and emerging spatial imbalances in the later periods. Meeting the basic civic service needs has been the core running theme of the entire urban development endeavours. Efforts were also made to develop necessary institutional arrangements at the national and state levels.

Though the need for formulating national level urban policies was well recognised and unsuccessful attempts were made, the governmental commitment has been missing. However, some national sectoral policies have been formulated after prolonged deliberations recently to shape the urban sector development viz., National Urban Housing and Habitat Policy, 2007, National Policy on Street Vendors, 2009 and the National Urban Transport Policy, 2014. These policies are guiding the legal and administrative reforms to a certain extent. The national level urban policies are of recent phenomenon, therefore, their impact is yet to be experienced. However, a policy strategy dealing with the growing urbanisation in the country is non-existent. The country badly needs an urban policy for managing the increasing urbanisation in the country. Moreover, the draft National Slum Policy awaits a final shape.

The urban India as a result of policy deficit has become a living paradox: lure of wealth amidst distressing poverty; enclaves of rich juxtaposing the mushrooming of slums; economic vibrancy as against marginalisation of labour and life threatening environmental pollution; self-contradicting governmental action as against policy intents and; crippling urban services, infrastructure and governance in the face of rapid urban expansion—all reflecting the present urban enigma. Cities of today face a critically complex challenge of balancing: the growth with equity; development with sustainability; micro development with macro development; and short-term goals with long-term goals. The urban policies and governance therefore, demand some out of the box thinking to get into totally a different approach and strategies to deal with the current issues and problems thereby achieving the critical balances.

Historically, cities and towns are intrinsically linked to economic prosperity of the countries. The economic significance of the cities and towns was emphasised by the National Commission on Urbanisation in 1988 itself in India. The economic reports brought-out that urban India contributes a major portion to country's GDP. But the concrete policies and measures to integrate economic growth with the urban development endeavours are yet to be crystallised. The integration has to begin at the planning stage itself where spatial and infrastructure development planning is effectively linked with the economic development planning at different levels—national, state, district and local. Ideally all the planning exercises must converge at the local level leading to comprehensive development. The urban planning exercises unfortunately were confined to master plans, development control and building regulation with no lateral and forward linkages to the development endeavours.

Emerging Policy Perspectives

In the context of global concerns of urbanisation and climatic change interfaces and, fast rising inequities in the developing countries, sustainable development and inclusive cities have become catchy policy discourses. These concepts are deliberated extensively across the globe especially since the World Summit on Social Development held in Johannesburg in 2002 seeking not only environmental sustainability but also economic and social.[6]

Developed and developing countries have taken a confrontational stance on the environmental policy issues as they are mostly related to energy and consumption patterns. The social inequities and exclusions which are socially sensitive and more pronounced in Asian countries like India received government attention.

Environmental Concerns and Sustainable Habitat

There is increased awareness in recent years about the environmental costs of fast and indiscriminate urban growth in the country in line with the global thinking. The deterioration in quality of city environment is partly attributed to the climatic change arising out of unregulated urban growth and metropolitanisation putting tremendous pressure on urban space. The spatially concentrated urbanisation has led to energy intensive growth patterns affecting global climatic cycles and consequential climate changes. The recent experiences across the globe and in India too indicate that they have damaging effect on urban infrastructure like water, sanitation, energy, transportation, health-care, fire services and other forms of emergency measures. The effects are interrelated having cascading effect in urban context. For instance, a change in water availability and supply affects sanitation and drainage systems. When water supplies reduce, sewerage systems also become vulnerable. Further, sewage treatment plants are vulnerable to floods or sea level rise, as these are often located near rivers or seas. Storm water drainage systems could become overloaded and cause flooding if heavy storms become more frequent due to climate change. The impact of inadequate drainage systems in cities like Mumbai, Chennai, Delhi, to quote a few, is already felt leading to flooding and huge economic losses. More frequent floods could also present a significant public health problem. Climate change may impact transport and other infrastructure too leading to significant economic losses. Rising sea-levels in coastal areas would increase the risk of flooding and increase the vulnerability of communities residing in these areas, especially the poor. The urban-heat island effects could get exacerbated due to increase in baseline temperatures, affecting climatic comfort of the urban populations and may consequently result in additional mitigation costs[7]. In the backdrop of these effects, policy-makers in the country started showing interest in the climate change related issues of urbanisation.

On the development sphere, it is sustainable development mantra which is stimulating the policy-makers and administrators. In the context of growing environmental and climatic change concerns across the globe, the concept of sustainable development has assumed prime significance. Sustainable development has received attention in India not only due to environmental concerns but also due to rising socio-economic inequities. The National Mission on Sustainable Habitat is one of the eight missions formed under the Prime Minister's National Action Plan for Climate Change (NAPCC) to look into the aspects of climate change from the sustainable development angle. The National Mission for Sustainable Habitat was approved by the Prime Minister's Council for Climate Change in June 2010. One of the deliverables of the Mission is the formulation of National Sustainable Habitat Standards. The National Mission on Sustainable Habitat was launched in order to draw necessary action plans to achieve sustainability of human habitats. The National Mission on Sustainable Habitat seeks to promote sustainability of habitats through:[8]

(i) Improvements in energy efficiency in buildings through extension of the energy conservation building code - which addresses the design of new and large commercial buildings to optimise their energy demand;

(ii) Better urban planning and modal shift to public transport - make long-term transport plans to facilitate the growth of medium and small cities in such a way that ensures efficient and convenient public transport;

(iii) Improved management of solid and liquid waste, e.g. recycling of material and urban waste management - special areas of focus will be development of technology for producing power from waste.

(iv) Improving ability of habitats to adapt to climate change by improving resilience of infrastructure, community-based disaster management and measures for improving advance warning systems for extreme weather events and to addresses sustainability concerns related to habitats, primarily in urban areas.

(v) Conservation by appropriate changes in legal and regulatory framework.

The key deliverables of the Mission include:[9] (a) development of sustainable habitat standards that lead to robust development strategies while simultaneously addressing climate change related concerns; (b) preparation of city development plans that comprehensively address adaptation and mitigation concerns; (c) preparation of comprehensive mobility plans that enable cities to undertake long-term, energy efficient and cost effective transport planning; and (d) capacity building for undertaking activities relevant to the Mission. National committees are constituted to formulate standards and measures in the specific areas like water supply and sewerage, solid waste management, urban planning, etc. The recommendations of these committees are yet to be considered seriously by the governments for incorporating into the relevant policies, laws, reforms and programmes.

Inclusive Cities

Inclusive growth, according to a paper of Asian Development Bank,[10] involves creation of equal access to opportunities as unequal opportunities arise from social exclusion associated with market, institutional, and policy failures. It is stated that a development strategy anchored on inclusive growth consists of two mutually reinforcing strategic pillars of high and sustainable growth to create economic opportunities, and social inclusion to ensure equal access to opportunities. This will enable to accomplish the agenda of eradicating extreme poverty and, at the same time, address the development challenge brought about by rising inequities. Inclusive growth primarily focuses on eradicating circumstance related to inequities and not personal effort related inequities. In this perspective, inclusive growth could be a more broad-based development paradigm compared to the strategy of eradicating mere income related inequities. It provides opportunities for optimum utilisation of human resources leading to higher productivity and more sustainable development.

The concept of inclusive growth basically has two dimensions--economic and social. The economic angle of the inclusive growth involves growth with sustainability. Though it is not realistic to aim at high growth, a moderate growth with sustainability could be the feasible policy objective in the context of high incidence of absolute and relative poverty in the developing countries like India. Measures like extending easy micro credit, promoting small enterprises, and several livelihood programmes are needed to address the income poverty issues. From the social perspective, inclusive growth involves investing in social services like education and health for building social capital of the disadvantaged people, forming social safety nets to absorb the transitional problems and address the issues like temporary job losses, employer exploitation in the informal sector, insurance, homelessness, street children and destitute, etc., and providing necessary policy and institutional structures for social and economic justice.

Framework of Inclusive Cities

India's urbanisation should be guided towards inclusive, equitable and sustainable growth of towns and cities (12th Five Year Plan)

Achieving the goal of Inclusive Cities to a large extent revolves around good governance characterised by principles of sustainability, subsidiarity, equity, efficiency, transparency, accountability, civic engagement and security which are interdependent and mutually reinforcing as summed-up by the UN.[11] The process involves: decentralisation of responsibilities and resources to local authorities based on the principles of subsidiarity and accountability; encouraging the participation of civil society, particularly women, in the design, implementation and monitoring of local priorities; using a wide-variety of partnerships, including with the private sector, to achieve common objectives; building capacity of all actors to contribute fully to decision-making and urban development processes; facilitating networking at all levels; and taking full advantage of modern information and communications technologies (ICTs) to support good urban governance and sustainable urban development.[12] In strategic terms, it starts with identifying factors of exclusion namely, who is excluded, from what excluded and how excluded and finally, devising mechanisms of inclusion on similar analogy.

Inclusive Cities implies improving the livelihoods of the urban working poor, most of who are employed in the informal economy, many as waste pickers, street vendors, and home-based workers. These workers, while numerous, have very little representation or "voice" in the municipal policy and urban planning processes that significantly impact their everyday lives and their ability to generate enough income to support their families. An inclusive city is primarily based on the shared belief that reducing urban poverty requires reversing the current exclusionary trend of many modernising cities. This requires fundamental shifts on policy front and reshaping of urban plans and regulations to incorporate the poor.

The challenge is to promote inclusive cities that embrace the informal economy as a vibrant and key component of the urban economy, and view it as a solution to reducing urban poverty and violence. Inclusive cities support home-based production through basic infrastructure and appropriate zoning regulations; allow street vendors and hawkers to operate alongside small kiosks and shops as well as large retail shops and malls; and incorporate waste pickers into modern solid waste management systems. Inclusive cities require an inclusive, rather than exclusive, approach to urban infrastructure and services; urban spatial planning and zoning; urban regulations and laws; and urban policies. To achieve this, representatives of the working poor must have a voice in urban planning processes. The Twelfth Five Year Plan has rightly captioned its urban policy thrust as faster, more inclusive and more sustainable growth reflecting the current global thinking. The Plan aims at the following[13]:

- Good urbanisation that ensure that towns and cities are free from slums and provides adequate opportunities for productive employment and a decent quality of life to all their inhabitants including the poor.
- The cities must provide world class infrastructure and services at affordable costs to give a competitive edge to the economic activities they host.
- Cities should be able to provide basic services to migrant workers, their families and other vulnerable sections of society including women and children.
- The future renewal of our cities should facilitate transition from 'informality' of large number of workers towards more formal livelihoods in line with their aspirations. They should address various vulnerabilities including residential, occupational and social vulnerabilities, associated with urban poverty.
- People should be brought to the heart of the urban agenda as an overriding principle, for both, deciding the vision of their city and for choosing the process of reaching that goal.
- All citizens should have access to basic services of clean water, sanitation, sewage, solid waste management, urban roads, safe and affordable public transport systems, affordable housing, and a clean and healthy environment.
- Indian cities should also meet the rising aspirations of people for a better quality of life besides creating avenues for gainful employment

- Future growth should be consistent with cities' natural endowments and the economic potential of the region in which they are situated. All cities should be efficient in using available resources particularly energy, water and land. Our cities must also preserve and foster their cultural and historical heritage and benefit from the tourism potential of their heritage and natural endowments.

The report of the McKinsey Global Institute (2010)[14] observed that cities of India continue to play significant role in India's economy and society while playing dominant role globally. But, the critical challenge is funding and managing fast urban growth. It warned that unless necessary measures are taken, the rapid urban growth devoid of good infrastructure and services will further lead to deterioration in the urban living conditions. The strategy recommended includes following critical elements:

- ***Providing for Adequate Funds to Build Infrastructure and Services in Urban Areas:*** The strategies suggested to raise funds are: land monetisation, property taxes and user charges, debt and private participation and formula-based grants from the government.[15]
- ***Strengthening of Governance:*** The focus should be on building effective administrative and technical talent at different levels to manage the proposed spending properly.
- ***Effective Planning***: Indian urban planning suffers from narrow perspective and ad hoc decisions. The focus need to be on developing high dense cities which could reduce service costs.
- ***Sector Policies:*** The sector specific policies should focus on job creation, public transport, affordable housing, and climate-change mitigation efforts.

Role of International Agencies

A pronounced strategic shift observed among the international agencies in recent past has been from 'project financing' to 'policy-oriented lending' in the urban sector. A review of on-going externally aided urban sector projects shows that majority of these are designed to bring about changes in urban sector institutions, policies and programmes aiming at structural impacts.[16] For instance, The World Bank[17] which has funded about 30 urban sector projects with lending commitments of Rs 15,100 crore in India has been focusing on policy reform and institutional change influencing regulation, pricing and incentive systems covering the entire range of governance issues viz., local government autonomy; resource mobilisation; and patterns of relationships among local government, the private sector, informal organisations and households. The new strategy of the Bank for the twenty-first century is a shift from mere financial intermediation to strategic intermediation which calls for a holistic approach to urban development with more democratisation, decentralisation, public-private partnership and collaboration among stakeholders.[18] The pronounced aim of the World Bank is to promote sustainable cities and towns that fulfill the promise of development for their inhabitants with a view to build national perspective in urban development.[19] The sustainability implies that the cities are livable ensuring decent quality of life and equal opportunities for all residents including the poorest. The cities must be competitive, well governed and efficiently managed and financially sustainable, in other words bankable to achieve this goal. Competitiveness means the cities require buoyancy with growing employment, incomes and investment coupled with appropriate urban equity and social safety nets. The Bank is supposed to act as a facilitator for city wide economic analysis and strategy development through a participative process involving a wide range of stakeholders. Good governance implies inclusion and fair representation to all groups in the urban society, accountability, integrity and transparency and management capacities at the municipal level. Bankability reflects the financial soundness of municipalities that decides credit worthiness to borrow and enter into the capital market. This requires specific and internally consistent local revenues and expenditures, transparent and credible intergovernmental transfers, prudent conditions for municipal borrowing and generally accepted financial accounting, asset management and procurement practices and, more importantly adopting a commercial approach to the service and administrative functions keeping in view the social concerns. The World Bank's strategy currently aims to achieve greater[20]:

- ***Replicability*** - promoting lasting improvements in raising the living standards of the urban poor on a scale commensurate with the demands.
- ***Relevance*** - mobilising skills and knowledge across a wide range of urban issues and offering assistance with flexible designs, realistic timeframes and appropriate forms of financing.
- ***Reach*** - ensuring more local governments can be reached directly or indirectly.

The current urban dilemmas reflect the struggles to move in the directions envisioned by the international agencies and the Twelfth Plan at one end, and the diverse pulls of social realities, political preferences and administrative inadequacies at the other end. We shall now proceed to discuss how fast the urban reforms initiated are moving and in which direction.

National Urban Reform Agenda

The urban reforms in India were isolated local or state level initiatives till recently. Most reforms were related to improving the system of municipal governance. Certain reforms were introduced in the areas like election and role of chairperson, property tax and community mobilisation. The local level reservation system for weaker sections viz., SCs and STs, BCs, and women began on experimental basis. A unique system called the Mayor-in-Council emphasising strong mayor was introduced in Kolkata Municipal Corporation which was followed by other corporations in West Bengal, Madhya Pradesh, Chhattisgarh, and Tripura. This system is similar to the cabinet system of governance designed to make the mayor more powerful and a real executive enjoying the powers as against the widely prevalent in other corporations of the country. But, the system did not find favour in other states.

Reforms were also made in respect of the election of the chairpersons. Certain states like Andhra Pradesh, Tamil Nadu, Uttar Pradesh, and Jharkhand have tried with direct election of the chairpersons/mayor as against the indirect election of the municipal chairpersons indirectly by the council. The direct election of the chairpersons without much change in the powers of the chairperson appears to be a cosmetic attempt than a real reform. The experience also proves this as there were frequent disruptions in the council meeting where majority of the councillors belonged to the party other than the chairperson.

Another most experimented reform in the country has been property tax reform which continues even today. Standardisation of the property tax assessment was the main objective which received people's acceptability in some states like Andhra Pradesh in the 1990s. Another reform area which received much significance relates to community mobilisation. With the launch of various slum improvement and urban poverty alleviation programmes in 1980s, involvement of the NGOs and formation of various community-based organisations like the Neighbourhood Committees has become a popular strategy of the governments. The role of these organisations in the provision of the various urban basic services in the slum areas has been found to be of immense help to the municipal authorities. The community involvement initiatives were made on experimental or pilot basis in cities like Hyderabad, Chennai and Mumbai. International institutions notably, UNICEF and ODA (now DFID) came forward to offer financial assistance to improve slum areas. These institutions promoted community participation by involving various community-based organisations in the slum improvement programmes. The success of these initiatives in Hyderabad, Vijayawada, Indore, Cuttack and other select cities and towns, led to a country-wide awareness for involving the local communities in the latter periods.

These ad hoc and isolated reform attempts could not yield the desired results due to structural deficiencies associated with low status of the local bodies in Indian federation. Major structural deficiencies in the local governance system have ultimately led to the thinking that unless local bodies are provided constitutional status, the situation is not going to improve in true sense. The untiring efforts of the urban think tank led to the Constitution 74th Amendment Act, 1992 and since then, efforts are being made to strengthen the municipalities in terms of powers, functions finances and capacities. It is dispiriting that even after two decades of the 74th CAA, the status of urban local governments has not changed much. The reluctant and half-hearted efforts of the state authorities to implement the mandatory provisions of the Constitution Amendment failed to yield the desired results especially, in the areas of devolution of powers, functions and finances to the municipalities and achieving the goal of decentralised, democratic and participatory governance remained a distant dream. The strong state level lobby is not allowing the true decentralisation process since Independence in India. The structural corrections could not be carried out due to inherent systemic influences in spite of the Constitution Amendment. This proves the widely prevalent perception that Indian administrative system is decentralised in vision but centralised in practice.

The latest urban reforms have a close linkage with the economic liberalisation reforms initiated in 1990s. The economic reforms in the first phase included the components like opening up of the economy to foreign direct investments (FDI), foreign institutional investors, import of technology and consumer goods and more significantly to privatisation. Evidence shows that the reform process has been very slow due to governance deficiencies and political constraints. The FDI which were expected of did not come through. Some thinkers[21] argue that though the foreign investments as a proportion of total capital accounts inflow in India increased from a mere one per cent in 1991-92 to 25 per cent in 2000-01, it was not the foreign investments but the loan accounts whose share increased from 38 per cent to 45 per cent. At the other end, the share of FDI, which is more robust, and a stable investment has remained quite low at five per cent in 2000-01 but the portfolio investment which is very unstable and can fly out of the country even at the hint of a small crisis in the economy was dominant with about 20 per cent in 2000-01. The most visible effects of economic reforms have been the flooding of foreign consumer durables and increased debt burden on the governments. In the urban front, reforms were mostly project-based where international agencies played significant role. It is no exaggeration to state that they were debt driven. Most of the urban reforms were focused on financial sustainability of the municipalities through opening-up of urban infrastructure to the private sector, accessing capital market, cost recovery and improving financial management by introducing accrual based double entry accounting system. The importance of land as a resource was slowly recognised.

The Government of India began to formulate national level reforms in 1996 through guidelines which provided a broad vision of urban reform strategies. The guidelines suggested that the traditional system of funding based on plan and budgetary allocations be reduced and ultimately withdrawn due to fiscal deficit. Subsidies need to be rationalised and urban development plans and projects need to be placed on a commercial format by designing commercially viable urban infrastructure services and area development projects. Other innovative resource mobilisation measures suggested include using land as a resource, increase in the non-property taxes and using Public-Private Partnership in service delivery. In order to increase the pace and incentivise urban reforms, the Urban Reforms Incentive Fund was formed by the Government of India and a package of reforms including abolition of Urban Land (Ceiling & Regulation) Act (ULCRA), reforming rent control legislation, reduction of stamp duty, levy of user charges, property tax and accounting reforms was linked to the funding by 100 per cent grant. The reform agenda formulated under JnNURM was the consolidated version drawing much from earlier attempts broadly focusing on financial sustainability of municipalities, governance improvement, shelter and poverty alleviation. However, new initiative Atal Mission for Rejuvenation and Urban Transformation (AMRUT) focuses on implementation of urban reforms agenda at grass-roots level. The approach is now changed from 'project based' to 'action based'.

Revisiting JnNURM

The spatial, economic and social implications of urbanisation are not properly articulated into comprehensive policies and plans in urban sector. But, urban sector funding has been increasing from the Central Government in vital fields of urban infrastructure and services, slum development and urban poverty alleviation through a series of centrally sponsored schemes like Environmental Improvement of Urban Slums (EIUS), Integrated Development of Small and Medium Towns (IDSMT), Nehru Rojgar Yojana (NRY), Urban Basic Services for the Poor (UBSP), National Slum Development Programme (NSDP), Accelerated Urban Water Supply Programme (AUWSP), Mega City Scheme (MCS), Swarna Jayanti Shahari Rojgar Yojana (SJSRY), etc. Yet, the level of outlay in the urban sector after Independence under Five Year Plans reveals a very disappointing picture.

The outlays for urban development in the Five Year Plans, for instance, did not go beyond three per cent. It is only from the Tenth Plan onwards the allocations for urban development have picked-up. For instance, the allocation in the Eleventh Plan was almost doubled to Rs 63,465 crore from the Tenth Plan outlay of Rs 31,719 crores mainly because of the launching of the JnNURM. The estimations[22] reveal that India needs investment of about $ 1.2 trillion (about 60 trillion rupees) in the next 20 years that is, from 2010 to 2030. Current annual spending on capital fronts is very low that is, mere $ 17 which is far less compared to China's spending of $116. Further, if large cities spend $130 per capita per year smaller cities and towns spend as low as $12-38. It is suggested that the investment has to be scaled-up to at least $134 per capita, almost eight times of the current spending. In other terms, the annual spending has to be at least two per cent of the GDP. Dr Ahluwalia Committee (2011)[23] which inquired into urban investment needs estimated that India needs Rs 39.2 lakh crores at 2009-10 prices for the next 20 years period from 2012 to 2031. The overall objective is to increase the level of investment from current 0.7 per cent of GDP to 1.1 per cent of GDP by 2031-32. Out of the total investment, 15 per cent that is, about Rs five lakh crore was to be allocated in the Twelfth Plan itself. As against this, the Twelfth Plan tentatively provided an outlay of Rs 1,64,078 crore which is 158 per cent more compared to the Eleventh Plan allocation. In terms of its share in the total sectoral allocation, the increase has been from 3.99 per cent to 4.60 per cent. All these estimations strongly indicate that the level of spending must be scaled up considerably.

The launching of JnNURM in 2005 was considered a landmark in terms of expectations and quantum of investment in urban sector. The JnNURM was a centrally sponsored scheme with collaborative funding from the central government in the form of Additional Central Assistance (ACA), state governments and municipalities including borrowings if any. The quantum of investments reached more than two lakh crores in the urban sector which was considered massive compared to earlier funding. The development emphasis was outlined as below:

(i) Integrated urban infrastructure development in cities and towns;
(ii) Securing effective linkages between asset creation and asset management so that the infrastructure services were maintained efficiently and become self-sustaining over a period;
(iii) Pumping adequate investments to bridge the deficiencies in urban infrastructure;
(iv) Improving delivery systems of civic services with emphasis on universal access to the urban poor;
(v) Renewed attention on inner city development and reducing congestion;
(vi) Alleviation of urban poverty;
(vii) Slum rehabilitation with affordable housing including security of tenure.

Urban reforms received big push to ensure that increased investments are properly utilised to achieve the desired objectives of sustainable urban development, inclusive growth and revitalised governance. As a strategically essential measure, the Government of India tied urban reforms with the

JnNURM, which provided better opportunity to push the urban reforms on fast track. A national urban reform agenda was formulated and efforts were made to implement the reforms on mission mode at the state and municipal levels. The urban reform agenda revolved around following themes:

- Financial sustainability with local level resource mobilisation through a combination of innovative strategies like land monetisation, accessing capital market, PPPs, setting up of development funds and traditional methods;
- Effective financial management by accounting and budgetary reforms;
- Good effective governance focusing on decentralisation, capacity building, peoples' participation, transparency and accountability;
- Efficient civic service delivery management;
- Sustainable urban development by way of inclusive growth, conservation and participatory planning.

Learnings from JnNURM

The reform focus of JnNURM was commendable demonstrating commitment of the Government of India to fast track urban reforms in the country toward developing a new statecraft. It was a major mission mode initiative in the urban sector launched for planned development of the urban areas through the City Development Plans and a bold attempt to incentivise and commit state governments and municipalities for implementing critical urban reforms. The state governments though reluctantly committed for the mandated reforms, JnNURM succeeded in creating a facilitating environment to implement the reforms by way of statutory amendments and, establishing financial and administrative institutional arrangements at the central, state and local levels.

The reforms under JnNURM were at various stages of implementation. The mid-term appraisal of the progress[24] revealed that about 66 per cent of the State level mandatory reforms and 82 per cent of the optional reforms were completed as against the sixth year commitments. The pace of progress in State level reforms and some optional reform implementation is considered satisfactory by the government appraisal. Key optional reforms which were moving fairly well as per mid-term assessment included revision of building byelaws to streamline building approval process, making rainwater harvesting mandatory, introduction of computerised process for registration of land and property. The municipal level reforms however, moved a bit slowly as the quantified national level achievement stood at 56 per cent. Key reforms that moved well included accounting reforms and earmarking of funds. In respect of property tax, many cities improved coverage to 85 per cent and improved collection efficiency to 90 per cent. Full cost-recovery of O & M through levy of user charges was however, found a difficult reform.

Excepting government reporting, no critical independent evaluation has been done in India to assess the performance of JnNURM in a comprehensive manner. The information available from government records and field observations provided certain broad trends on the lines of progress, problems and issues. Certain strategic lapses and basic deficiencies were observed which should be considered seriously. The sharply voiced criticism against JnNURM was the mission city or large city bias and glaring inter-sector variation in terms of funding. The focus bestowed on the mission cities by the central and state governments since beginning speaks of the mission city bias. In fact, the state authorities and other implementing agencies felt that the JnNURM was for mission cities and the other cities were of lesser importance. The impression gathered from officers of the small and medium towns reveal that JnNURM was only for mission cities and UIDSSMT and IHSDP programmes were on different platform. It is not surprising therefore to find that the mission city performance was better compared to the smaller cities and towns. There is a strong opinion among a section of the thinkers and authorities that the mission city focus was needed and a better strategy in the context of increasing metropolitanisation centred around million plus cities more particularly mega cities.

The erstwhile Planning Commission did some exercise to evaluate the JnNURM performance. According to its assessment, the major defect noticed was absence of any long-term city level planning based on participatory planning process preventing any holistic development of cities in tune with the local economic base and resource endowments.[25] Another major lapse was exclusion of the peri-urban areas where intense urban activities are occurring taking advantage of the tax incentives, availability of cheap land and liberal administrative regulations, etc. These areas are growing in haphazard manner due to planning lapses devoid of proper civic facilities, infrastructure and connectivity. If absence of participatory planning led to ownership problem, exclusion of peripheral areas is expected to lead to expensive retrofits in future. Ultimately, these deficiencies raise sustainability debate in view of intensification of spatial imbalances, inequities and delining quality of life. Absence of convergence was another problem identified. For example, the Government of India launched two schemes namely, JnNURM for infrastructure development and Service Level Benchmarking (SLB) for improving civic service levels. Both aimed at improving the status of urban infrastructure and services. In practice, these two schemes were not linked so that each supplements the other for achieving the common goal. Since Independence, lack of adequate expertise at the municipality level has been the main reason to float autonomous urban institutions. It is surprising that the same approach was adopted though not by design but by excessive dependence on external consultants while ignoring the capacity building of municipalities. It is paradoxical that JnNURM which mandated for implementing all the 74th CAA provisions for strengthening of urban local governments adopted this strategy.

Urban Reforms—Fast Track but Slow Motion

The major strategic defect observed is adoption of "one size fits all" approach without taking into consideration capacities and needs of various categories of municipalities—big and small. This impacted the reform implementation adversely in the smaller cities and towns which lacked adequate capacities compared to the large cities. As there are considerable differences in local conditions and capacities among different sizes of municipalities in bringing-out reforms, there is need for proper calibration of the reforms as well as norms to suit the ULGs.

Further insight brings out certain interesting trends in the implementation of reforms. In respect of the legal amendments, the state authorities amended the Acts but the rules were unduly delayed. In Karnataka, for example, the ward committee rules were not made though legislative provisions were made for a long time. As a result, the ward committees could not be formed though elections were conducted. In Uttar Pradesh, the U.P. Municipal Corporation Act, 1959 was amended in 2009 providing the constitution of ward committee in each ward (earlier the provision was to constitute a wards committee on every ten wards), but the rules were drafted in 2012 and finalised in 2014. The ward committees are still to see the light of the day in this most populous state having second highest network of municipalities in the country.

Further, the functions were transferred through orders but no subsequent measures were taken to implement them in some cases. For instance, in Andhra Pradesh, urban planning was transferred to municipalities through orders but in practice it was not. The urban planning continues with the state level town planning department and urban development authorities. Interestingly in Uttar Pradesh, functions were transferred to the municipalities by amending both the Acts, but through government orders many of these mandatory functions are being performed by the parastatals as usual. Legal provisions have been made for forming Metropolitan Planning Committees in many of the states but they are not actually constituted and made functional even after a lapse of more than two decades. The story is more or less the same as regards District Planning Committees. Such is the fate of the supreme law of the land. The provisions of the 74th CAA have been implemented in letter (the respective laws have been amended) but not in true spirit. This shows the disinterestedness of the state governments in strengthening the urban local governments. The matter needs to be revisited and the local governments deserve their independent status (entry) in the Constitution. Other ad hoc arrangements will never make the Indian Federation three tier—Union, State and Local.

Another major deficiency has been inability to attract private sector funding either through PPPs or accessing capital market. Though all the mission cities were rated by the rating agencies, mobilisation of private investment was very discouraging. The municipalities also failed to adopt innovative methods like land monetisation to mobilise funds for critical sectors. In respect of resource mobilisation, there has been increased tendency to depend on the external funding from the central and state governments. It is in this context that the suggestion made by the High Power Committee (2011) of the Ministry of Urban Development on estimating financial needs of urban services assumes significance. The Committee recommended strengthening of the municipalities by way of devolution of financial powers and development of the planning and management capabilities. It is pertinent to mention here the expectations of the government while approving the JnNURM in 2005. The Prime Minister stated that 'the core reforms proposed at the municipal level for urban governance should be insisted upon both as a method of improving urban governance and a way of locating financial resources internal to the city to make it undertake a programme of urban renewal. This will also help increase the ability of the local body to mobilise institutional finance.'[26]

An overview of the implementation of state and municipal level reforms reveals that less controversial reforms like e-governance reforms, introduction of accrual based double entry accounting system, citizen charters, reservation of funds for weaker sections in the municipal budgets, reduction of stamp duty, abolition of Urban Land (Ceiling & Regulation) Act (ULCRA), etc., were implemented with ease relatively. Though legal provisions were made to implement water harvesting reform, its implementation is suffering from lack of people's awareness and municipal commitment. In regard to mandatory provisions of the 74th CAA, excepting provisions like constitution of the metropolitan committees, transfer of planning and fire services and, devolution of funds and powers to municipalities, others were implemented in some form or other. The problem mainly relates to some complicated or difficult reforms. Implementation of some difficult reforms like district planning committees, metropolitan planning committees, cost recovery, people's participation, ward committees and area sabhas, reservation of land for the weaker sections in the private lay-outs, resource mobilisation by innovative methods and PPPs lagged behind due to several institutional and political reasons. Even in this disappointing scenario, some states like Andhra Pradesh, Gujarat, Maharashtra, Tamil Nadu, Karnataka, Madhya Pradesh and Chhattisgarh have been performing relatively better which speaks of the positive local environment and commitment of the state government. In many states the reforms under JnNURM were implanted on papers for getting the financial assistance from the Central Government.

Reshaping Strategy

There are difficulties in formulating national urban policy agenda in federal systems where states or sub-national governments are equal partners in the power structure. In diversified countries like India, it is still challenging because urban problems are vast and amorphous collection of different physical and social conditions existing in society and translating all of them into a common policy at national level would be a herculean task for the policy makers.[27] The difficulties experienced by the municipalities and the state governments in implementation of the nationally formulated policies and reforms envisaged under JnNURM reflect such a predicament.

Based on the lessons learned from the earlier efforts, the Government of India has reshaped the strategy by shifting from 'one size fits all' approach to 'calibrated strategy' to suit different types of municipalities and local situations. The Union Government launched, in 2015, a comprehensive package of mission-mode schemes viz., Atal Mission for Rejuvenation and Urban Transformation (AMRUT), Smart Cities Mission, and Housing for All by 2022-Prime Minister's Awas Yojana (PMAY), besides many other innovative initiatives including Swachh Bharat Mission (SBM), dreaming to make the country open defecation free within a fixed time-frame. A detailed discussion about PMAY has been made in Chapter- Nine. We are not throwing light on SBM as it has no direct linkage with the reform process, though a most important aspect of urban sanitation and environment.

Atal Mission for Rejuvenation and Urban Transformation (AMRUT)[28]

The Atal Mission for Rejuvenation and Urban Transformation (AMRUT) is a reform-based mission mode scheme focusing on creation of infrastructure having a direct impact on the real needs of urban people. The focus is on infrastructure that has a direct link to provision of better services to people. However, the pursuit of better outcomes will not stop with the provision of universal coverage of taps and sewerage connections (in earlier scheme the emphasis was on creating water supply and sewerage infrastructure and not providing individual connections). Further, the new initiative is linked with Service Level Benchmarking (SLB) and an incremental process is being adopted for achieving SLBs gradually according to National Priorities. This new strategy replaces the earlier approach of project based sanctions and rather adopts approving the State Annual Action Plan (SAAP) once a year and the states have to sanction the projects at their end. Thus, AMRUT makes state equal partners in planning and implementation of projects.

Contrary to JnNURM strategy to cover selected 65 metro cities, AMRUT covers five hundred cities. These cities include all class-I towns (towns having a population of one lakh and more), all capital cities, all cities classified as Heritage Cities under HRIDAY scheme, 13 cities on the stem of the main rivers with a population above 75,000 and less than one lakh, and ten cities from hill states, islands and tourist destinations. Thus, the AMRUT follows holistic approach in coverage.

Under AMRUT, each municipality has to prepare a Service Level Improvement Plan (SLIP) for bridging the gap in the provision of service delivery. At the state level, the SLIPs of all Mission Cities will be aggregated and a State Annual Action Plan (SAAP) will be prepared by the state governments. The SAAP is basically a state level service improvement plan indicating the year-wise improvements in water supply and sewerage connections to households. Implementation of Reforms is an important objective of SLIP. The municipalities have to prepare a road-map for Reforms. The state government will consolidate and include the same as a part of SAAP.

Under AMRUT, creation of a sound institutional structure is conceived as the foundation. Therefore a focused Capacity Building approach and a set of Reforms have been incorporated in the Mission. It is aimed that Reforms will lead to improvement in service delivery, mobilisation of resources and making municipal functioning more transparent and functionaries more accountable, while Capacity Building will empower municipal functionaries and lead to timely completion of projects. As regards capacity building, the state governments have to submit an annual plan for the same as part of the SAAP. The realigned Capacity Building Plan consists of two strategic interventions—Individual Capacity Building and Institutional Capacity Building. The purpose of individual training is to enhance the functional knowledge, improve the job related skills and change the attitude of municipal functionaries. The aim of the institutional capacity building is to improve institutional outcomes, as set out in the Reform Agenda. The effectiveness of the strategy, however, depends on the training institutions imparting training and capacity building specialists engaged at the state level to monitor and guide the whole exercise. The customary approach of imparting class-room training and patronising the training institutions will not help in any way. In many cases the institutions, trainers and capacity building specialists are found lacking in appropriate capacity.

The main goals of AMRUT may be enumerated as under:

(i) Improving service delivery standards in urban areas including basic services for urban poor;

(ii) Empowering municipalities and municipal functionaries;

(iii) Facilitating participatory governance;

(iv) Managing land resources effectively;
(v) Fostering sustainable, inclusive and faster growth;
(vi) Alleviating urban poverty.

The focus under AMRUT is on Action-oriented development rather than project-based development. The action areas are made more focused and well directed to achieve these goals. Within this blanket provision, the willingness of the municipalities to implement the reforms has been given priority in sanctioning of the projects. Reinforcing the emphasis on reforms, the element of incentivisation is introduced in the fund outlay. Ten per cent of the total fund will be allocated for providing incentives for better performing municipalities in implementation of difficult reforms. The reform package under this new initiative has been reworked on calibrated model. We have discussed in details about Reforms Agenda in Chapter- Two. The Twelfth Plan has given some indications in this regard. The reform initiatives are more focused on specific action areas to achieve the development goals.

Smart Cities Mission[29]

Another mission mode scheme launched by the Union Government in 2015 is Smart Cities Mission, proposing to develop some 100 cities as Smart City in five years. The purpose of the Mission is to drive economic growth and improve the quality of life of people by enabling local area development and harnessing technology, especially technology that leads to Smart outcomes. Area-based development will transform existing areas (retrofit and redevelop), including slums in to better planned ones, thereby improving livability of the whole city. New areas (Greenfield) will be developed around cities in order to accommodate the expanding population in urban areas. Application of Smart Solutions will enable cities to use technology, information and data to improve infrastructure and services. Comprehensive development in this way will improve quality of life, create employment and enhance incomes for all, especially the poor and disadvantaged, leading to inclusive cities. The strategic components of area-based development are thus city improvement (retrofitting), city renewal (redevelopment) and city expansion (Greenfield development). This aims to adopt a holistic approach of city development by bridging the gaps as left in the earlier efforts of creating infrastructure and services.

Future Urban Vision and Way Forward

The future of urban India as envisioned by the Twelfth Plan is planned, inclusive and sustainable urban development which is to be founded on five basic enablers viz., affordable housing, sustainable livelihood and enterprise, universal access to water and sanitation, quality and affordable public transport and clean and healthy environment.[30] The deliverables are slum redevelopment, reducing urban poor, universal access to water and sanitation, increasing share of public transport, and all new commercial buildings to be energy efficient. These ambitious goals require legal and administrative interventions in the core sectors of governance, planning, financing and capacity building. The on-going reforms in these areas should be implemented with more commitment, innovation and sharpness.

The success of urban reforms ultimately depends on devising and implementing proper governing systems. Improving urban governance is the prerequisite for moving forward on the development front. The experience so far in India depicts a picture of oscillation between centralisation and decentralisation based on the changing situations and contexts. One school of thought, presumably dominant, propagating centralised system for efficient and effective results while the other fighting for decentralisation to meet diverse needs and aspirations of the people by strengthening of local governing systems. The need of the hour is to break across the silos and create convergence.[31] This implies targeting a devolved structure with power closer to where results are required with different parts of the system working collaboratively with faster adaptations and learning. This is taken care of in new initiatives.

Ensuring social accountability, financial sustainability of civic services, institutional convergence and better relations between the deliberative and administrative wings in municipalities are the major concerns of urban governance today. The key reforms suggested by the Planning Commission to address these concerns are vesting executive authority in the mayor with fixed five year term, setting-up municipal service regulator, implementing citizen charters effectively, corporatisation of service delivery in metro cities, role specification of parastatal organisations and municipalities and setting-up Ombudsman or Lok Ayukta.[32] Excepting the citizen charters to some extent, the other reforms are difficult riddled with complex social and political problems. Ensuring financial sustainability without necessary cost recovery is next to impossible and municipalities are struggling to maintain services with meagre funds available. To resolve this issue, the municipalities should improve efficiency in the delivery of quality services (as envisaged in AMRUT) and at the same time, the local communities who utilise the services should understand the financial aspects of delivering better services should get convinced to bear the costs. The sustainability aspect cannot be solved unless efforts are made on both the fronts. The strained relationship between the deliberative and administrative wings has been the critical factor plaguing the functioning of the municipal bodies. Several solutions are proposed since Independence but no long-term solution has been found so far to resolve this chronic problem. Introduction of the strong mayor system is considered a better option and some states like West Bengal, Madhya Pradesh, Chhattisgarh, and Tripura have already introduced this system. The Government of India is suggesting the states to adopt the strong mayor system in some form or other to improve the municipal functioning while giving due importance to the democratic functioning. In view of widely prevalent corruption in the public administration, introduction of the Ombudsman to deal with the municipal corruption cases is considered very essential and the Government of India has already made its commitment by including it as one of the conditions for getting performance based grant under the Thirteenth Finance Commission grant-in-aid to the urban local governments. Another critical unresolved issue is decentralisation by devolution of powers and functions. It appears that unless the capacities of the municipalities in terms of staff, financial and administrative powers are strengthened, the aspect of decentralisation will continue to be a debatable and unresolved issue. In this respect two important reforms viz., Administrative Reforms and Institutional Reforms proposed under erstwhile JnNURM need proper attention even today by the state governments.

People centred planning is essential to formulate need-based plans and ensuring their successful implementation. The 74th CAA has already mandated to transfer main urban planning function to the municipalities. Unfortunately, state governments are reluctant on this reform. Involving people in the local planning process continues to be an important reform and thus has been recognised as an important focal point under AMRUT. Strengthening of the District Planning Committees and Metropolitan Planning Committees with urban development authorities as the technical wings and making the ward committee system effective are essential to achieve this objective. The Twelfth Plan and its working groups suggested for strategic densification, Guaranteed Land Title Act, including peri-urban areas in city development plans and entrusting strategic spatial planning responsibility to the state planning boards. The states should take necessary legal and administrative measures to implement these reforms for making urban planning development oriented and to fructify the benefits as envisaged under AMRUT.

Though funding is being increased in the successive plan periods, there is still considerable need for increasing the urban sector outlay in view of the increasing urbanisation in the country. Another neglected area is local resource base. There is a school of thought which believes that unless a Local Government Finance List on the lines of the State List is provided in the Constitution itself, the financial status of municipalities cannot change radically. It is therefore need of the hour that a new Local List is incorporated in the country's Constitution. This is the only way to make the country federal with three tier system of governance—Union, State and Local. As a result of 74th CAA a 'federation within the federation' has been created. The country has two 'federations'—one, Union and States; and the other, States and Local Governments (Urban and Rural). The Second Administrative Reforms Commission has rightly recommended making 'District' as a unit of Local Government. The earlier experiences of District Government in Karnataka (during Ram Krishna Hegde regime) and Zila Sarkar (District Government) in Madhya Pradesh (during Digvijay Singh regime) need to be explored. The real strength of local government depends on granting autonomy and independence to these grass-roots governance units under the supreme law of the land. Local government deserves its own status as the state governments are reluctant in decentralising the powers and devolving the funds despite the lapse of more than two decades of promulgation of historical 74th CAA.

Ensuring regular revenue flows to the municipalities from the states is essential. Except in few states like Karnataka and Tamil Nadu, the transparent devolution is not institutionalised and implemented. The State Finance Commissions (SFCs) should be made more effective by synchronising them with the Central Finance Commission and by fixing specific time-lines for implementation of recommendations. Innovative resources like pricing FSI, land monetisation, imposing user charges to cover O & M costs of core services, and setting-up city level development funds to facilitate preparation of financial plans by the municipalities are some reforms proposed rather, reinforced. Though the Government of India emphasised these reforms earlier, not much progress was made so far. It is a pious hope that the proposed incentivisation mechanism of the AMRUT funding would achieve some results. The foremost need today is to make municipalities financially strong and it should be part of the institutional strengthening programme.

Capacity building of the municipalities is the most neglected aspect today. Creating municipal professional cadres in core areas like accounting, planning, engineering, environment, etc., is the essential step in this regard. At present, municipalities are grossly understaffed and underutilised. Unfortunately, the outsourcing and privatisation methods are given more importance than capacity building at the municipal level. Neglect of the municipalities practiced since Independence will continue if no serious efforts are made in this regard. Devising and implementing staffing standards and competency development by necessary training should be given top priority to make municipalities stronger and capable. The AMRUT has truly recognised its importance and emphasis is laid of focused training of functionaries. But the effectiveness depends on its implementation. The 'desk training' as has been the usual practice will not cut the ice much.

It is apt to conclude by summing-up the current urban scenario in India. The existing urban system is demanding for reforms and some reforms are initiated in a piece meal way without addressing the core governance issues seriously. That is why paradoxical situations are being experienced. If the centralised governance systems demand decentralisation, the efficiency and convenience factors dominate the scene and; if small enterprises are to be encouraged, the interests of the big business dominate. On another plane, subsidy-financial sustainability interfaces pull in opposing directions. All these undercurrents make whatever reforms being pursued ineffective. The system is becoming more turbulent and oscillating due to strongly influencing socio-political undercurrents. In some areas, reforms pick-up which raises hopes and expectations and at the same time, the other areas lag behind or move in opposing direction. In the process, people become impatient and lose trust in the governance which in turn leads to blame game and political logjam. The solution is right orchestration of the core institutions and new statecraft built on collaborative and participative principles and practices.

References

1. Sivaramakrishnan, K. C. (2013): Revisiting the 74th Constitutional Amendment for Better Metropolitan Governance, Economic and Political Weekly, XLVIII, No. 13, March, 30, p. 86.
2. Sood, Ashima (2013): Urban Multiplicities Governing India's Megacities, Economic and Political Weekly, XLVIII, No. 13, March, 30, p. 100.
3. Mohanty, P.K., Misra, B.M., Goyal, Rajan and Jeromi, P.D.(2007): Municipal Finance in India: An Assessment, Development Research Group, Study No. 26, RBI, p. 139.
4. UNHABITAT (2006): World Urban Forum 3, Municipal Finance: Innovation and Collaboration, Report of the Third Session of the World Urban Forum, June 19-23.
5. As quoted in Joshi, Rutul (2003): An Overview of Urban Sector Reforms in India: Changes in the Paradigms of Urban Development, N-AERUS Annual Seminar, p. 5.
6. UNHABITAT (2006): op.cit, p. 44.
7. Ministry of Urban Development, Government of India, National Mission on Sustainable Habitat, p. 6.
8. National Sustainable Habitat Parameters on Urban Storm Water Management, p. 2.
9. *Ibid.*, p. 3.
10. Ali, Ifzal and Zhuang, Juzhong (2007): Inclusive Growth toward a Prosperous Asia: Policy Implications, Asian Development Bank, July.
11. UNHABITAT, United Nations Human Settlements Programme, Global Campaign on Good Governance, p. 19.
12. *Ibid.*, p. 14.
13. 12th Five Year Plan, op. cit., pp. 324-340.
14. McKinsey Global Institute (2010): India's Urban Awakening: Building Inclusive Cities, Sustainable Urban Growth, April.
15. India needs investment of about $ 1.2 trillion in next 20 years that is, from 2010 to 2030. Current annual spending on operational and capital fronts is very low

that is, mere $ 50 which is far less compared to China's spending of $362. Further, if large cities spend $130 per capita per year smaller cities and towns spend as low as $12-38. At these current levels of spending rates, it is estimated that India could spend $300 billion in next 20 years which is very inadequate compared to the demand. Hence, the level of spending must be scaled up considerably.

16. Joshi, Rutul (2003): An Overview of Urban Sector Reforms in India: Changes in the Paradigms of Urban Development, N-AERUS Annual Seminar, p. 5.
17. *Ibid.*
18. The World Bank (2000): Cities in Transition-World Bank, Urban and Local Government Strategy, Washington.
19. *Ibid.*
20. *Ibid.*
21. Mahadevia, Darshini (2003): Globalisation, Urban Reforms and Metropolitan Response, India, Manak Publications Private Limited, New Delhi, pp. 68-73.
22. Mckinsey Global Institute (2010): op.cit, pp. 19-20.
23. Government of India, Ministry of Urban Development (2011): Report on Indian Infrastructure and Services, pp. 79-80.
24. Planning Commission, Report of the Working Group on Urban Governance, p. 4.
25. Report of the Committee on JnNURM II, Planning Commission, op. cit, p. 6.
26. *Ibid.*, pp. 9-10.
27. Lawrence, Eric, Stoker, Robert and Wolman, Harold (2010):, Crafting Urban Policy in Urban Affairs Review, Vol. 45, No. 3, January, p. 414.
28. Government of India (2015): Guidelines of AMRUT.
29. Government of India (2015): Guidelines of Smart Cities Mission.
30. Report of the Steering Committee on Urbanisation, Planning Commission, *op. cit.*, p. 3.
31. Government of India: Planning Commission (2013): Scenarios: Shaping India's Future, July, p. 5.
32. *Ibid.*, pp. 4-16.

Index